Functional Inorganic Materials for Tumor Diagnostics and Therapy

Xinyue Song • Min Gao
Lishang Liu • Dan-Dan Wang • Shusheng Zhang
Editors

Functional Inorganic Materials for Tumor Diagnostics and Therapy

Springer

Editors
Xinyue Song
School of Medicine
Linyi University
Linyi, China

Lishang Liu
School of Medicine
Linyi University
Linyi, China

Shusheng Zhang
School of Medicine
Linyi University
Linyi, China

Min Gao
School of Chemistry and Chemical Engineering
Linyi University
Linyi, China

Dan-Dan Wang
School of Medicine
Linyi University
Linyi, China

ISBN 978-981-95-0527-2 ISBN 978-981-95-0528-9 (eBook)
https://doi.org/10.1007/978-981-95-0528-9

This Springer imprint is published by the registered company Springer Nature Singapore Pte Ltd.
The registered company address is: 152 Beach Road, #21-01/04 Gateway East, Singapore 189721, Singapore

This book is dedicated to the scientific community working to develop advanced materials for biomedical applications, especially those working to address complex challenges of cancer diagnosis and treatment.
We extend our deepest respect to the pioneers who have made the field possible through their foundational contributions to inorganic and nanomaterials, and to the clinicians and translational researchers who have worked to translate laboratory innovations into clinical practice.
We thank our colleagues and Springer Nature—your intellectual support and collaboration were crucial to the success of this work.
Finally, we dedicate this book to the next generation of researchers who will continue to explore, innovate, and push the boundaries of cancer nanomedicine.
May this book contribute to the burgeoning field of oncology nanomedicine and provide inspiration to those who seek to advance it.

Foreword

In the evolving field of cancer diagnostics and treatment, the intersection of materials science and biomedicine has emerged as a critical and transformative area. The complexity and heterogeneity of tumors require innovative, precise, and personalized therapeutic strategies to overcome the limitations of conventional treatments. In this context, functional inorganic materials show extraordinary promise in addressing some of the most pressing challenges in oncology—from early detection and targeted therapy to multi-modal imaging and controlled drug delivery.

The integration of inorganic nanomaterials with advanced biomedical technologies not only expands our understanding of tumor biology, but also lays the foundation for next-generation diagnostic and therapeutic platforms. These materials have unique and tunable properties, such as optical responsiveness, catalytic activity, high surface area, and biocompatibility, which can be engineered to achieve specific biomedical goals. Therefore, functional inorganic materials have become central to the development of precision medicine approaches in oncology.

This book, *Functional Inorganic Materials for Cancer Therapy*, brings together six chapters written by experts in the field. The editors—Xinyue Song, Min Gao, Lishang Liu, Dan-Dan Wang, and Shusheng Zhang—bring together expertise from multiple disciplines including chemistry, nanotechnology, and biomedical engineering to provide a timely and comprehensive overview of this rapidly evolving field. Each chapter explores a specific class of inorganic nanomaterials, including noble metals, upconversion particles, carbon nanostructures, nanozymes, and liquid crystal-based platforms. The book also pays special attention to the cutting-edge integration of nanomaterials with microfluidic systems, highlighting the translational potential of these innovative technologies in clinical oncology. Chapter 1 introduces noble metal nanomaterials, demonstrating their photothermal, imaging, and drug delivery capabilities in tumor precision targeting. Chapter 2 delves into upconversion nanoparticles, which are activated by near-infrared light and have unique advantages for deep tissue imaging and minimally invasive treatment. Chapter 3 explores liquid crystal-based nanostructures, focusing on their applications in tumor sensing and responsive therapeutic devices. Chapter 4 explores nanozymes—engineered nanomaterials with enzyme-mimicking properties—and their applications in catalytic tumor therapy and multimodal imaging. Chapter 5 focuses on carbon nanomaterials such as graphene and carbon nanotubes, emphasizing their versatility in cancer diagnosis and therapeutic delivery. Chapter 6 introduces the combination of functional nanomaterials with microfluidic chip technology, emphasizing their potential in real-time tumor analysis and precision treatment.

The distinctive feature of this book is its interdisciplinary perspective and practical significance. The authors not only summarize the existing literature, but also critically examine the opportunities and limitations of each material class and propose future research and clinical translation directions. This makes the book particularly valuable for graduate students, early career researchers, and professionals dedicated to developing material-based biomedical applications. In addition, the book emphasizes the importance of collaboration across scientific fields. Breakthroughs in cancer nanomedicine are often the result of joint efforts of chemists, material scientists, bioengineers, and clinicians. The content of this book reflects this spirit of collaboration and innovation, and the insights provided are not only scientifically rigorous, but also in line with the evolving needs of precision oncology.

As the field continues to mature, the need for accessible, authoritative resources becomes increasingly urgent. *Functional Inorganic Materials in Diagnosis and Treatment of Tumor* fills this gap, providing a fundamental and forward-looking guide to one of the most promising areas of contemporary biomedical research. We hope it will serve not only as a reference but also as a source of inspiration for continued innovation in cancer nanomedicine.

School of Chemistry and Chemical Engineering
Nanjing University,
Nanjing, China

Hongyuan Chen

Preface

We are pleased to present this book, *Functional Inorganic Materials in Diagnosis and Treatment of Tumor*, which is a comprehensive and timely exploration of the emerging role of inorganic nanomaterials in advancing cancer diagnostics and therapeutics. This book represents the collective efforts of researchers working at the forefront of nanotechnology, materials science, and biomedical engineering, and reflects a shared commitment to address the persistent challenges in modern oncology.

Cancer remains one of the most significant threats to global public health, with global incidence and mortality rates continuing to rise. Although conventional treatment modalities—such as chemotherapy, radiotherapy, and surgery—have yielded remarkable outcomes, their efficacy is often limited by systemic toxicity, poor selectivity, and the inability to detect tumors at an early stage. These challenges have driven the scientific community to explore interdisciplinary strategies, with materials science emerging as a particularly promising frontier in the fight against cancer. In this field, functional inorganic materials offer a promising paradigm shift due to their tunable physicochemical properties, superior structural stability, and intrinsic imaging or therapeutic capabilities. These materials—ranging from noble metals and carbon-based nanostructures to nanozymes and upconversion nanoparticles—have shown great potential in improving the precision and efficacy of cancer diagnosis and treatment. Their applications cover advanced imaging, targeted drug delivery, photothermal and photodynamic therapy, biosensing, and more.

This book comprises six comprehensive chapters that showcase the latest progress and emerging trends in this dynamic field of functional inorganic materials in diagnosis and treatment of tumor. Each chapter focuses on a distinct class of inorganic nanomaterials, presenting their fundamental principles, fabrication strategies, functional mechanisms, and biomedical applications. This book aims to provide readers with a clear, up-to-date, and technically sound understanding of how these materials are shaping the future of oncology. Chapter 1 examines noble metal nanomaterials and their applications in photothermal therapy, bioimaging, and targeted drug delivery. Chapter 2 discusses upconversion nanomaterials, emphasizing their unique optical properties and utility in deep tissue imaging and light-activated therapy. Chapter 3 explores liquid crystal-based materials for tumor sensing and the development of responsive therapeutic devices. Chapter 4 focuses on nanozymes, artificial enzyme-like nanomaterials, and their role in catalytic tumor treatment and multimodal imaging strategies. Chapter 5 highlights the diverse functionalities of carbon-based nanomaterials in cancer detection, imaging, and targeted therapeutics. Chapter 6 presents the integration of functional inorganic nanomaterials with microfluidic chip platforms for precise, real-time diagnostic and therapeutic applications.

This book is intended for researchers, graduate students, and professionals in the fields of nanomedicine, materials science, biomedical engineering, and oncology. We hope that it will not only serve as a reference for current methodologies and applications, but also as a catalyst for future research and clinical translation.

We would like to express our sincere appreciation to all the contributors for their dedication and expertise. We also extend our gratitude to Linyi University and fund committee (National Natural Science Foundation of China and Nature Science Foundation of Shandong Province) that supported this work. It is our hope that this book will stimulate further research and inspire new ideas in the ongoing battle against cancer.

Linyi, China Xinyue Song
Linyi, China Min Gao
Linyi, China Lishang Liu
Linyi, China Dan-Dan Wang
Linyi, China Shusheng Zhang

Acknowledgments

It is a great honor to extend our sincere appreciation to all those who have given us great support in the preparation of the book *Functional Inorganic Materials in Tumor Diagnosis and Therapy*.

First and foremost, we are profoundly grateful to the Fund Committee. This work was supported by the National Natural Science Foundation of China and Nature Science Foundation of Shandong Province. Their sustained funding and encouragement have played a critical role in facilitating the research presented in this book.

We would also like to extend special thanks to the colleagues and mentors who offered insightful discussions, constructive feedback, and valuable guidance throughout the preparation of this book. Their expertise and encouragement helped refine our ideas and deepen the scientific rigor of each chapter. To the chapter authors—Xiangfang Lin, Changmin Hou, Wanqing Teng, Lishang Liu, Jinrong Chen, Wang Dan-Dan, Zhang Chuangli, and Min Gao—we express our deepest appreciation for your dedication, professionalism, and scholarly contributions. Your hard work and commitment to excellence have enriched this book with depth and clarity.

Finally, special thanks to the editorial and production teams of Springer Nature whose professionalism and guidance throughout the publication process have been exceptional.

Contents

Chuangli Zhang Shandong Provincial Key Laboratory of Tumor Imaging Equipment Development and Tumor Diagnosis and Treatment Integration Technology, School of Medicine, Linyi University, Linyi, Shandong, People's Republic of China

Shusheng Zhang Shandong Provincial Key Laboratory of Tumor Imaging Equipment Development and Tumor Diagnosis and Treatment Integration Technology, School of Medicine, Linyi University, Linyi, Shandong, People's Republic of China

Chapter 1
Application of Noble Metal Nanomaterials in Tumor Diagnosis and Treatment

Xiangfang Lin

Abstract Noble metal nanomaterials, particularly gold (Au), silver (Ag), and platinum (Pt), and many more, have garnered significant attention in the fields of tumor diagnosis and treatment due to their unique optical properties, excellent biocompatibility, and high stability in physiological environments. In this chapter, the various preparation methods for noble metal nanomaterials are firstly introduced. Then, the optical properties of noble metal nanomaterials are described. Finally, we focus on the latest progress of noble metal nanomaterials in biomarker sensing and tumor treatment, such as photothermal therapy, photodynamic therapy, and many more. We hope that this review will guide the further development of noble metal nanomaterials and promote their practical applications in cancer diagnosis and treatment.

Keywords Noble metal nanoparticles · Preparation method · Optical properties · Biosensing · Cancer treatment

1.1 Introduction

With the rapid development of nanomaterials and technology, noble metal nanomaterials have garnered significant attention because of their excellent physical and chemical properties as well as their surface plasma behavior (Arvizo et al. 2012; Jiang et al. 2018). Since the first synthesis of colloidal gold nanoparticles (Au NPs), noble metal nanomaterials have found extensive applications across various fields, including optoelectronics, catalysis, biosensing, drug delivery, and clinical diagnosis and treatment. By altering the composition, size, shape, microstructure, and dielectric environment of these nanoparticles, the properties and characteristics of noble metal nanomaterials can be easily adjusted, establishing a foundation for their

X. Lin (✉)
Shandong Provincial Key Laboratory of Tumor Imaging Equipment Development and Tumor Diagnosis and Treatment Integration Technology, School of Medicine, Linyi University, Linyi, Shandong, People's Republic of China

X. Song et al. (eds.), *Functional Inorganic Materials for Tumor Diagnostics and Therapy*, https://doi.org/10.1007/978-981-95-0528-9_1

functionality in diverse application scenarios. Noble metal nanomaterials primarily consist of Au, Ag, Pt, and other elements, with their most notable feature being the ability to generate surface plasmons in the visible and near-infrared regions (Beik et al. 2019; Liu et al. 2020; Cui et al. 2019; Garcia-Peiro et al. 2022). Various nanostructures, such as nanospheres, nanorods, nanotubes, nanoshells, hollow structures, and heterostructures, exhibit unique physical and chemical properties, such as localized surface plasmon resonance (LSPR), which has sparked significant research interest in cancer diagnosis and treatment (Sau and Rogach 2010). More importantly, noble metal nanomaterials can accumulate in tumors more effectively due to the enhanced permeability and retention effects (EPR). Meanwhile, the surfaces of noble metal nanomaterials can be easily functionalized with active targeting moieties, such as antibodies, proteins, peptides, and small molecules, to minimize nonspecific uptake and achieve tumor-specific targeting (Goddard et al. 2020). In addition, noble metal nanomaterials can absorb light energy to generate high-energy hot electrons, enhance electromagnetic fields, and release heat energy through electron–phonon relaxation, exhibiting excellent photothermal conversion performance, making these materials suitable photothermal therapy (PTT) (Xiong et al. 2023; Lv et al. 2021). Some research has reported that noble metal nanomaterials can generate cytotoxic reactive oxygen species (ROS) when exposed to light and ultrasound irradiation, making them suitable for enhancing photodynamic therapy (PDT) and sonodynamic therapy (SDT), respectively (Wu et al. 2022a). Based on this, noble metal nanomaterials have shown broad application prospects in the field of tumor diagnosis and treatment.

1.2 Preparation of the Noble Metal Nanomaterials

1.2.1 Physical Method

Various physical methods, including pyrolysis (Gao et al. 2020), physical vapor deposition (Habibullah et al. 2021), pulsed laser ablation (Riedel et al. 2020), and radiation-induced techniques (Kim et al. 2020), are commonly employed to synthesize noble metal nanomaterials. Pyrolysis is a widely used technique that decomposes chemical precursors by breaking their bonds, resulting in the formation of nanoparticles through subsequent chemical reactions (Odularu 2018). Physical vapor deposition involves the conversion of raw gases into nanoparticles under high-temperature and vacuum conditions (Harish et al. 2022). By carefully controlling the vacuum environment and employing processes such as electron beam excitation, ion bombardment, and evaporation, metals, alloys, and compounds are transformed from solid to gaseous states and subsequently deposited onto a target surface to form thin films or coatings. Pulsed laser ablation, a technique developed by Cotton et al., involves the generation of noble metal nanoparticles, including gold, silver, and platinum, from bulk metal particles (Lee et al. 2022). This technique consists of two main steps: intense ablation to produce small metal seeds and mild irradiation to encourage the growth of nanoparticles. The advantages of this approach include rapid processing times, precise control over particle size and shape, and high yields. Radiation-induced synthesis employs ionizing radiation, including ultraviolet (UV) light, X-rays, and gamma rays, to generate nanomaterials. This technique can produce high-purity, fully reduced metal nanoparticles, representing an advancement over traditional synthesis methods. However, these processes often demand substantial energy to sustain the required high temperature and pressure conditions, resulting in increased costs. Furthermore, nanoparticles produced through physical methods may display imperfect surface structures, which can negatively impact their physicochemical properties.

1.2.2 Chemical Reduction Method

Chemical reduction is one of the most commonly employed methods for synthesizing noble metal-based nanomaterials. Taking typical Au NPs as an example, spherical Au NPs can be easily synthesized using various reducing agents such as trisodium citrate, sodium borohydride ($NaBH_4$), and ascorbic acid, to reduce the gold salt precursor, chloroauric acid ($HAuCl_4$) (Arvizo et al. 2012; Zhao et al. 2013). The NPs are then stabilized in the growth medium by introducing stabilizers, such as citrate, cetyltrimethylammonium bromide (CTAB), thiolate, polymer, etc. In some cases, the reducing and capping agents are the same. Among all the methods for the in situ synthesis of Au NPs by $HAuCl_4$ reduction, the Turkevich-Frens method, which utilizes trisodium citrate as both a reducing agent and stabilizer, has been widely applied and further improved via several research groups (Zhao et al. 2013). This method involves the rapid addition of trisodium citrate to a boiling $HAuCl_4$ solution under vigorous stirring, resulting in a wine-red colloidal suspension. By controlling the ratio of trisodium citrate to Au, Au NPs in a large size range (from 15 to 150 nm) can be obtained; however, when the size exceeds 20 nm, the particles tend to become polydisperse. The Brust–Schiffrin method, which employs thiolates as stabilizers, further improves this situation. The thiolates bind to the surface of Au NPs through relatively strong Au-S bonds, significantly inhibiting the growth process and forms smaller and monodispersed Au NPs. Additionally, the rapid addition of $NaBH_4$ and cooling of the solution also yield smaller and more monodisperse AuNPs. Similar strategies are employed to synthesize other noble metal, such as silver, platinum, and palladium in different size and shapes.

In addition to spherical noble metal nanoparticles, anisotropic noble metal-based nanomaterials, such as nanorods, nanocubes, and nanostars, have also attracted extensive attention due to their enhanced physical and chemical properties compared to spherical noble metal nanoparticles (Sau and Rogach 2010). The seed-mediated growth method is widely used to synthesize anisotropic noble metal-based nanomaterials due to its simplicity, precise size control, and versatility. This method involves utilizing metal nanoparticles as seeds, which are subsequently introduced

attention for its environmentally friendly nature compared to microbial-based methods (Paiva-Santos et al. 2021; Nasrollahzadeh et al. 2020). This approach eliminates the need for developing and maintaining costly cell cultures and bypasses the complex and time-consuming processes of surface modification, microbial screening, and carrier construction. So far, various plant extracts, including those derived from leaves, flowers, fruits, and roots, have been utilized to synthesize noble metal nanoparticles (Jadoun et al. 2020). For example, Adak et al. used leaf extract from purple-colored rice to green-synthesize Ag NPs, where secondary metabolites played a stabilizing role in nanoparticle formation (Adak et al. 2021). Similarly, Pradeep et al. utilized St. John's wort [Hypericum perforatum (H. perforatum) L.] extract for the synthesis of Ag NPs, with phenolic acids and flavonoids contributing to the reduction of Ag^{+} ions, while xanthones and phloroglucinols acted as capping agents, and naphthodianthrones were involved in both processes (Pradeep et al. 2021). Ahn et al. explored the synthesis of Ag NPs using 58 plant extracts sourced from Vietnam and Indonesia (Ahn and Park 2020). Among these 58 Ag NPs samples, they selected six Ag NPs samples synthesized from the extracts of Areca catechu, Hypotrachyna laevigata, Ardisia incarnata, Maesa calophylla, Maesa laxiflora, and Adinandra poilanei to study their antitumor effects. The results demonstrated that the Ag NPs samples prepared from Ardisia incarnata, Maesa calophylla, and Maesa laxiflora exhibited relatively high cytotoxicity and excessive reactive oxygen species (ROS) generation among the six selected Ag NPs samples. In addition, Fahmy et al. reported the green synthesis of platinum and palladium nanoparticles (Pt NPs and Pd NPs) using the alkaloid fraction of Peganum harmala seeds (Fahmy et al. 2021). These Pt-Pd NPs demonstrated significant cytotoxic activities against lung cancer (A549) and breast adenocarcinoma (MCF-7) cells.

Microbial-based synthesis is another commonly used biological method, although the number of studies on the bacterial synthesis of noble metal nanoparticles is significantly lower than that on plant-mediated synthesis (Ali et al. 2019). Both extracellular and intracellular syntheses are possible in bacteria. Intracellular synthesis primarily occurs through a specialized transport system in microorganisms, where positively charged metal ions bind to the negatively charged cell wall via electrostatic interactions. After being transported into the cell, the ions are reduced by enzymes in the periplasmic space, while the nanoparticles can be removed from the cell and subsequently attached to the cell surface, preventing the metal from re-entering the periplasm and acting as a catalyst for further metal reduction. For example, Capeness et al. demonstrated that *D. alaskensis* G20 can synthesize nickel (Ni), palladium (Pd), and Pt NPs. Zhao et al. used DH5α strain of *Escherichia coli* (*E. coli*) as a bio-factory to synthesize Au NPs (Hao et al. 2023). As shown in Fig. 1.2, Au^{3+} ions were absorbed by the extracellular polymeric substance and reduced by reducing agents, including enzymes, present in the extracellular polymeric substance. After a specific incubation period, a significant number of Au NPs accumulated in the submarginal area and tended to be secreted from the bacterial cells, dispersing in the extracellular environment. The Au NPs prepared by this method consist of a gold core and an outer layer of bacterial membrane. The Au core can induce a high-temperature effect, while the bacterial membrane elicits

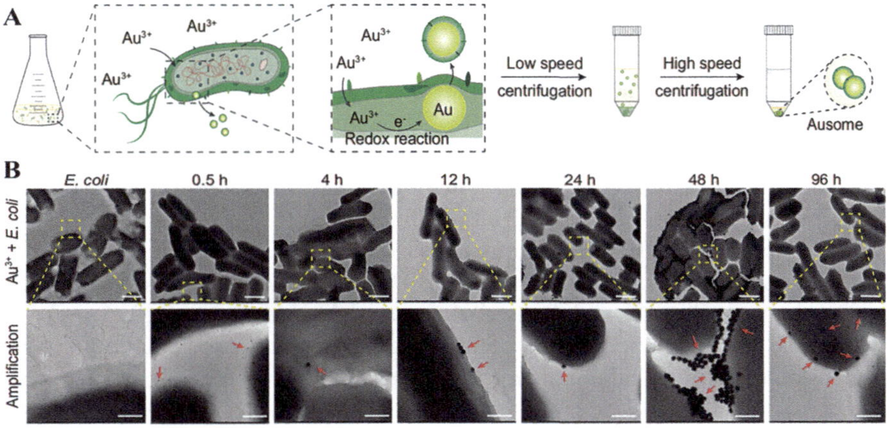

Fig. 1.2 (**a**) Schematic illustration of bacteria-synthesized Au NPs. (**b**) Transmission electron microscopy (TEM) of images of *E.coli* and the synthesized Au NPs at different time (Hao et al. 2023)

multiple immune responses, thereby selectively enhancing local antitumor immune function. Extracellular synthesis is facilitated by compounds secreted by bacterial cells that reduce noble metal ions to zero-valent metals, along with biomolecules that stabilize the nanoparticles involved in the process. Yu et al. employed fungi for the biosynthesis of Au NPs in solution (Yu et al. 2022). They found that the biosynthetic Au NPs exhibited a uniform size and an approximately spherical shape, whether produced by Saccharomycetales or Eurotiales. Additionally, they tested ten different species of bacteria from three orders—Bacillales, Enterobacteriales, and Actinomycetales—as candidates for synthesizing Au NPs. In comparison to the Au NPs biosynthesized by fungi, those mediated by bacteria displayed a larger hydrodynamic size, ranging from 100 to 1000.

1.3 Properties of the Noble Metal-Based Nanomaterials

The physical and chemical properties of noble metal nanoparticles are primarily determined by their size, shape, and composition. Based on their distinctly different characteristics, noble metal nanoparticles can be broadly classified into two categories: plasmonic nanoparticles and molecular-like nanoparticles. These two types of nanoparticles exist in different size regions. Plasmonic noble metal nanoparticles are typically larger than 2 nm and exhibit semicontinuous electronic structures, while noble metal nanoparticles smaller than 2 nm display discrete electronic states due to strong quantum confinement effects. In the following content, we mainly focus on the LSPR properties of plasmonic nanoparticles and the photoluminescent properties of molecular-like nanoparticles or nanoclusters.

1.3.1 Localized Surface Plasmon Resonance (LSPR)

When incident light strikes a conductive nanostructure at a specific angle, it interacts with free carriers (electrons or holes) in the metal or semiconductor, inducing coherent oscillations known as surface plasmon resonance (SPR) (Yan et al. 2023). This vibration will induce the displacement of the surface electron cloud relative to the nucleus, however, the Coulomb attraction between the nucleus and the electron will produce a restoring force, leading to oscillation of the electron cloud within the confines of the nucleus (Zhao et al. 2021). This phenomenon is called LSPR. The size and shape of noble metal-based nanomaterials significantly influence LSPR properties. For example, Huang et al. successfully synthesized ultrathin, two-dimensional (2D) palladium nanosheets with a uniform hexagonal shape (see Fig. 1.3a) (Huang et al. 2011). These palladium nanosheets exhibited size-dependent LSPR peaks as the edge length increased from 21 to 27, 41, and 51 nm. Similarly, Espinosa et al. synthesized Au NSs of varying sizes and observed that as the size increased, the absorbance band gradually shifted to the NIR region at 790, 800, 900, and 950 nm, respectively (Fig. 1.3b) (Espinosa et al. 2016). This shift confirmed that the plasmon resonances of Au NSs are strongly dependent on size and morphology, particularly the length and number of spikes. Li et al. discovered that the LSPR band of porous cubic AuAg alloy nanoshells can be easily adjusted from 787 to 1540 nm by varying the pore number and particle size (Fig. 1.3c) (Li et al. 2022).

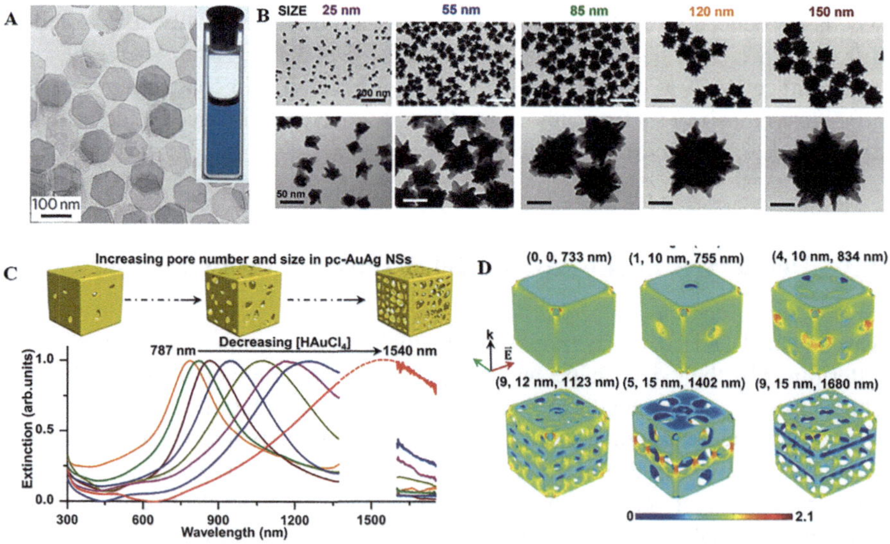

Fig. 1.3 (**a**) TEM image of the palladium nanosheets (Huang et al. 2011). (**b**) TEM images of Au NSs with different size (Espinosa et al. 2016). (**c**) The LSPR band of porous AuAg alloy nanoshells can be adjusted by controlling the pore size and number. (**d**) near-field distribution at the plasmon resonance wavelength (Li et al. 2022)

They found that these nanoshells exhibit pore size- and number-dependent plasmonic properties, affecting both intensity and plasmon resonance wavelength (Fig. 1.3d).

The LSPR effect is significantly influenced by the shape of NPs. In comparison to spherical noble metal nanoparticles, which typically exhibit LSPR peaks around 520 nm, anisotropic noble metal-based nanoparticles with branched shape, such as nanorods, nanobipyramids, and nanostars, exhibit LSPR peaks in the longer wavelength region, especially the NIR region. This shift is attributed to the increased absorption cross-section resulting from their multiple branches, which ultimately enhances laser absorption. For instance, Ye et al. synthesized Au NRs with well-defined lengths and dimensions (Ye et al. 2013). They demonstrated that the longitudinal LSPR peaks can be tuned between 800 and 1200 nm by altering the aspect ratios. Similarly, Abadeer et al. reported that the LSPR peaks of Au NRs could be tuned between 530 and 850 nm by changing the aspect ratios (Fig. 1.4a) (Nardine et al. 2014). Yang et al. prepared a series of Au NRs with different aspect ratios and

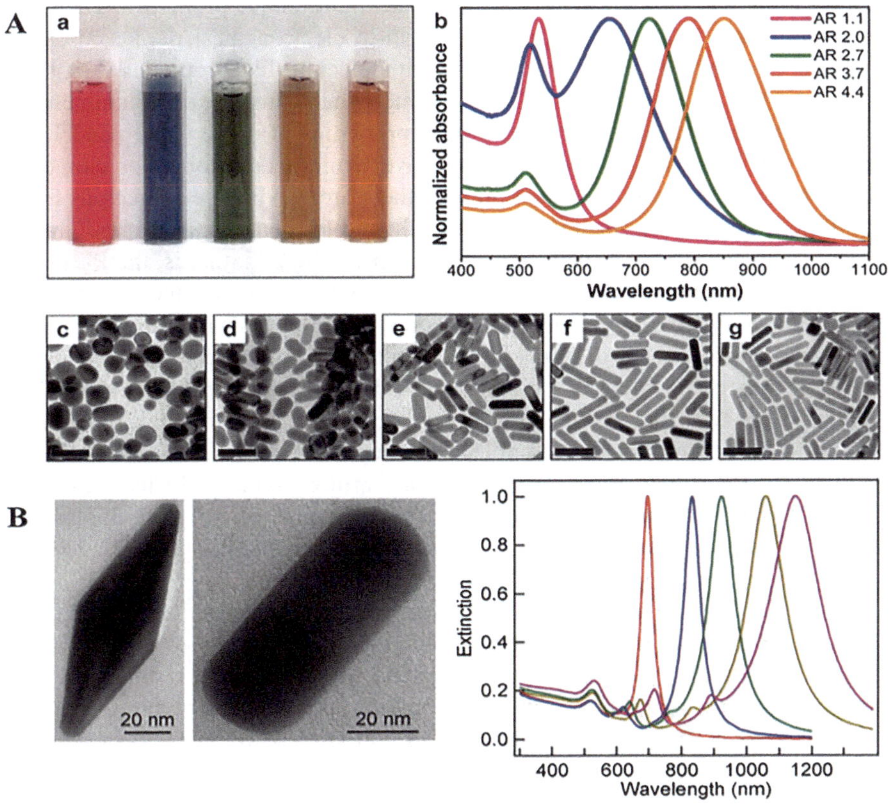

Fig. 1.4 (**a**) The influence of aspect ratio on LSPR peak of Au NPs (Nardine et al. 2014). (**b**) TEM images of a single Au NBPs (left) and a single Au NRs (right), along with the extinction spectra of five differently sized Au NBPs (Chow et al. 2019)

longitudinal LSPR peaks, exploring how size and surface properties affect tumor cell internalization (Yang et al. 2020). Their results indicated that Au NRs with moderate width, length, and lactoferrin modification exhibited significant tumor-targeting effects, effectively destroying xenograft tumors under 980-nm laser irradiation. As an alternative to Au NRs, gold nanobipyramids (Au NBPs) with sharp tips at both ends have been developed (Fig. 1.4b). One of the most notable features of Au NBPs is their ability to generate exceptionally large local electric field enhancements due to their sharp tips (Chow et al. 2019). The full width at half maximum (FWHM) of Au NBPs is nearly half that of Au NRs with similar longitudinal dipolar plasmon wavelengths.

1.3.2 Photoluminescence Property

Unlike plasmonic noble metal nanoparticles, noble metal nanoparticles smaller than 2 nm, commonly referred to as nanoclusters, exhibit unique properties. Their sizes are comparable to the de Broglie wavelength of electrons near the Fermi energy, which endows noble metal nanoclusters with molecular-like properties. Among the various molecular-like properties of noble metal nanoclusters, photoluminescence (PL) is the most attractive because the PL properties of noble metal nanoclusters provide a new avenue for designing effective luminescent probes for biomedical applications (Qiao et al. 2021). At present, there are two main explanations for the origin of noble metal nanoclusters PL: ligand-to-metal-metal charge transfer (LMCT); ligand-to-metal-metal charge transfer (LMMCT), that is, the PL properties of noble metal nanoclusters are affected by both metal core and surface ligand (Goswami et al. 2016). In terms of metal cores, nanoclusters with different numbers of metal core atoms have different light absorption and fluorescence emission properties. Kamat et al. found that glutathione-stabilized gold nanoclusters (Au NCs) exhibited excited state and electron transfer characteristics that depended on the number of gold core atoms (Stamplecoskie and Kamat 2014). As the number of gold core atoms increased, the absorption starting point gradually red-shifted. Zheng et al. found that the excitation light, emission light, Stokes shift, fluorescence lifetime, and quantum yield of Au NCs depended largely on the number of gold core atoms (Huang et al. 2018).

In terms of surface ligands, changes in the size and conformation of Au NCs can be achieved by changing the ligands, providing an opportunity to regulate the PL properties of noble metal nanoclusters (Wu and Jin 2010). Chen et al. reported that the special effect of methylthiolbenzene ligands on the PL properties of Au NCs. Au_{130}, Au_{104}, and Au_{40} NCs with different PL emission were prepared by adjusting the methyl group at ortho-, meta-, and para sites (Chen et al. 2015). Ligand containing electron-rich atoms (such as N, O) or functional groups (such as –COOH, –$CONH_2$) directly affect the delocalized electron density of ligand-to-metal, significantly improving the PL quantum yield of noble metal nanoclusters. Jin et al. found that the fluorescence intensity of Au_{25} NCs gradually increased with the enhancement of the electron-donating ability of the ligands, that is, the enhancement of PL is directly related to the electron-donating ability of the surface ligands (Jin 2015). Besides, the rigid environment around the surface of NCs has been shown to reduce non-radiative relaxation of NCs with minimal structural changes. Xie's groups reported for the first time the controlled aggregation of Au(I)—thiolate complexes on in situ generated Au(0) cores to form aggregation-induced emission (AIE) type Au NCs (Luo et al. 2012). The enhancement of PL emission is mainly attributed to strong intermolecular and intramolecular interactions (e.g., van der Waals force and aurophilic interactions) that cause ligand aggregation on the surface of Au NCs, increasing the restraints on the intramolecular vibrations and rotations of the complexes, thereby reducing the probability of non-radiative relaxation of the excited states. By changing the solvent polarity (Zhang et al. 2018), phase (coagulation or gelation) (Nirmal et al. 2016), and/or ionic condition (acidity or metal ion) (Chang et al. 2020) could restricted the intramolecular motion, achieving high quantum yield.

1.4 Application of Noble Metal Nanomaterials in Cancer Diagnosis

1.4.1 Small Molecular Detection

Small molecule metabolites are essential components of cancer biomarkers and effectively reflect the condition of cells and tissues. ROS are highly reactive molecules containing oxygen, such as hydrogen peroxide (H_2O_2), superoxide radicals (O_2^-), hydroxyl radicals (·OH), peroxynitrite anion ($ONOO^-$), and hypochlorite (ClO^-), which are closely related to the occurrence, development, and apoptosis of tumors. Excessive accumulation of highly reactive oxygen species (hROS) can induce oxidative stress, leading to cell damage. Therefore, monitoring the concentration of ROS is of great significant for the diagnosis and treatment of tumors. Due to ROS being a type of molecule or ion with high oxidative activity, it can oxidize and destroy the structure of nanoclusters, thereby altering their fluorescence properties. Based on the changes in fluorescence signals, ROS detection can be achieved. For example, Quan et al. designed a ratiometric fluorescent probe based on the cellulose nanocrystal (CNC)-stabilized Au NCs for imaging of hROS, including ClO^-, $ONOO^-$, and ·OH (Fig. 1.5) (Quan et al. 2022). In this ratiometric fluorescence probe, o-phenylenediamine serves as an anti-interference internal standard, which can be oxidized by hROS to form 2,3-diaminophenazine, producing yellow fluorescence. Based on this, the coexistence of o-PD with CNC-stabilized Au NCs facilitates hROS detection with excellent selectivity and sensitivity for fluorescence imaging of cells and zebrafish. ClO^- as one of hROS can be endogenously produced by neutrophils. Abnormal levels of ClO^- in vivo can be associated with arthritis, cardiovascular diseases, lung injury, neuron degeneration, and even cancers. Lin

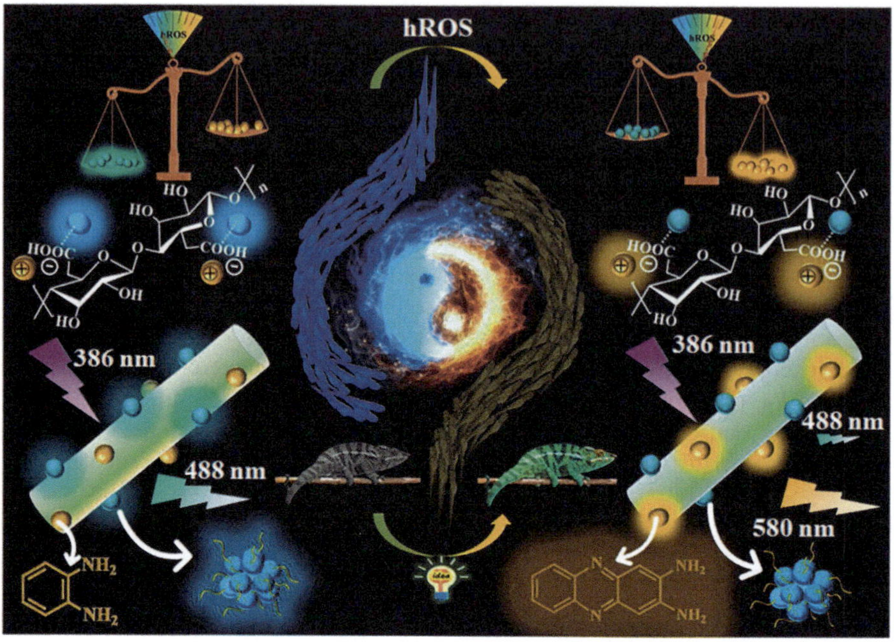

Fig. 1.5 Schematic diagram of CNC-stabilized Au NCs for hROS detection (Quan et al. 2022)

et al. synthesized strongly PL and water-soluble Au(I)–Ag(I)-cysteine complexes nanoplatelets for intracellular ClO^- detection (Lin et al. 2021). In the presence of ClO^-, the metal-thiolate ligand structure of the Au(I)–Ag(I)-Cys nanoplatelets can be destroyed by oxidation, which will lead to PL quenching, thus exhibiting a biological recognition function for ClO^- anions. Similarly, they reported that the Ag core of amphiphilic copolymer-modified luminescent Ag NCs can be oxidatively damaged by ClO^- (Lin et al. 2024). Based on it, a fluorescent nanoprobe can be constructed for detecting exogenous and endogenous ClO^- in living cells. H_2O_2 is a significant ROS in biological systems, playing a crucial role in cell proliferation, apoptosis, and intracellular signal transduction. In comparison to normal cells, cancer cells demonstrate an increased rate of H_2O_2 production (0.5 nmol/10^4 cells/h), leading to higher levels of H_2O_2 in tumors. Jain et al. synthesized bovine serum albumin (BSA)-coated Au NCs for the detection of H_2O_2 in mammalian cells (Jain et al. 2021). The bright red fluorescence intensity of internalized BSA-Au NCs was significantly quenched in cells exposed to 3-amino-1,2,4-triazole, which can indicate the accumulation of H_2O_2. Therefore, the developed biocompatible BSA-Au NCs can be effectively utilized to detect H_2O_2 levels in the cytoplasm.

1.4.2 Protein Detection

The analysis of protein biomarkers is highly significant in the diagnosis of cancer. Noble metal NPs have excellent conductivity, which can improve and enhance the original conductivity of electrodes. By appropriate modification, noble metal NPs can be prepared into a series of biosensors for use in electrochemical catalytic reactions. Alpha-fetoprotein (AFP) is a broad-spectrum tumor marker that can be detected utilized noble metal nanomaterials-based biosensing. Quan et al. developed a label-free electrochemical aptamer-based sensor for highly selective and sensitive detection of AFP (Quan et al. 2022). A screen-printed graphene-carbon paste electrode (SPGE) modified with carboxylated graphene oxide (Pt NPs/GO-COOH) was utilized as the immobilization platform, with AFP aptamer serving as the bio-recognition element. Compared to bare SPGE, Pt NPs/GO-COOH modified SPGE significantly enhanced the current response of the hydroquinone electrochemical probe. Then, BSA was modified on the electrode surface to avoid non-specific binding. In the presence of AFP, this aptasensor showed high sensitivity to AFP with detection limit of 1.22 ng/mL. Chen et al. designed an innovative sandwich-based electrochemical immunosensor, as shown in Fig. 1.6, for the ultrasensitive determination of AFP. On the one hand, DNA-functionalized Au NPs (AuNPs@DNA) serve dual purposes: they act as crosslinkers to immobilize primary antibodies (anti-AFP antibody 1, Ab1), resulting in the formation of AuNPs@DNA-Ab1, and they also function as templates for the synthesis of Ag NCs on AuNPs to the creation of AuNPs@DNA-AgNCs. On the other hand, p-sulfonated calix aromatic hydrocarbon (pSC4)-modified Au was selected to immobilize the secondary antibody (anti-AFP antibody 2, Ab2) by master-guest recognition

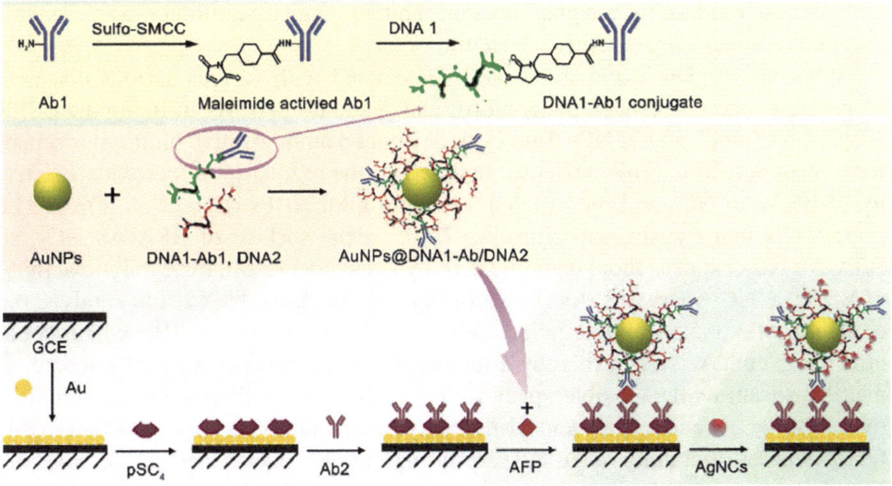

Fig. 1.6 Schematic of the preparation procedure for the sandwich-type electrochemical immunosensor for AFP detection (Chen et al. 2023a, b)

between Ab2 and pSC_4. In the presence of AFP, the immune response signal can be significantly amplified by the electrochemical reduction of Ag NCs (Chen et al. 2023a, b).

As another key broad-spectrum tumor marker, monitoring the levels of carcinoembryonic antigen (CEA) is highly significant for the early diagnosis and treatment of cancer. Lan et al. developed a highly sensitive, label-free electrochemical immunosensor for detecting CEA based on novel Pt NPs-decorated reduced graphene oxide@polystyrene nanospheres (PtNPs@rGO@PS NSs) (Lan et al. 2020). The PtNPs@rGO@PS NSs were prepared by the electrostatic self-assembly of negatively charged GO sheets onto positively charged PS nanospheres, followed by hydrazine reduction of GO and in situ deposition of Pt NPs. The nanospheres were further biofunctionalized with streptavidin and a biotinylated antibody, enabling their use as a highly sensitive electrochemical immunosensor for the label-free determination of CEA. Similarly, Niu et al. designed terahertz toroidal metal-biosensor by integrating with functionalized AuNPs (Niu et al. 2023). The designed 8-shaped biosensor, characterized by a high Q-factor, benefits from the presence of a confined closed-loop magnetic field. Meanwhile, the as-synthesized Au NPs were capable of improving the sensitivity of the toroidal meta-biosensor due to their high refractive index. Consequently, this biosensor can achieve highly sensitive and specific detection of CEA. Chen et al. developed an innovative photoelectrochemical immunosensor utilized a PdPt nanozymes and CdS NPs for the high-sensitivity detection of CEA (Chen et al. 2021). The PdPt nanozymes not only showed exceptional peroxidase-like activity but also significantly improves the photoelectrochemical performance of the pure CdS NPs. PdPt nanozymes can catalyze 4-chloro-1-naphthol (4-CN) to generate insoluble precipitation in the presence of H_2O_2, triggering the biological etching of CdS NPs, resulting in synergistic signal amplification. Furthermore, the combination of the CdS/PdPt/H_2O_2/4-CN system with glucose oxidase (as a signal labeling) and a secondary antibody (as a recognition unit) enables ultra-sensitive detection of CEA.

Abnormal trypsin levels are strongly associated with various serious diseases, including pancreatic cancer, pancreatitis, and cystic fibrosis. By utilizing the LSPR properties of noble metal NPs, Luo et al. developed an innovative multicolor sensor for trypsin detection. This biosensor is based on the regulation of peroxidase activity of BSA-Au NCs and the efficient etching of Au NBPs (Luo et al. 2023). The principle is that trypsin can hydrolyze BSA on the surface of BSA-Au NCs, so BSA-Au NCs exposes more catalytic activity sites. As a result, the catalytic activity of BSA-Au NCs increases. As shown in Fig. 1.7, BSA-Au NCs could catalyze the oxidation of 3,3′,5,5′-tetramethylbenzidine (TMB) to produce TMB^+, which could effectively etch Au NBPs, thereby affecting the aspect ratio of Au NBPs accompanied by an ultraviolet–visible spectra blue shifting. According to the quantitative relationship between trypsin and TMB^+, varying amounts of trypsin can lead to different degrees of etching of Au NBP, resulting in rich color changes in the solution. Based on this, a multicolor sensor could be developed for the detection of trypsin. Similarly, He et al. developed a novel strategy to induce precisely tunable LSPR and fluorescence-composited dual-mode signals by initially using mono-DNA as an

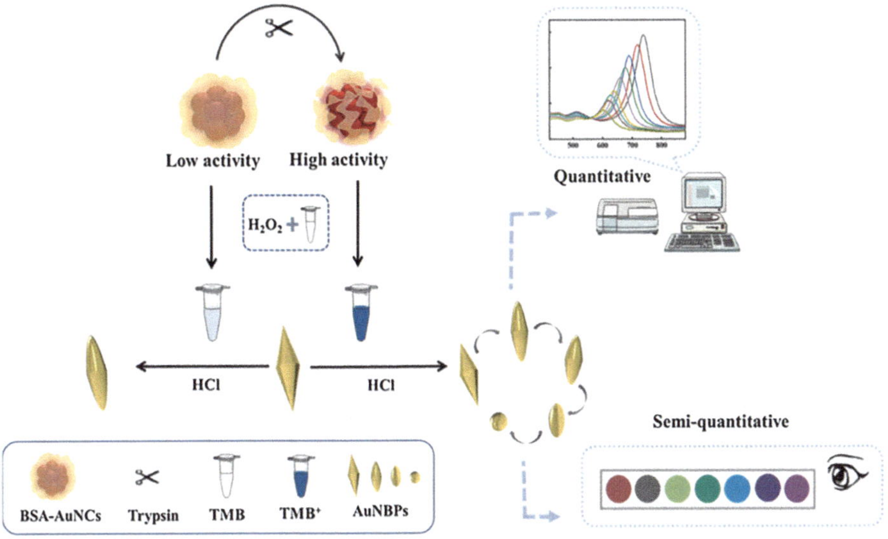

Fig. 1.7 Schematic of the multicolor sensor for trypsin detection based on the regulation of peroxidase activity in Au NCs and the etching of Au NBPs (Luo et al. 2023)

etching agent to modify the morphology of Ag nanotriangles and subsequently as a template for synthesizing fluorescent Ag NCs (He et al. 2023a, b). Utilizing the unique LSPR shift of Ag nanotriangles and the fluorescence of Ag NCs, a biosensor mediated by BSA was developed for the ultra-sensitive detection of trypsin activity in a label-free and instrument-free manner. Additionally, fluorescence detection provides supplementary information, which avoided the influence of factors that may compromise morphological stability.

In addition, Chen et al. developed an electrochemiluminescence (ECL) biosensor using glutaraldehyde-cross-linked chitosan confined Ag NCs (Glu-CS-Ag NCs) as ECL emitters and X-shaped DNA as recognizers for detecting Apurinic/apyrimidinic endonuclease 1 (APE1), which abnormal expression is closely associated with malignant tumors (Chen et al. 2024). Zhang et al. developed ECL biosensor for the detection of exosomes and their surface proteins by the in-situ decoration of AuNPs on aptamer modified MXenes (Au NPs-MXenes-Apt) (Zhang et al. 2020a). The exosomes were efficiently captured on a sodium alginate-poly(acrylamide) electrode interface. Additionally, the in situ-formed Au NPs-MXenes-Apt hybrid provided a bare catalytic surface with high electrocatalytic activity due to the predominance of (111) facets in the Au NPs, which significantly enhanced the ECL signal of luminol. Therefore, the ECL sensing platform has high selectivity for extracellular vesicles and surface proteins derived from various tumor cell lines, including HeLa, OVCAR, and HepG2 cells. It can sensitively and accurately detect extracellular vesicles in human serum. Xu et al. developed a proportional antifouling electrochemical biosensor using a two-dimensional nanomaterial MXene loaded with multifunctional peptides and loaded with AuNPs and methylene blue (MB)

(Xu et al. 2021). MXene nanocomposites loaded with Au NPs showed excellent electrochemical properties and enhanced electrical conductivity. In addition, Au NPs are able to capture biomolecules containing sulfhydryl termini, providing a large number of reaction sites. Thus, the model target prostate-specific antigen was detected with high selectivity and sensitivity.

1.4.3 Nucleic Acid Detection

The occurrence of cancer is often accompanied by abnormal gene expression, and these abnormally expressed genes produce specific nucleic acid markers, such as microRNA (miRNA), messenger RNA (mRNA), and so on. By detecting the presence and levels of these nucleic acid markers, early diagnosis and monitoring of cancer can be carried out. For example, Yang et al. developed a terahertz biosensor based on metamaterials coupled with nanoparticles and strand displacement amplification (SDA) for the detection of miRNA (Yang et al. 2021). As shown in Fig. 1.8a, the miRNA-21 first hybridizes with the template DNA, triggering DNA strand extension using Klenow fragment polymerase. NPs are initially modified with a thiol probe to form the nanoprobe. Following the SDA process, the nanoprobes and biotin probes hybridize with their complementary segments of the trigger DNA chains. The complex formed by the miRNA-21 SDA products and nanoparticles exhibits a significant terahertz frequency shift in metamaterials, making it suitable for the highly sensitive detection of miRNA-21. Pothipor et al. developed a label-free electrochemical biosensor utilizing a Au NPs/graphene quantum dots/graphene oxide (AuNPs/GQDs/GO) modified three-screen-printed carbon electrode (3SPCE) array for the simultaneous detection of miRNA-21, miRNA-155, and miRNA-210 (Pothipor et al. 2021). Three miRNA probes are assembled on the modified three-working area electrodes. Redox dyes (AQ, MB, and PDA) serve as signal indicators, producing distinguishable electrochemical signals. With high sensitivity and multiplexing capability, the fabricated biosensor enhances diagnostic accuracy for early breast cancer detection. In addition, Zoughi et al. introduced a novel biosensor based on Ag NCs to detect miRNA-21 (Zoughi et al. 2024). The conformation of the hairpin sequence is changed in the presence of miRNA-21, allowing the G-rich sequence to pair with it and interact with Ag NCs, thereby increasing their fluorescence emission intensity. The designed probe responds to miRNA-21 concentration with a detection limit of 2 pM. Furthermore, it was successfully applied to detect miRNA-21 in human serum samples as well as in human lung and ovarian cancer cells.

CircRNAs are a group of circulars, blocked, single-stranded non-coding RNAs that are important biomarkers for cancer diagnosis and prognosis. This study proposes an isothermal amplification system based on dual-catalyzed hairpin self-assembly (CHA) and chameleon DNA-templated silver nanoclusters (DNA-AgNCs) for label-free ratio detection of circRNAs (Yang et al. 2022). The dual-cascade amplification platform begins with upstream CHA 1, where the target catalyzes two

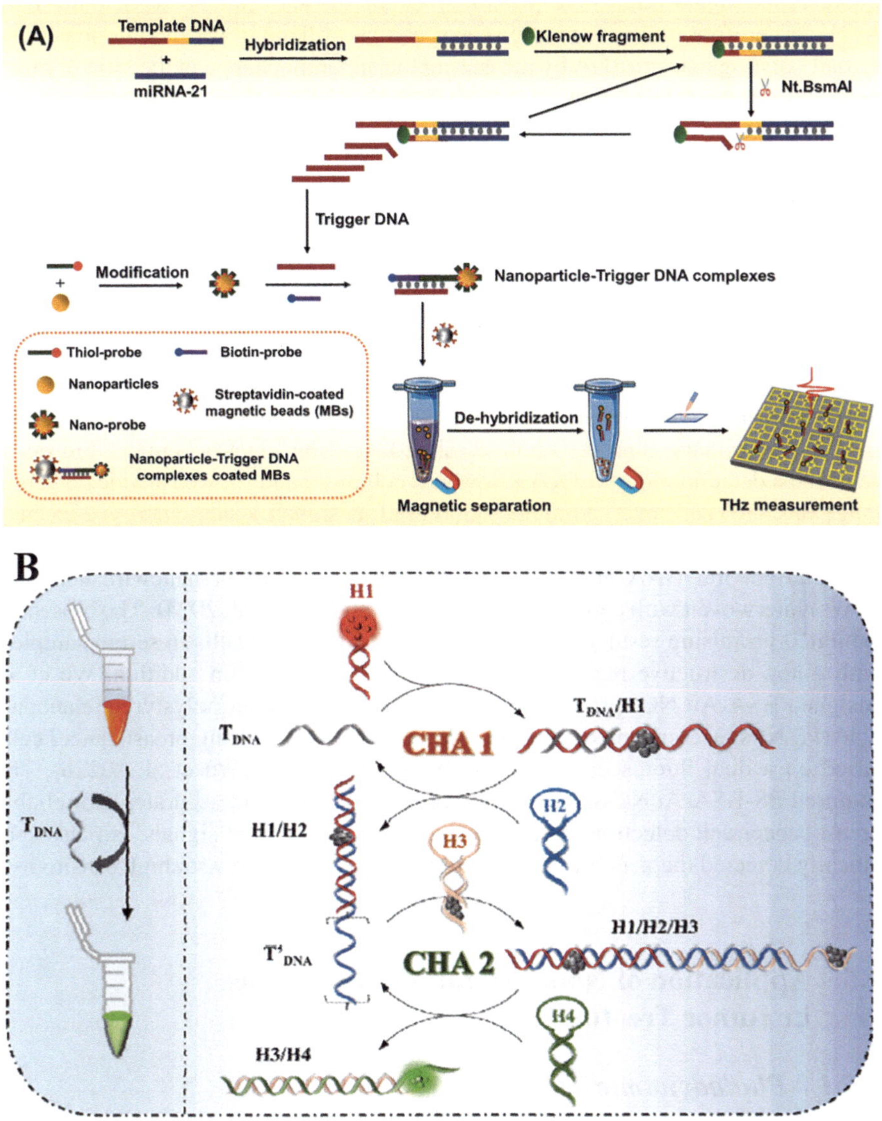

Fig. 1.8 (**a**) Schematic description of THz metamaterial biosensor for miRNA detection. (**b**) Schematic illustration of Ag NCs for detecting circRNA

hairpin substrates to generate a substantial amount of dsDNA product, which subsequently triggers the downstream CHA 2 loop. During this process, the hairpin structure in upstream CHA 1 is disrupted, leading to the quenching of the red fluorescence emitted by AgNCs. The product of upstream CHA 1 drives the self-assembly of the downstream CHA 2 hairpin reactant, resulting in a significantly enhanced green fluorescence signal due to the proximity of the G-rich sequence to another dark AgNC. Therefore, this ratiometric sensor can be utilized to sensitively detect and visually distinguish circRNA by measuring the apparent change in the ratio of green to red fluorescence intensities.

1.4.4 Cancer Cell Detection

In addition to the aforementioned, noble metal nanomaterials can also be applied for detecting cancer cells. Circulating tumor cells (CTCs) that are shed from the primary tumor site into the bloodstream are closely associated with tumor-related mortality. For example, Zhou et al. constructed a PdRu/Pt heterostructure working with heme/G-tetrad DNA enzyme to specifically detect circulating tumor cell (CTC) shed from primary tumors (Zhou et al. 2021). The electrochemical biosensor achieved a detection limit (LOD) as low as 2 cell mL^{-1} and showed a wide detection range of 2–10^6 cell mL^{-1}. Similarly, Guo et al. designed a sandwich-type electrochemical aptasensor composed of holothurian-shaped AuPd nanoparticles (AuPd HSs), tetrahedral DNA nanostructures (TDNs), and CuPdPt nanowire networks (NWs) interwoven with a graphdiyne (GDY) sheet (Guo et al. 2024). The biosensor exhibited promising results when applied to detect MCF-7 cells in serum samples, with a non-destructive recovery rate of 95.98%–102.39%. In addition, Wu et al. designed BSA-Au NCs protein corona stabilized fluorescent polystyrene nanoparticle (PS NPs) accompanied with EpCAM aptamer for detecting breast cancer cells based on a dual fluorescence signal approach (Fig. 1.9) (Wu et al. 2022b). The acquired PS-BSA-AuNCs-Apt not only performed sensitive, accurate, and reliable breast cancer cell detection in vitro by dual-color fluorescent signals, but also specifically detected the xenografted breast tumor tissue ex vivo with high sensitivity.

1.5 Application of Noble Metal Nanomaterials in Tumor Treatment

1.5.1 Photodynamic Therapy

PDT utilize light-excited photosensitizers (PSs) that can undergo type I (electron transfer) or type II (energy transfer) photochemical mechanisms to activate endogenous O_2, leading to the production of highly cytotoxic ROS, which in turn induces apoptosis for cancer treatment. Noble metal complexes nanomaterials (such as Ir, Ru) are highly efficient PSs for PDT due to their high ROS production. Although noble metal complexes have good ROS production ability, their hydrophobic structure often limits their biological application. Therefore, integrating noble metal complexes with hydrophilic functional components (such as polymer) can provide

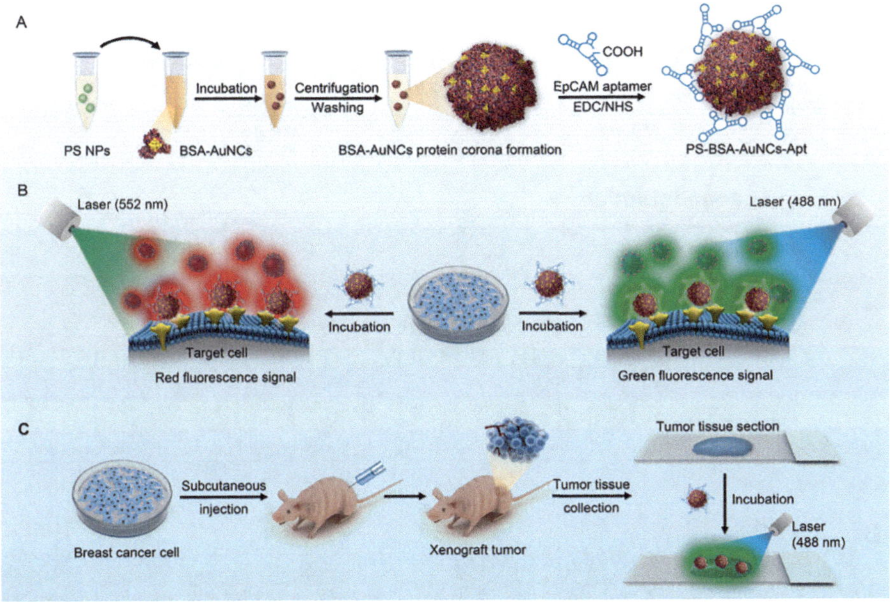

Fig. 1.9 Schematic illustration of the workflow for the detection of EpCAM-positive breast cancer cells by PS-BSA-AuNCs-Apt. (**a**) Fabrication of PS-BSA-AuNCs-Apt. (**b**) Detection of breast cancer cells based on dual-color fluorescent signals. (**c**) Detection of xenografted breast tumor tissue (Guo et al. 2024)

complementary functions such as water-solubility, biocompatibility and stability. For instance, Kuang et al. reported a novel mitochondrial-targeting iridium (III) peroxide prodrug (2-O-IrAn), which could be facility encapsulated by DSPE-PEG polymer to form DSPE-PEG-Biotin@2-O-IrAn (Fig. 1.10a) (Kuang et al. 2022). The DSPE-PEG-Biotin@2-O-IrAn exhibited excellent water solubility, good biocompatibility, tumor-targeting ability, and highly efficient ROS production ability. During the process of cancer treatment, DSPE-PEG-Biotin@2-O-IrAn internalized and reached the acidic lysosome environment, and then 2-O-IrAn was released and further accumulated in the mitochondria. 2-O-IrAn could generate toxic singlet oxygen ($^1O_2$) from the covalently bound oxygen molecules and producing alkoxy radicals, which could lead to the loss of mitochondrial membrane potential and ultimately triggering apoptosis. In vitro and vivo experiments have shown that DSPE-PEG-Biotin@2-O-IrAn almost completely eradicate tumors in mouse models through two-photon irradiation in the NIR (Fig. 1.10b). Similarly, Huang et al. designed an iridium (III) photocatalyst (Ir3) with excellent photo-physical and anticancer effect. Ir3 exhibited a high excited state lifetime generates $^1O_2$ and also catalyzes photooxidation of reduced nicotinamide adenine dinucleotide 2′-phosphate (NADPH) to $NADP^+$ which in turn leads to the generation of O_2^- and H_2O_2 as ROS. It is evident from zebrafish and mouse cancer models that Ir3 exhibits high in vivo biocompatibility and photocatalytic anticancer efficiency. In addition, Zhang

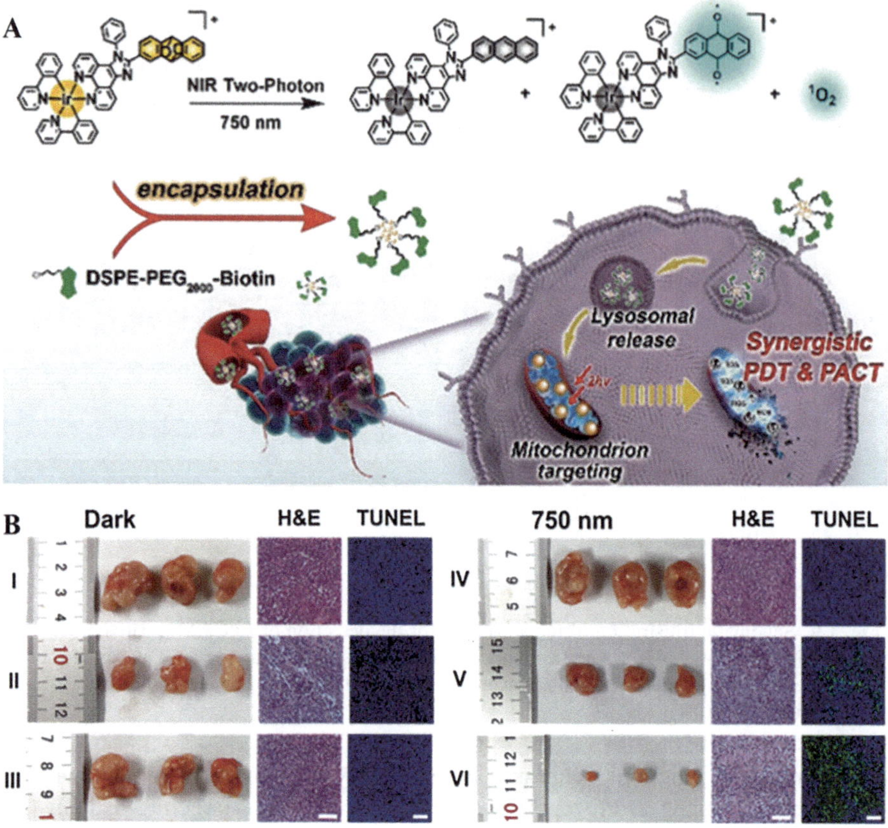

Fig. 1.10 (**a**) Schematic illustration of the mechanism of the DSPE-PEG-Biotin@2-O-IrAn for PDT. (**b**) The tumor therapy of different groups. (I) (IV) Physiological saline; (II) (V) cisplatin; (III) (VI) DSPE-PEG-Biotin@2-O-IrAn (Kuang et al. 2022)

et al. designed NP2 containing a photodynamic polymer and an anti-metastatic ruthenium complex (NAMI-A) for breast cancer treatment (Zhang et al. 2024). NP2 can generate reactive ROS under 808-nm wavelength irradiation to kill tumor cells, reduce the expression of matrix metalloproteinases (MMP2/9) in cancer cells, and decrease the invasive and migratory capabilities of cancer cells, ultimately eliminating them. In vivo experiments, NP2 combined with laser irradiation effectively inhibited the growth of primary tumors, suppressed the metastasis of primary tumor cells, and reduced the formation of metastatic lesions in the liver and lungs. Moreover, this therapeutic approach activated the systemic immune response in mice, providing protection against tumor recurrence.

Au NCs containing several to dozens of Au atoms have emerged as promising candidates for PSs in PDT due to their high intersystem crossing efficiency and long lifetime of electronically excited states. For instance, Santhakumar et al. prepared two highly fluorescent silver-doped gold nanoclusters (TPGNCs) using custom-synthesized tripeptides DCG (Asp-Cys-Gly) and DCS (Asp-Cys-Ser) as reducing agents and stabilizers (Santhakumar et al. 2023). TPGNCs could produce ROS for PDT under laser irradiation. This may due to the longer lifetimes of gold nanoclusters under laser irradiation, which transfer energy from nanoclusters to O_2, thereby generating ROS. In vitro and vivo experiments indicated that TPGNCs exhibit excellent anticancer activity, as they could increase the production of ROS in cancer cells through photoinduced, and then activate the mitochondrial apoptotic pathway in cancer cells. Zhang et al. designed a pH-responsive AuNCs@MOFs-DOX nanocomposites for enhancing PDT/chemotherapy, in which GSH-stabilized Au NCs were used as PS for PDT under 670-nm irradiation (Zhang et al. 2020b). The encapsulation of ZIF-8 prevents the release of Au NCs and DOX in neutral media, thus avoiding the threats to normal tissues. More importantly, the collapse of pH-responsive structural ZIF-8 in acidic environment accelerated the release of Au NCs and DOX in tumor cells, thus enhancing the efficacy of PDT and chemotherapy. Under 670-nm laser irradiation, the prepared AuNCs@MOF-DOX nanoprobes generate a substantial amount of $^1O_2$ for PDT. In addition, the loading and release of DOX could for highly effective chemotherapy. Based on this, this pH-responsive nanocomposite could achieve a synergistic effect in PDT/chemotherapy.

In addition to being as PSs for PDT, noble metal nanoparticles can also serve as carriers for PSs loading, which can prevent the interference of the intracellular environment. Typically, noble metal nanoparticles can form bio-conjugated metal nanohybrids with PSs such as Chlorin e6 (Ce6) or Indocyanine Green (ICG) to enhance the efficacy of PDT or to promote synergistic combination therapies. For example, Sun et al. developed a multifunctional nanotherapeutic platform for synergistic near-infrared-I (NIR-I) PDT and near-infrared-II (NIR-I) PTT by combining gold nanodendrites (AuND) with mitochondria-targeting compound (triphenylphosphine, TPP) and ICG (Fig. 1.11a) (Sun et al. 2021). The AuND-TPP-ICG NPs were further encapsulated by macrophage cell membranes (MCM), which can efficiently capture ICG in the interbranch space of AuND and selectively interact with MDA-MB-231 cells to facilitate endocytosis, and achieve selective accumulation within tumors. Upon laser irradiation, the excited ICG converts adjacent molecular oxygen (O_2) into highly cytotoxic singlet oxygen ($^1O_2$) (Fig. 1.11b). Meanwhile, modified TPP could enable AuND-TPP-ICG@MCM NPs target mitochondria, limiting the production of $^1O_2$ within mitochondria to induce targeted damage, and thus maximize the cytotoxicity of PDT. It is worth mentioning that the combination of AuND-TPP-ICG@MCM with 1064-nm and 808-nm laser irradiation resulted in the highest tumor ablation observed, with the tumor nearly completely disappearing after 16 days (Fig. 1.11c). Similarly, Wu et al. successfully developed a multifunctional system (PDA-Ce6-GSH-AuNFs) based on gold nanoflowers (Au NFs) for synergistic PDT and PTT (Wu et al. 2020). Au NFs bound to Ce6 through glutathione (GSH) linkage and was enveloped by a layer of polydopamine (PDA). With the assistance of GSH, the loading capacity of Ce6 reached up to 14.0 wt.%, with $^1O_2$ production approximately 91.0% that of free Ce6. In vitro and in vivo studies have demonstrated that PDA-Ce6-GSH-AuNFs exhibited excellent phototoxicity and synergistic effects in killing cancer cells.

respectively. Under the combined action of PTT, PDT, and catalytic therapy using BSA@Au/Pt-IR808, the cell viability was significantly reduced to 12.8%, and the MCF-7 tumor was significantly inhibited, demonstrating excellent antitumor activity (Fig. 1.12c). Similarly, Da et al. introduced Pt NPs with CAT-like activity and catalytic hydrogen (H_2) generation capability to enhance photocatalytic NADPH oxidation of Ru complexes (Da et al. 2023). By co-encapsulating Ru complexes and Pt NPs in BSA within tumor-targeting ability, the resulting Ru/Pt@BSA exhibited highly efficient selective anticancer activity both in vitro and in vivo.

Noble metal nanomaterials can couple with semiconductors to enhance ROS production by extend the lifetime of photogenerated energetic charges. Pu et al. synthesized Au NR/TiO_2 hybrid nanoparticles for PDT under irradiation with NIR light (Pu and Pons 2023). As shown in Fig. 1.12d, they found that Au NR/TiO_2 hybrid nanoparticles could the generation of •OH through the transferring the hot electrons generated by plasmon excitation to the titanium dioxide (TiO_2) tips, and then reducing dioxygen molecules. More importantly, even in the absence of O_2, the residual photogeneration of •OH still existed, which could prove beneficial for PDT in hypoxic tumor microenvironment. Yu et al. designed a hybrid Au NR@aTiO_2@UCNPs nanocomposites for combined PDT and PTT therapy, which composed of anatase titania-coated plasmonic Au NRs decorated with upconversion nanoparticles (UCNPs) (Yu et al. 2021). The near-field enhancement generated by the LSPR increased the upconversion emission threefold, which was subsequently reabsorbed by the adjacent TiO_2 and Au NRs. Due to the efficient formation of free radicals and heat generation induced by LSPR, the hybrid nanostructures exhibited significant in vitro anticancer effects under near-infrared light. In vivo animal experiments demonstrated a cytotoxicity of up to 86.2% against cancer cells, further confirming the ability of Au NR@aTiO_2@UCNPs to inhibit tumor growth through the combined effects of PDT/PTT. In another study, Jang et al. utilized low-temperature hydrothermal reactions to direct deposition of anatase TiO_2 onto sharp-tipped Au NBPs to synthesize Au NBP/a-TiO_2 (Jang et al. 2022). Under NIR irradiation, high-energy LSPR-derived hot electrons are generated on the Au NBPs surface, which are then directly injected into the conduction band of the deposited anatase TiO_2 nanoclusters, resulting in the reduction of nearby O_2 to O_2^- and ultimately to •OH. In vitro and in vivo experiment indicated that Au NBP/a-TiO_2 can achieve synergistic PTT and PDT therapy.

1.5.2 Photothermal Therapy

PTT involves the induction of localized hyperthermia to ablate tumors using photothermal agents (PTAs) under NIR light. It has garnered significant attention due to its non-invasive, high spatiotemporal resolution, excellent selectivity, and minimal damage to healthy tissues. The ideal PTAs should have high photothermal conversion efficiency (PCE), strong absorption in the NIR region, good photostability and excellent biocompatibility. Noble-based nanomaterials (e.g., nanoparticles, nanorods, and nanostars) have emerged as one of the most widely used PTAs due to their good biocompatibility, low toxicity, and LSPR effect. As mentioned earlier, the LSPR can be modified by adjusting the aspect ratio, size, shape, and aggregation of the nanoparticles. Compared to anisotropic noble metal-based nanomaterials, small-sized noble metal nanoparticles exhibit a longer blood residence time and a shorter biological half-life, making them more suitable for in vivo applications. However, small-sized noble metal nanoparticles often exhibit low PCE, limited absorption in the NIR region, and inadequate penetration into deep tissues. Recently, the in situ aggregation of small-sized noble metal nanoparticles has attracted increasing interest in addressing the aforementioned issue. For example, Chen et al. developed a furin-responsive strategy for intracellular aggregation of Au NPs to enhance the efficacy of PTT for cancer treatment (Chen et al. 2020). As shown in Fig. 1.13a, the furin-responsive platform (AuNP@1) consists of 5-nm Au NPs modified with a furin specific recognition peptide sequence. When AuNP@1 was internalized by cancer cells, it undergoes a condensation reaction facilitated by intracellular GSH and the furin protein, followed by cross-linking to form Au NPs aggregates (Fig. 1.13b). The aggregation of Au NPs showed a red shift in the LSPR wavelength and high photothermal efficiency under 808-nm laser irradiation (Fig. 1.13c). Furthermore, animal experiments have demonstrated that AuNP@1 can effectively accumulate in MDA-MB-468 cells at the tumor site. Under laser irradiation, it significantly raises the temperature at the tumor site more rapidly and to a greater extent than other groups. Similarly, Zhang et al. developed a pH-triggered self-assembly Au NPs for photoacoustic (PA) imaging-guided PTT (Zhang et al. 2020c). The Au NPs were coated with a zwitterionic surface that possesses mixed charges, exhibiting pH-sensitivity and long blood circulation time. After reaching the tumor site, Au NPs rapidly aggregate into larger particles within the acidic tumor microenvironment. Aggregated Au NPs exhibited enhanced LSPR, making them suitable not only for effective PTT but also for enhancing PA imaging in tumors upon NIR laser irradiation.

Anisotropic noble metal-based nanomaterials, which exhibit the highest PCE, are widely used as PTAs for PTT. Valle et al. reported a platinum-modified Au NSs (Pt-AuNS) for the treatment of multidrug-resistant tumors (del Valle et al. 2020). Pt-AuNS exhibits high PCE and a strong capacity to absorb near-infrared light, allowing it to effectively ablate localized tumors through efficient photothermal heating. Meanwhile, Pt-AuNS can release toxic Pt and Au species under NIR irradiation, resulting in increased levels of lipid peroxidation associated with oxidative stress-induced lethal injuries. In addition, they found that Pt-AuNS can enhance cell damage through ferroptosis by inducing more effective depletion of GSH and inactivation of glutathione peroxidase 4. Compared to conventional drug-resistant cancer therapies, Pt-AuNS, which respond to NIR light as ferroptosis-inducing agents, demonstrate significant potential for treating chemo-resistant cancers. Liu et al. synthesized three-dimensional Ag nanocubes@branched Au shell nanostructure (Ag@Au) using an amino acid-encoded Stranski-Krastanov mode for tumor therapy (Liu et al. 2023). Due to the highly branched structure, Ag@Au has significant PCE

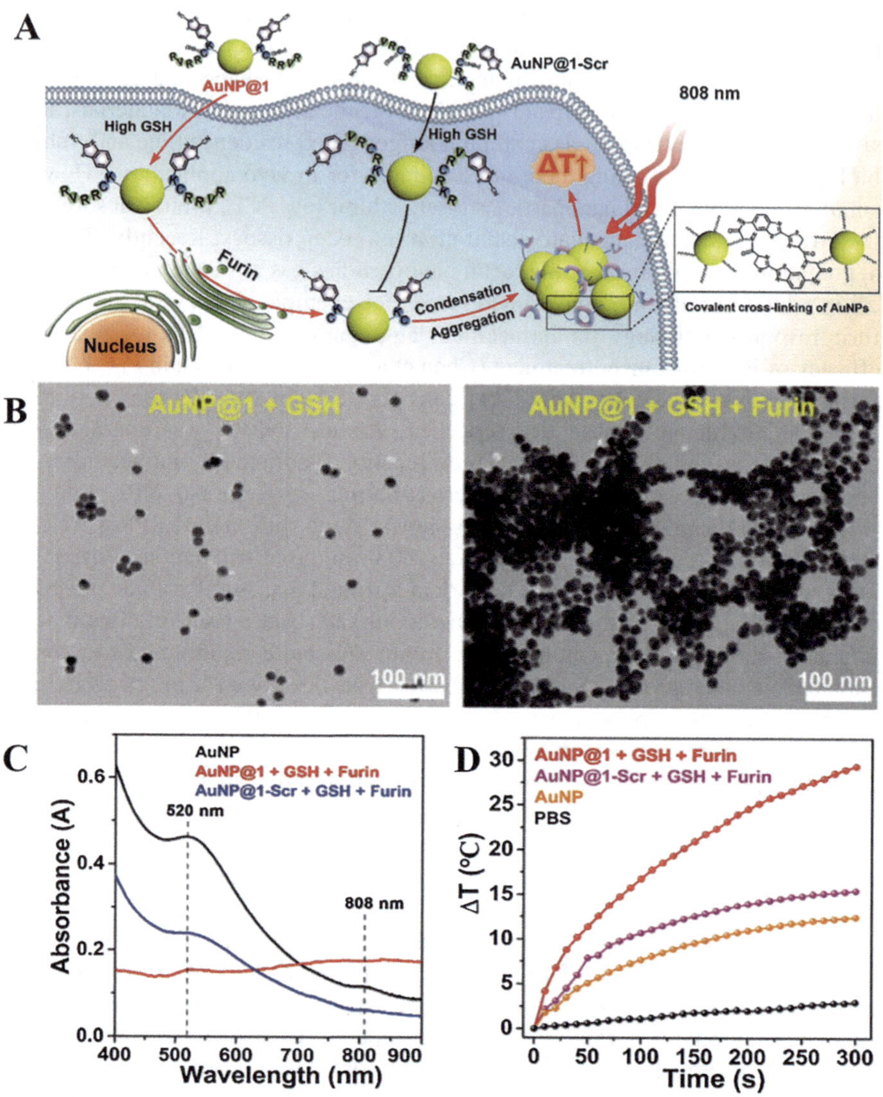

Fig. 1.13 (**a**) Schematic illustration of intracellular furin-induced Au NPs aggregation for PTT. (**b**) TEM images of Au NPs aggregation induced by furin. (**c**) Absorption and (**d**) temperature changes of Au NPs after aggregation (Chen et al. 2020)

(87.28%). In vitro and in vivo experiments demonstrate that Ag@Au exhibits significant antitumor therapeutic efficacy through laser irradiation.

Although anisotropic noble metal-based nanomaterials exhibit good PCE; however, their cellular uptake is lower than that of conventional nanoparticles. Anisotropic noble metal-based nanomaterials can be functionalized to selectively target lesion tissues for cancer diagnosis and treatment. By coating noble metal-based nanomaterials with various types of cell membranes, such as leukocytes, red blood cells, and platelets, researchers have developed diverse nanoplatforms that exhibit distinct features and functions. For example, Liu et al. developed a biomimetic nanoplatform (GMCMs) composed of Au NRs and manganese porphyrin, encapsulated in 4T1 cell membranes for surface-enhanced Raman scattering (SERS) and PA imaging-guided synergistic PTT and PDT (Fig. 1.14a) (Liu et al. 2024). The encapsulation of 4T1 cell membranes allows GMCMs to specifically target 4T1 cells through homologous targeting, while also protecting manganese porphyrin from detaching from the Au NRs. The biomimetic GMCMs demonstrate excellent biocompatibility, stability, water dispersibility, and strong NIR-I absorption. In addition, under 808-nm laser irradiation, GMCMs can produce significant thermal effects and generate a substantial amount of $^1O_2$. Furthermore, the presence of manganese ions not only quenches the fluorescence of porphyrin, enhancing the SERS imaging effect, but also depletes cellular GSH, thereby increasing the yield of $^1O_2$. GMCMs also function as high-performance contrast agents for in vivo PA imaging, facilitating subsequent animal therapies. In vitro and animal experiments demonstrate that GMCMs effectively inhibit the growth of 4T1 cells and tumors through SERS/PA imaging-guided PTT/PDT. Similarly, Guo et al. developed an Au NRs-based nanoplatform for highly efficient PTT/gene combination therapy (Guo et al. 2023). This nanoplatform was synthesized by loading siRNA-KRAS into the polyethyleneimine-coated Au NRs and incorporating gadolinium ions onto the surface of the Au NRs through electrostatic interactions. Then, the red blood cell membrane encases the Au NRs to improve biocompatibility and promote long-term systemic circulation. Additionally, the nanoplatform was functionalized with Plectin-1 antibodies, which are often overexpressed in pancreatic tumors, to enhance targeting. The results indicated that this Au NRs-based nanoplatform shows good antitumor efficacy for pancreatic cancer.

In addition, peptides play a crucial role in enhancing the biocompatibility, stability, and targeting capabilities of noble metal-based nanoparticles. Ma et al. designed mitochondria-targeting Pt NPs (TPP-Pt) composed of lipophilic triphenylphosphine (TPP) peptides, which serve as highly effective mitochondrial ligands with improved biocompatibility (Ma et al. 2020). The ultra-small Pt NPs as the core to effectively induce cellular hyperthermia under NIR-II irradiation. The membrane potential of tumor cells and their mitochondria is approximately 60 mV lower than that of normal cells. This difference allows TPP-Pt to preferentially localize in tumor mitochondria, enhancing the proton coupling effect between particles. Subsequently, localized NIR-II irradiation can induce high temperatures in situ within thermosensitive mitochondria, effectively ablating tumors while preserving low systemic cytotoxicity. Furthermore, due to their ultra-small size, TPP-Pt can be efficiently eliminated from the body through the kidneys, thereby minimizing concerns about long-term toxicity. Yin et al. reported a NIR-responsive multifunctional nanoplatform for tumor-targeting and SERS imaging-guided chemo-PTT for cancer therapy (Yin et al. 2023). As shown in Fig. 1.14b, the nanoplatform (pAS@AuNCs) is composed of Au NSs as the core and an outer shell of Au NCs. To enhance the tumor-targeting ability, pAS@AuNCs was modified with PEG terminated with an RGD

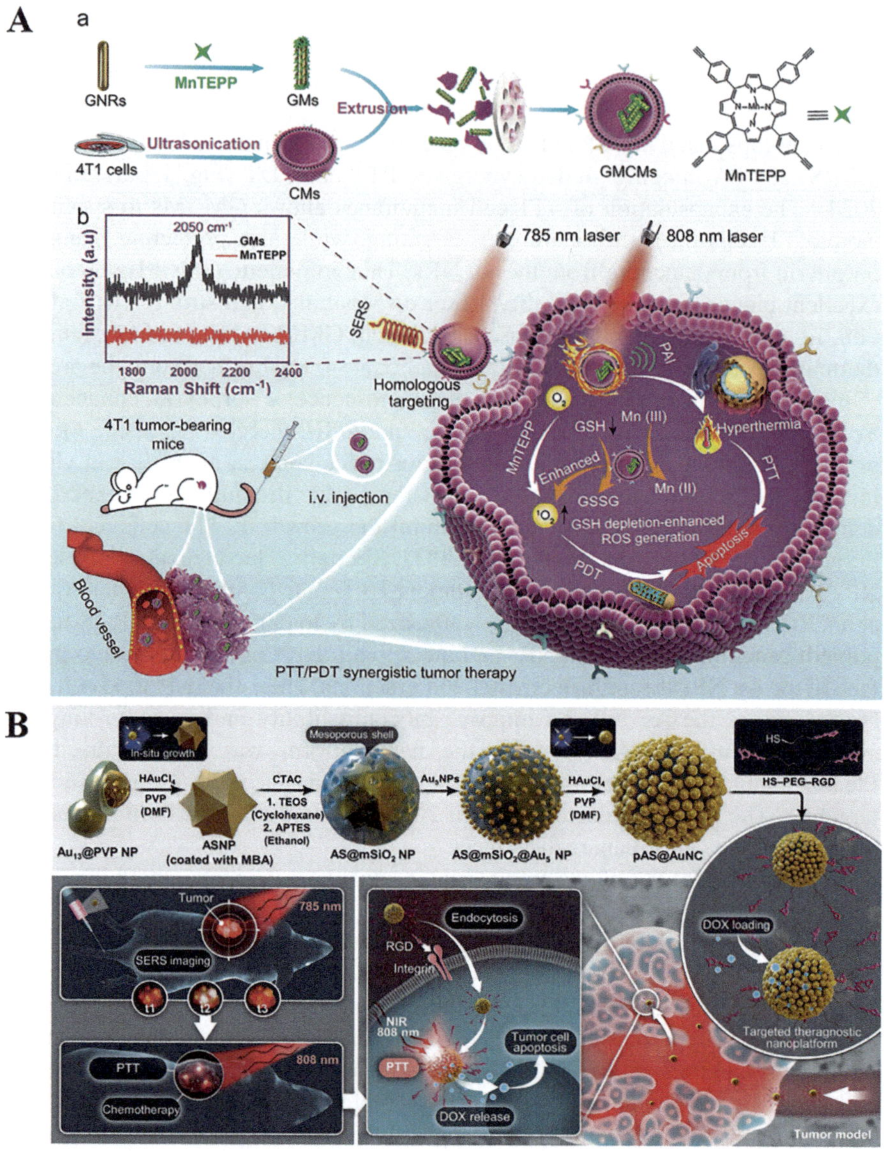

Fig. 1.14 (**a**) Schematic illustration of the main synthesis process, the SERS effect and the mechanism underlying the PTT/PDT antitumor properties of GMCMs (Liu et al. 2024). (**b**) Schematic illustration of the synthesis route and application of the multilayered mesoporous gold nanoarchitecture (RGD/DOX-pAS@AuNCs) for SERS imaging-guided synergistic (Yin et al. 2023)

peptide. Meanwhile, this surface modification can improve biocompatibility, enhance water-solubility and colloidal stability. Importantly, the presence of multiple hotspots between adjacent nanoparticles in pAS@AuNCs, along with the closely coupled Au NCs shell and Au NS core, is expected to facilitate effective plasmonic coupling. This significantly enhances the SERS signal and improves the NIR induced photothermal conversion efficiency to 85.5%. Furthermore, the LSPR peak of pAS@AuNCs is centered around 808 nm in the NIR range. This characteristic facilitates deep tissue penetration while minimizing overheating of non-target tissues, thereby enabling non-invasive localized chemotherapy-PTT. It is important to note that through SERS imaging, pAS@AuNCs can effectively treat tumors within 1 month without recurrence.

Although PTT exhibits significant therapeutic effects on cancer cells, the application of high laser power and extended exposure can induce stress in the cells, resulting in the production of excessive ROS. These ROS can interact with neighboring healthy tissues and off-target sites, leading to unintended cell death and side effects. Therefore, the development of noble metal nanoparticles with intrinsic reactive oxygen species scavenging properties is essential for improving the efficacy of PTT. For instance, Wang et al. synthesized polyethylene glycol-modified iridium tungsten oxide (IrWOx-PEG) nanoparticles to modulate hypoxia, exert anti-inflammatory effects, and treat metastatic breast cancer through PTT (Fig. 1.15a) (Wang et al. 2022). $IrWO_x$-PEG exhibits good absorption in the NIR range, allowing it to selectively destroy in situ breast cancer cells through PTT. In addition, $IrWO_x$-PEG exhibits multienzyme-like activity by decomposing H_2O_2 to produce O_2, thereby alleviating the hypoxic tumor microenvironment. This action diminishes the inflammatory response induced by PTT and regulates the expression of proteins associated with metastatic pathways in breast cancer cells, thereby inhibiting tumor metastasis to the lungs. Notably, under the guidance of dual-phase imaging, the PTT induced by $IrWO_x$-PEG completely eradicated in situ breast cancer in 4T1 tumor-bearing mice without causing significant systemic toxicity (Fig. 1.15b). Recently, Singh et al. developed a novel bimetallic palladium nanocapsule (Pd Ncap) that features a solid gold bead as the core and a thin palladium shell, intended

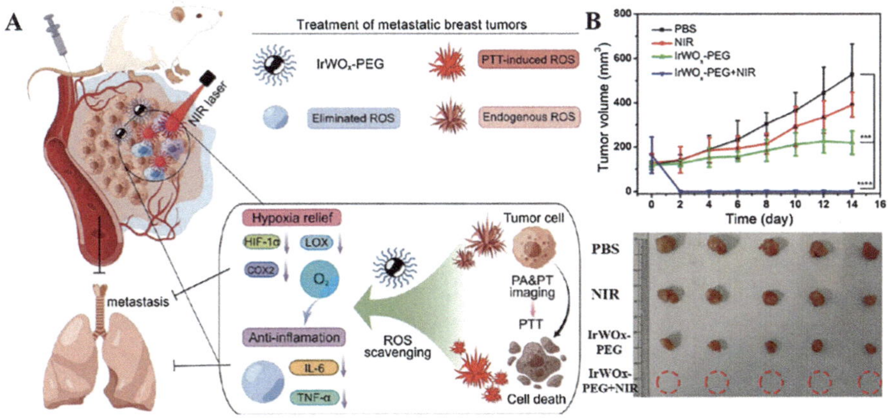

Fig. 1.15 (**a**) Schematic illustration of IrWOx-PEG-mediated hypoxic regulation for duplex imaging-guided PTT. (**b**) Tumor volume curves and digital tumor photographs (Wang et al. 2022)

for use in plasmonic photothermal therapy (PPTT) (Singh et al. 2023). The Pd Ncap exhibited excellent photothermal stability and notable anticancer activity at very low laser power densities. When utilizing a 1064-nm laser, the photothermal conversion efficiency is approximately 49%. Additionally, palladium has the capacity to scavenge reactive oxygen species, thereby safeguarding healthy cells and tissues from the harmful effects of excessive reactive oxygen generated during the PPTT process. Notably, at extremely low concentrations of nanoparticles and laser power densities, this nanostructure demonstrates high photothermal conversion efficiency and lethal effectiveness, making it a potent phototherapeutic agent for the treatment of various types of cancer, while also providing protective capabilities against ROS.

In addition, one of the primary concerns during photothermal therapy (PTT) is the overexpression of heat shock proteins (HSPs), such as heat shock protein 70 (HSP70) and heat shock protein 90 (HSP90), which enhance the thermal resistance of cancer cells. This protective effect may necessitate an increase in the dosage of photothermal agents to achieve higher temperatures capable of effectively killing cancer cells. Such adjustments could potentially impact the efficacy of PTT, leading to thermal damage in normal tissues and an increased risk of recurrence and metastasis. With advancements in gene therapy (GT) technologies, the clustered regularly interspaced short palindromic repeats (CRISPR)/CRISPR-associated protein 9 (CRISPR/Cas9) gene editing system has emerged as an effective tool for precisely regulating gene expression, achieving significant suppression of heat-sensitive protein (HSP) levels. Noble metal-based nanomaterials have garnered considerable attention as targeted carriers for delivering the CRISPR-Cas9 system. For instance, Li et al. developed hypoxia-responsive Au NRs-based nanocomposites for the targeted delivery of CRISPR-Cas9, facilitating mild PTT through tumor-targeted gene editing (Li et al. 2021). As shown in Fig. 1.16a, the Cas9/sgRNA ribonucleoprotein complexes (Cas9) are covalently cross-linked to Au NRs through a

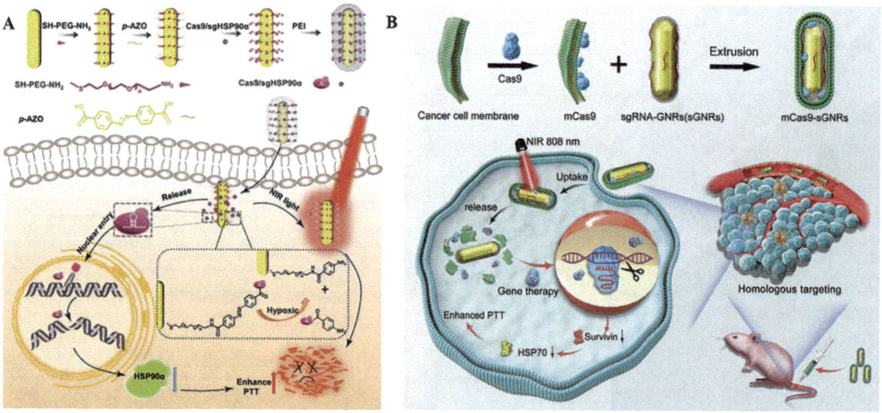

Fig. 1.16 (**a**) Schematic illustration of the hypoxia-responsive CRISPR-Cas9 system for gene editing and mild PTT (Li et al. 2021). (**b**) Schematic illustration of mCas9-sGNRs for combined PTT/gene therapy

hypoxia-responsive azobenzene linker (p-AZO) under normal oxygen pressure. In the hypoxic tumor microenvironment, the p-AZO group is selectively reduced by overexpressed reductases, leading to the release of Cas9. The Cas9 can specifically knock out the HSP90α gene, significantly reducing the thermal tolerance of tumor cells. Under mild photothermal therapy conditions, APACPs significantly inhibit tumor cell proliferation, achieving effective tumor ablation both in vitro and in vivo. In another study, Huang et al. proposed a cancer membrane-derived nanocarrier (mCas9-sGNRs) for the effective delivery of Au NRs and CRISPR/Cas9 for synergistic photothermal and gene therapy (Huang et al. 2022). In this approach, the Cas9 protein can be effectively incorporated into the cell membrane (denoted mCas9) through electrostatic interactions (Fig. 1.16b). Then, mCas9 was further used to coat siRNA loaded Au NRs to prepared mCas9-sGNRs. With the aid of cancer cell membranes, mCas9-sGNRs not only possess the ability to target tumor cells and evade the immune response, but also effectively encapsulate Cas9 for gene editing. The mCas9-sGNRs can achieve significant antitumor effects through the gene editing of survivin, which enhances anticancer activity and reduces the heat tolerance of cancer cells induced by Au NRs during PTT due to the downregulation of HSP70.

1.5.3 Sonodynamic Therapy

Sonodynamic therapy (SDT) is a treatment method that employs ultrasound (US) with a frequency of 20 kHz or higher to activate sonosensitizers, resulting in the generation of ROS that induce cellular damage. This approach is a promising non-invasive and site-directed strategy for the eradication of solid tumors. Compared to PDT and PTT, SDT shows greater potential for treating deep-seated solid tumors due to the low tissue attenuation coefficient of ultrasound and its non-radiative nature, which allows it to penetrate deeper tissues (up to 11 cm). However, the efficacy of SDT is significantly influenced by the biological potency of the sonosensitizers. Therefore, there is a critical need for the development of biocompatible, high-efficiency ROS-generating, and easily synthesized noble metal-based nanosenosensitizers to facilitate the translation of SDT from laboratory research to clinical applications.

The tunable optical and thermal properties, low toxicity, excellent cellular uptake, and anti-angiogenic characteristics of gold-based nanomaterials render them appealing candidates for SDT. Recently, Loke et al. reported a promising SDT nanomaterial, alginate-coated Au NRs (Au NRs^{ALG}), for the effective SDT of human breast cancer cells (Loke et al. 2023). The Au NRs^{ALG} exhibited stability under ultrasound irradiation and preserved its structural integrity across three irradiation cycles. As shown in Fig. 1.17a, Au NRs^{ALG} as potent nanosonosensitisers and Fenton-like agents, can produce a high amount of ROS under ultrasound irradiation. It was found that Au NRs^{ALG} significantly enhanced the cavitation effect during ultrasound irradiation, resulting in $^1O_2$ production that was 3–8 times higher than those of other commercially reported TiO_2 nanosonosensitisers. In vitro studies on

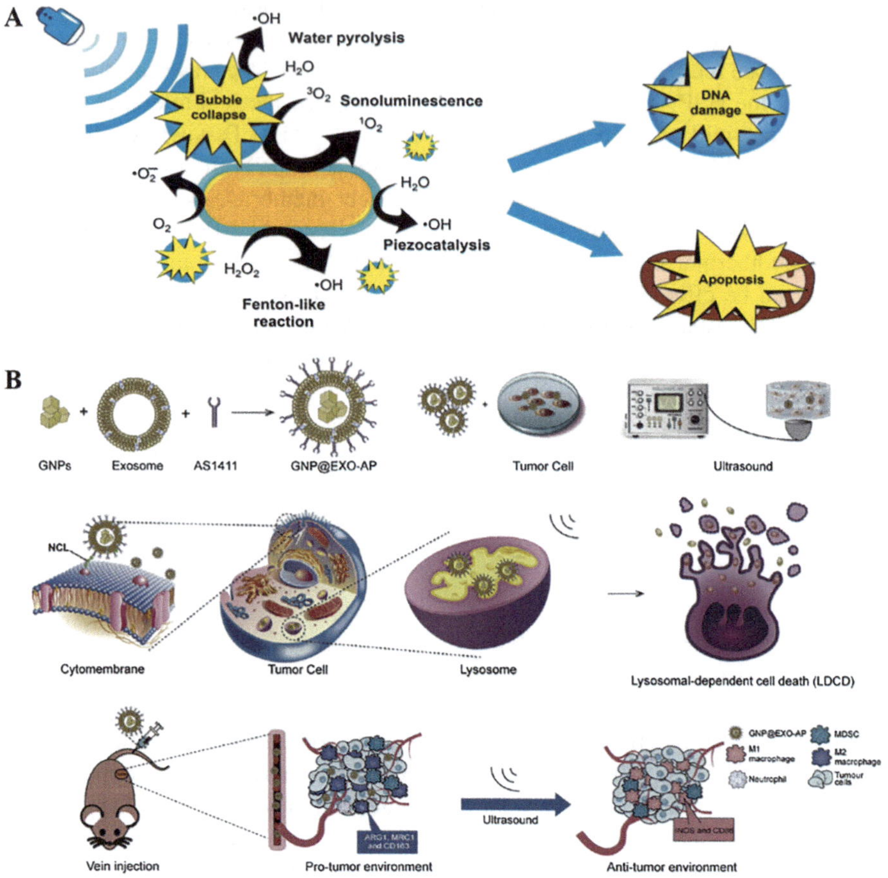

Fig. 1.17 (**a**) Schematic illustration for the antitumor mechanism of Au NRs^{ALG} under ultrasound irradiation (Loke et al. 2023). (**b**) Schematic illustration of the GNP@EXO-Aps for SDT (Tang et al. 2023)

human MDA-MB-231 breast cancer cells revealed that Au NRs^{ALG}-mediated SDT induces cancer cell death primarily through apoptosis. In addition, protein expression analysis demonstrated that Au NRs^{ALG}-mediated SDT can also induce DNA damage in cancer cells and may trigger apoptosis via the mitochondrial pathway. In another study, Tang et al. developed an acoustic-actuated, tumor-targeted nanomachine (GNP@EXO-APs) utilizing Au NPs encapsulated with macrophage-derived exosomes modified by AS1411 aptamers for cancer SDT (Fig. 1.17b) (Tang et al. 2023). Under ultrasound stimulation, GNP@EXO-APs generate significant mechanical stress in cancer cells, resulting in lysosome-dependent cell death. It is important to note that ultrasound irradiation can effectively stimulate the antitumor activity of macrophages within the tumor microenvironment, reprogramming tumor-associated macrophages into antitumor phenotypes in the presence of ultrasound. This innovative ultrasound tumor treatment method demonstrates remarkable tumor-suppressing effects.

In recent years, iridium(III) complexes have garnered significant attention due to their exceptional stability, straightforward synthesis, and adjustable photophysical properties. Several sonosensitizers based on iridium(III) complexes have been developed for ultrasound-activated cancer therapy. For instance, Tang et al. developed a NIR fluorescent dinuclear iridium(III) nanoparticle for immunogenic sonodynamic therapy (ISDT) (Fig. 1.18a) (Tang et al. 2024). The dinuclear iridium(III) complex, Ir-DPP-Ir, consists of a diketopyrrolopyrrole core and two iridium (III) complexes that incorporate phenylpyridine ligands. Then, Ir-DPP-Ir was assembled with the commercially available amphiphilic polymer DSPE-PEG_{2000} to form NanoIr. Under ultrasound irradiation, NanoIr can generate $^1O_2$ and ·OH, effectively inducing immunogenic cell death (ICD) in vitro. After intravenous injection, the accumulation of NanoIr at the tumor site was observed, which can be monitored using fluorescence imaging for ultrasound irradiation. Under ultrasound irradiation, NanoIr effectively eliminated tumor cells and induced ICD. Furthermore, the dead cells can release pro-inflammatory factors that promote the activation of $CD8^+$ T cells and facilitate ISDT. Similarly, Zhu et al. developed a novel sonosensitizer (AuNPs@Ir1) by combining an Ir(III) complex with ultrasmall Au NPs for effective SDT (Zhu et al. 2022). AuNPs@Ir1 exhibit excellent biocompatibility and served as both a potent sonosensitizer for generating intracellular $^1O_2$ and an efficient sonocatalyst for the oxidation of intracellular 1,4-dihydronicotinamide adenine dinucleotide (NADH). The production of $^1O_2$ and the oxidation of NADH induced by US disrupt the redox balance in cancer cells, resulting in significant in vitro anticancer activity and ultimately leading to cancer cell death. Due to the effective passive retention of AuNPs@Ir1 in tumors following intravenous injection, there was a

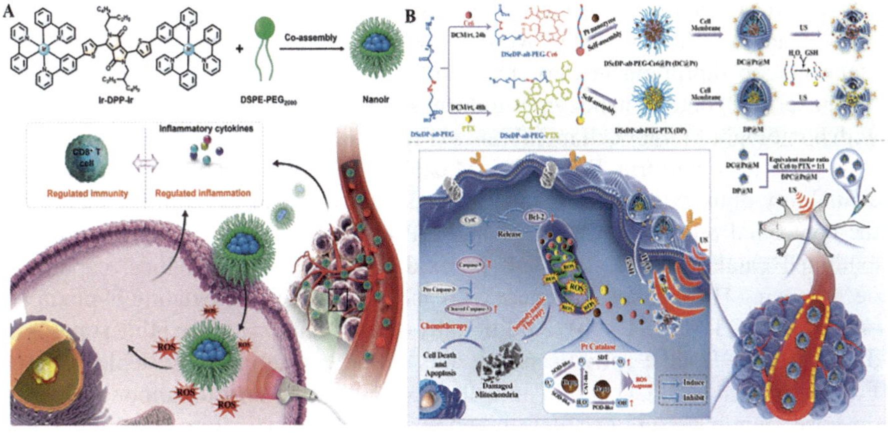

Fig. 1.18 (**a**) Schematic illustration of dinuclear iridium(III) nanoparticle (NanoIr) for ISDT (Tang et al. 2024). (**b**) Schematic diagram of DPC@Pt@M for synergetic sonochemotherapy on colon cancer (Zhao et al. 2023)

further inhibition of tumor growth in vivo under ultrasound stimulation, achieving an inhibition rate of approximately 70% without causing long-term toxicity to other organs.

Some studies have demonstrated that the interference of SDT with tumor physiology and microenvironmental conditions enhances the thermal sensitivity of cancer cells, thereby improving the therapeutic efficacy of PTT. Furthermore, during PTT treatment, blood flow significantly increases, which correspondingly elevates oxygen supply and enhances the sensitivity of tumor cells to oxygen-dependent SDT. By leveraging this effect, the synergistic combination of PTT and SDT can overcome the inherent limitations of single treatment modalities while improving overall treatment efficiency. For example, Shen et al. reported a nanoplatform (Ir-B-TiO_2@CCM) composed of black titanium nanoparticles functionalized with an Ir(III) complex and cancer cell membranes, designed for effective NIR-II PTT and SDT (Shen et al. 2021). Ir-B-TiO_2@ CCM can generate effective heat under NIR-II irradiation and catalyze the formation of ROS when exposed to ultrasound. In addition, Ir-B-TiO_2@ CCM have been demonstrated to selectively localize in mitochondria and preferentially accumulate in cancer cells rather than non-cancer cells. After exposure to NIR-II irradiation and ultrasound, particularly when used in combination, nanoparticles have shown the capability to identify tumor sites through multiphoton, photothermal, or photoacoustic imaging in various in vitro and in vivo models. Additionally, they have demonstrated the ability to completely eradicate tumors through the synergistic effects of NIR-II photothermal and sonodynamic therapy.

Due to the hypoxic conditions of the complex tumor microenvironment, the therapeutic efficacy of SDT in cancer is often limited. Increasing the production of ROS can fundamentally enhance the sensitivity of SDT. With advancements in nanotechnology, gold-based and platinum-based nanomaterials can catalyze the production of O_2 from H_2O_2, alleviating tumor hypoxia, and can be utilized in conjunction with ultrasound to increase ROS generation. For example, Zhao et al. reported a biomimetic nanoplatform (DPC@Pt@M) that involves loading the prodrugs 3,3′-polyethylene glycon dipropionate-Ce6 (DC) and 3,3′-polyethylene glycon dipropionate paclitaxel (DP) onto Pt nanozymes to form stable micelles, which are then modified with homologous tumor cell membranes (Fig. 1.18b) (Zhao et al. 2023). As a redox-responsive nanoplatform, DPC@Pt@M exhibits excellent water solubility and bioavailability, along with effective in vitro and in vivo targeting capabilities, facilitating the controlled release of Ce6, DC, and DP at the tumor site. The Pt nanozymes exhibited catalase (CAT)-like and peroxidase (POD)-like activities, capable of decomposing H_2O_2 in situ to generate O_2 and ·OH, thereby alleviating hypoxia and enhancing SDT (Fig. 1.17b). Additionally, the Pt nanozymes exhibit superoxide dismutase (SOD)-like activity, which aids in the consumption of depleted H_2O_2. Furthermore, the DPC@Pt@M micelles exhibit a homologous tumor membrane disguise, demonstrating excellent biocompatibility and targeting capabilities both in vitro and in vivo. Under ultrasound irradiation, the DPC@Pt@M significantly increase the production of ROS and effectively induce apoptosis, thereby enhancing the efficacy of sonochemotherapy for colon cancer.

1.5.4 Radiotherapy

Radiotherapy (RT) is a non-invasive treatment that targets cancer cells and effectively reduces the risk of tumor recurrence. RT can induce the decomposition of water molecules into ROS under ionizing radiation, which subsequently attack DNA molecules, leading to anti-proliferative effects and apoptosis of tumors. The current efficacy of RT requires high doses of X-rays or γ-ray beams. While these treatments have demonstrated effectiveness against cancer cells, they also produce significant toxic side effects on normal tissues. Therefore, radiosensitizers can enhance the sensitivity of tumors to radiation through physical, chemical, and biological interactions, thereby reducing the required radiation dose. Many high-Z element-based nanomaterials (such as gold and silver) have been used as radiosensitizers. Research has shown that nanomaterials containing high atomic number (Z) elements and hollow structures can improve the deposition efficiency of X-rays and increase the energy deposition from ionizing radiation, thereby triggering the Compton effect, which involves the emission of photoelectrons such as Compton electrons and Auger electrons, ultimately inducing cell apoptosis and necrosis.

Gold-based nanomaterials have been extensively utilized as radiosensitizers to enhance the production of ROS under X-ray radiation due to their high Z, direct photoelectric interactions, inert chemical properties, and Auger effects. For instance, Luo et al. developed prostate cancer-targeted Au NPs (Au-Gd(III)-PSMA NPs) for magnetic resonance (MR) -guided RT by conjugating prostate-specific membrane antigen (PSMA) targeting ligands and Gd(III) complexes to the surface of the Au NPs (Fig. 1.19a) (Luo et al. 2020). Both Au atoms and Gd(III) atoms possess a high Z number, which allows them to function as effective radiosensitizers. Meanwhile, the Gd(III) complex attached to the surface of Au NPs can enhance magnetic resonance sensitivity by approximately fourfold. Under the influence of PSMA, Au-Gd(III)-PSMA NPs exhibits excellent selectivity for prostate cancer cells that express PSMA, significantly enhancing magnetic resonance contrast and in vitro radiosensitization effects. Notably, upon systemic administration, Au-Gd(III)-PSMA NPs exhibited excellent tumor accumulation, enhanced magnetic resonance contrast, and significant amplification of in vivo radiation doses (Fig. 1.19b, c). Moreover, these particles can shield surrounding healthy tissues from radiation damage while facilitating the destruction of cancer cells, thereby reducing the necessary therapeutic X-ray dose. Scaramuzza et al. synthesized polymer-coated Au-B NPs for the combination of boron neutron capture therapy (BNCT) and X-ray RT, enabling the localization and quantification of radiosensitizers in tissues through computed tomography (CT) imaging (Scaramuzza et al. 2023). The Au-B NPs were prepared through a laser ablation process in a liquid medium and subsequently coated with biocompatible polymers, such as dexamethasone (DEX) or thiolated polyethylene glycol (PEG). The effective absorption of X-rays by Au-B NPs can be utilized for X-ray radiosensitization under irradiation conditions. In vitro study using a cancer cell model demonstrated the superior ability of Au-B NPs as radiosensitizers. Due to boron is present in high abundance in the Au-B NPs, they were

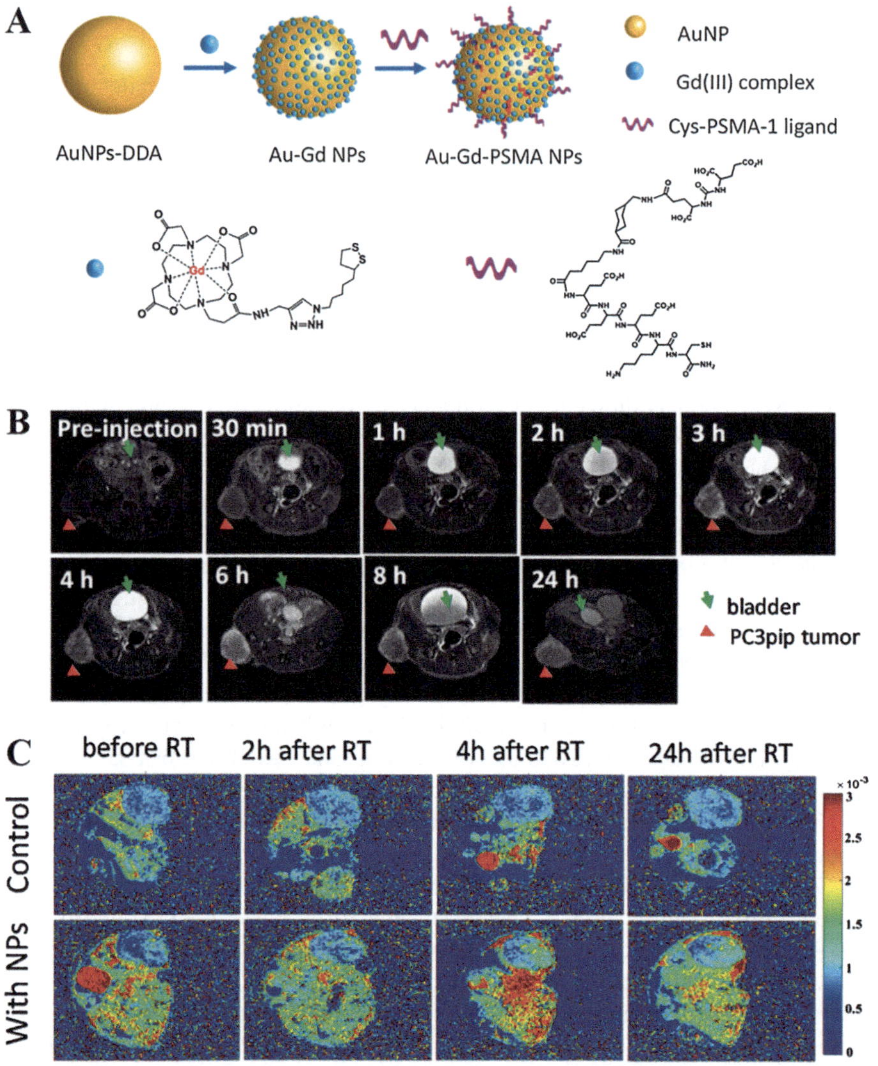

Fig. 1.19 (**a**) A schematic illustration of the synthesis process of Au-Gd(III)-PSMA NPs. (**b**) In vivo tumor targeting of Au-Gd(III)-PSMA NPs and MR imaging of mice with PC3pip tumor. (**c**) Apparent diffusion coefficient maps of mice injected with PBS or Au-Gd(III)-PSMA NPs (Luo et al. 2020)

used for BNCT. The results demonstrate that a fivefold improvement in BNCT performance can be easily achieved by utilizing a composition enriched with ^{10}B. By combining CT imaging, Au-B NPs can achieve effective synergistic BNCT and RT.

As a type of ultra-small gold nanomaterial (size <2 nm), Au NCs not only maintain the unique radiotherapy-sensitizing properties associated with gold-based nanomaterials but also evade clearance by the reticuloendothelial system, thereby minimizing toxic side effects. For instance, Cao et al. reported an X-ray-driven carbon monoxide (CO)-releasing nanoplatform (HA@AuNC@CO) for gas and RT synergistic therapy in tumor treatment (Cao et al. 2024). As shown in Fig. 1.20a, HA@AuNC@CO was synthesized through self-assembly involving cyclodextrin-protected gold nanoclusters (CD-Au NCs), adamantane-modified hyaluronic acid (HA-ADA), and adamantane-modified iron-CO complexes (Ada-CO). Under X-ray irradiation, the encapsulated CD-Au NCs effectively generate cytotoxic ROS, which not only directly inactivate cancer cells but also induce the in-situ generation of CO gas from the Ada-CO, thereby inhibiting key enzymes in the glycolytic pathway. Furthermore, the incorporation of CD-Au NCs enhances the nanoplatform's NIR II imaging capability, allowing for the monitoring of its biodistribution throughout the treatment process. In vitro and in vivo studies demonstrate that HA@AuNC@CO can specifically target CD44-expressing pancreatic cancer cells, effectively inhibiting tumor growth under low-dose X-ray radiation. In another study, Wu et al. developed a nanoplatform (denoted as AuNC-ASON) that utilizes pH-responsive Au NCs loaded with antisense oligonucleotides targeting biotin mRNA to enhance cancer RT and gene therapy (Wu et al. 2021). The AuNC-ASON exhibited structural disassembly when exposed to acidic pH, transitioning from Au NPs aggregates (approximately 80 nm) to Au NCs (around 2 nm). This size transformation in response to the tumor microenvironment facilitated the release of the loaded antisense oligonucleotides, thereby enabling gene silencing. In vitro experiments demonstrated that, due to the synergistic effect of convertible gold nanocluster radiosensitizers and surviving gene interference, AuNC-ASON could amplify and enhance the radiosensitivity of tumor cells, achieving a sensitization ratio of approximately 1.81. Notably, the size conversion capability facilitated high tumor retention and permeability, as well as renal metabolism of AuNC-ASON in vivo. Additionally, the interference with the surviving gene further enhanced the

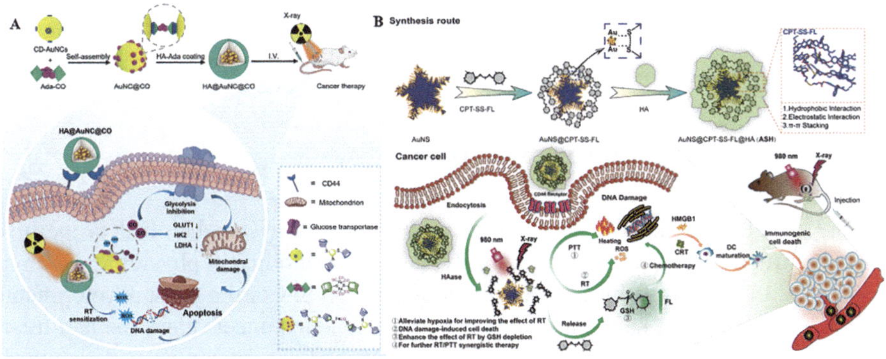

Fig. 1.20 (**a**) Schematic illustration of HA@AuNC@CO for RT/gas synergistic therapy (Cao et al. 2024). (**b**) Schematic illustration of the synthesis process of ASH nanoplatform and its synergetic PTT and RT (Hou et al. 2023)

radiosensitivity of cancer cells, collectively leading to improved outcomes in tumor RT and gene therapy.

In addition, various studies have shown that noble metal-based nanomaterials, which integrate RT and PTT, exhibit a remarkable synergistic effect in suppressing tumor growth and inhibiting metastasis. For example, Hou et al. designed a nanocomposite platform (ASH) for synergistic PTT and RT by loading a disulfide-linked camptothecin-fluoroborate prodrug (CPT-SS-FL) onto Au NSs and subsequently encapsulating it with hyaluronic acid (HA) (Fig. 1.20b) (Hou et al. 2023). The ASH effectively delivers the prodrug and targets the CD44 receptors on the surface of tumor cells. Endogenously overexpressed GSH in tumor cells can disrupt disulfide bonds, thereby activating the prodrug and releasing the radiosensitizer camptothecin (CPT) along with the fluorescent imaging agent rhodamine derivatives. The released rhodamine derivatives enable real-time tracking of the precise release location of the radiosensitizer CPT within tumor cells. Additionally, due to Au has high atomic number ($Z = 79$), Au NSs demonstrates strong X-ray absorption and deposition capabilities. Au NSs can also alleviate hypoxia in the tumor microenvironment through mild photothermal therapy. Therefore, by leveraging GSH depletion, the high atomic number of Au, and alleviating hypoxia, the designed ASH nanoplatform can effectively induce robust immunogenic cell death at the tumor site. Chen et al. constructed a core-shell nanoparticle, UiO-66-NH2@AuNS. Subsequently, the solid gold shell was etched to form hollow gold nanoshells (HAuNS), which were further modified with biotin-PEG-SH (PEG-bio) to obtain HAuNS@PEG-bio (Chen et al. 2023a, b). HAuNS@PEG-bio exhibits effective PTT performance in the near-infrared II region, where the temperature increase at the tumor site enhances blood circulation and alleviates the hypoxic conditions of the tumor microenvironment. Additionally, due to the high atomic number of gold and its hollow structure, HAuNS possesses strong X-ray absorption and deposition capabilities. The dual radiotherapy sensitization effect of high atomic number gold and PTT alleviating hypoxia can induce immunogenic cell death (ICD) in breast cancer cells. Notably, HAuNS@PEG-bio can improve tumor tissue hypoxia while enhancing X-ray deposition for amplified radiotherapy, achieving effective tumor treatment through the combination of NIR-II PTT and active targeted therapy.

1.6 Conclusions and Perspectives

Noble metal-based nanomaterials exhibit intriguing physicochemical properties, diverse structures, good biocompatibility, and high stability in physiological environments. In summary, this chapter provides an overview of the latest developments in noble metal-based nanomaterials for applications in cancer diagnosis and treatment. The preparation methods for noble metal-based nanomaterials include physical, chemical, and biological methods. Subsequently, the optical properties of these nanomaterials are discussed. Finally, their applications in biosensing and tumor therapy are examined in detail. Although noble metal-based nanomaterials have achieved significant advancements, several key issues and challenges remain unresolved and need to be addressed. Firstly, the main research focus so far has been on gold and silver due to their ease of functionalization, stability, and biological properties. Other noble metal nanoparticles, such as Pt and Pd, have been mixed with gold for various biomedical applications, including cancer treatment. In contrast, Pd, Ir, Rh, Os, and Ru are still catching up. In addition, precisely controlling the surface functionalization of noble metal-based nanomaterials is another important area of research. In many instances, additional ligands, such as active targeting molecules, drugs, and PSs, bind to these nanoparticles, inevitably influencing bio-nano-interactions and ultimately affecting the efficacy of nanomedicine. Therefore, the development of noble metal-based nanomaterials enables precise control over their surface modifications is very important. This capability is not only significant for optimizing nanomedicine but also ensures high reproducibility, which is crucial for clinical translation. Finally, before the application of noble metal-based nanomaterials in clinical and medicinal settings, more systematic research is required to assess their in vivo biocompatibility, biological distribution, and pharmacokinetic behavior. We are hopeful that this chapter will contribute to the future advancements of noble metal-based nanomaterials in cancer diagnosis and treatment.

References

Adak T, Swain H, Munda S et al (2021) Green silver nano-particles: synthesis using rice leaf extract, characterization, efficacy, and non-target effects. Environ Sci Pollut Res Int 28(4):4452–4462

Ahn EY, Park Y (2020) Anticancer prospects of silver nanoparticles green-synthesized by plant extracts. Mater Sci Eng C Mater Biol Appl 116:111253

Ali J, Ali N, Wang L et al (2019) Revisiting the mechanistic pathways for bacterial mediated synthesis of noble metal nanoparticles. J Microbiol Methods 159:18–25

Arvizo RR, Bhattacharyya S, Kudgus RA et al (2012) Intrinsic therapeutic applications of noble metal nanoparticles: past, present and future. Chem Soc Rev 41(7):2943–2970

Barbosa S, Agrawal A, Rodriguez-Lorenzo L et al (2010) Tuning size and sensing properties in colloidal gold nanostars. Langmuir 26(18):14943–14950

Beik J, Maziar K, Zohreh K et al (2019) Gold nanoparticles in combinatorial cancer therapy strategies. Coord Chem Rev 387:299–324

Cao L, Yating Y, Yanlin Z et al (2024) X-ray-triggered CO-release from gold nanocluster: all-in-one Nanoplatforms for cancer targeted gas and radio synergistic therapy. Adv Mater 36(25):2401017

Chakraborty I, Parak WJ (2019) Protein-induced shape control of noble metal nanoparticles. Adv Mater Interfaces 6(6):1801407

Chang H, Karan NS, Kwangsoo S et al (2020) Highly fluorescent gold cluster assembly. J Am Chem Soc 143:326–334

Chen Y, Chenjie Z, Douglas KR et al (2015) Tuning the magic size of atomically precise gold nanoclusters via isomeric methylbenzenethiols. Nano Lett 15:3603–3609

Chen J, Yinchu M, Wei D et al (2020) Furin-instructed intracellular gold nanoparticle aggregation for tumor Photothermal therapy. Adv Funct Mater 30(50):2001566

Chen G, Qin Y, Jiao L et al (2021) Nanozyme-activated synergistic amplification for ultrasensitive photoelectrochemical immunoassay. Anal Chem 93(17):6881–6888

Chen H, Li Y, Song Y et al (2023a) A sandwich-type electrochemical immunosensor based on spherical nucleic acids-templated ag nanoclusters for ultrasensitive detection of tumor biomarker. Biosens Bioelectron 223:115029
Chen Y, Wei M, Ming C et al (2023b) Biotin-decorated hollow gold nanoshells for dual-modal imaging-guided NIR-II photothermal and radiosensitizing therapy toward breast cancer. J Mater Chem B 11(41):10003–10018
Chen QL, Zhou XM, Zhao ML et al (2024) Nanoconfined silver nanoclusters combined with X-shaped DNA recognizer-triggered cascade amplification for electrochemiluminescence detection of APE1. Anal Chem 96(28):11383–11389
Chow TH, Li N, Bai X et al (2019) Gold nanobipyramids: an emerging and versatile type of plasmonic nanoparticles. Acc Chem Res 52(8):2136–2146
Cui M-L, Yi-Song C, Qing-Fan X et al (2019) Synthesis, properties and applications of noble metal iridium nanomaterials. Coord Chem Rev 387:450–462
Da X, Xiulian L, Chao L et al (2023) Ru/Pt@BSA nanoparticles for efficient photo-catalytic oxidation of NAD(P)H and targeted cancer treatment under hypoxic conditions. Chin Chem Lett 34(12):108317
Del Valle AC, Chih-Kuang Y, Yu-Fen H (2020) Near infrared-activatable platinum-decorated gold nanostars for synergistic photothermal/ferroptotic therapy in combating cancer drug resistance. Adv Healthc Mater 9(20):e2000864
Espinosa A, Silva AK, Sanchez-Iglesias A et al (2016) Cancer cell internalization of gold nanostars impacts their photothermal efficiency in vitro and in vivo: toward a plasmonic thermal fingerprint in tumoral environment. Adv Healthc Mater 5(9):1040–1048
Fahmy SA, Fawzy IM, Saleh BM et al (2021) Green synthesis of platinum and palladium nanoparticles using Peganum harmala L. seed alkaloids: biological and computational studies. Nano 11(4):965
Gao S, Hao S, Huang Z et al (2020) Synthesis of high-entropy alloy nanoparticles on supports by the fast moving bed pyrolysis. Nat Commun 11(1):2016
Garcia-Peiro JI, Bonet-Aleta J, Santamaria J et al (2022) Platinum nanoplatforms: classic catalysts claiming a prominent role in cancer therapy. Chem Soc Rev 51(17):7662–7681
Geng X, Leng W, Carter NA et al (2016) Protein-aided formation of triangular silver nanoprisms with enhanced SERS performance. J Mater Chem B 4(23):4182–4190
Goddard ZR, Marin MJ, Russell DA et al (2020) Active targeting of gold nanoparticles as cancer therapeutics. Chem Soc Rev 49(23):8774–8789
Goswami N, Yao Q, Luo Z et al (2016) Luminescent metal nanoclusters with aggregation-induced emission. J Phys Chem Lett 7(6):962–975
Guo K, Ren S, Zhang H et al (2023) Biomimetic gold nanorods modified with erythrocyte membranes for imaging-guided photothermal/gene synergistic therapy. ACS Appl Mater Interfaces 15(21):25285–25299
Guo H, Fu Y, Chen S et al (2024) Electrochemical cytosensor utilizing tetrahedral DNA/bimetallic AuPd holothurian-shaped nanoparticles for ultrasensitive non-destructive detection of circulating tumor cells. Mikrochim Acta 191(6):298
Habibullah G, Viktorova J, Ruml T (2021) Current strategies for Noble metal nanoparticle synthesis. Nanoscale Res Lett 16(1):47
Hao Q, Yang C, Zeming W et al (2023) Biosynthesized gold nanoparticles that activate toll-like receptors and elicit localized light-converting hyperthermia for pleiotropic tumor immunoregulation. Nat Commun 14(1):5178
Hao R, Gangwan Z, Jiahe Z et al (2024) Ultrasmall au/Pt-loaded biocompatible albumin nanospheres to enhance photodynamic/catalytic therapy via triple amplification of glucose-oxidase/catalase/peroxidase. J Colloid Interface Sci 654:212–223
Harish V, Ansari MM, Tewari D et al (2022) Nanoparticle and nanostructure synthesis and controlled growth methods. Nanomaterials 12(18):3226
He MQ, Ai Y, Hu W et al (2023a) Recent advances of seed-mediated growth of metal nanoparticles: from growth to applications. Adv Mater 35(46):e2211915

He MQ, Ai Y, Hu W et al (2023b) Dual-functional capping agent-mediated transformation of silver nanotriangles to silver nanoclusters for dual-mode biosensing. Anal Chem 95(14):6130–6137

Hou Y, Bin S, Rongtian L et al (2023) GSH-activatable camptothecin prodrug-loaded gold nanostars coated with hyaluronic acid for targeted breast cancer therapy via multiple radiosensitization strategies. J Mater Chem B 11(41):9894–9911

Huang X, Tang S, Mu X et al (2011) Freestanding palladium nanosheets with plasmonic and catalytic properties. Nat Nanotechnol 6(1):28–32

Huang Y, Fuksman L, Zheng J (2018) Luminescence mechanisms of ultrasmall gold nanoparticles. Dalton Trans 47(18):6267–6273

Huang L, Mengyang Z, Ghulam A et al (2022) A cancer cell membrane-derived biomimetic nanocarrier for synergistic photothermal/gene therapy by efficient delivery of CRISPR/Cas9 and gold nanorods. Adv Healthc Mater 11(16):e2201038

Iqbal M, Yong-Il C, Giyoong T (2007) An enhanced synthesis of gold nanorods by the addition of Pluronic (F-127) via a seed mediated growth process. J Mater Chem 17(4):335–342

Jadoun S, Rizwan A, Kumari JN et al (2020) Green synthesis of nanoparticles using plant extracts: a review. Environ Chem Lett 19(1):355–374

Jain V, Stuti B, Sanjay S (2021) Bovine serum albumin decorated gold nanoclusters: a fluorescence-based nanoprobe for detection of intracellular hydrogen peroxide. Sensors Actuators B Chem 327:128886

Jang D, Subin Y, Kyungwha C et al (2022) Direct deposition of anatase TiO2 on thermally unstable gold nanobipyramid: morphology-conserved plasmonic nanohybrid for combinational photothermal and photocatalytic cancer therapy. Appl Mater Today 27:101472

Jiang X, Du B, Huang Y et al (2018) Ultrasmall noble metal nanoparticles: breakthroughs and biomedical implications. Nano Today 21:106–125

Jin R (2015) Atomically precise metal nanoclusters: stable sizes and optical properties. Nanoscale 7(5):1549–1565

Kim H, Im SW, Cho NH et al (2020) Gamma-glutamylcysteine- and cysteinylglycine-directed growth of chiral gold nanoparticles and their crystallographic analysis. Angew Chem Int Ed Engl 59(31):12976–12983

Kuang S, Fangmian W, Johannes K et al (2022) Photodecaging of a mitochondria-localized iridium(III) Endoperoxide complex for two-photon Photoactivated therapy under hypoxia. J Am Chem Soc 144(9):4091–4101

Lan Q, Chuanli R, Alexander L et al (2020) Platinum nanoparticle-decorated graphene oxide@ polystyrene nanospheres for label-free electrochemical immunosensing of tumor markers. ACS Sustain Chem Eng 8(11):4392–4399

Lee SJ, Lee H, Begildayeva T et al (2022) Nanogap-tailored au nanoparticles fabricated by pulsed laser ablation for surface-enhanced Raman scattering. Biosens Bioelectron 197:113766

Li X, Yongchun P, Chao C et al (2021) Hypoxia-responsive gene editing to reduce tumor thermal tolerance for mild-photothermal therapy. Angew Chem Int Ed 60(39):21200–21204

Li L, Jiang R, Shan B et al (2022) Near-infrared II plasmonic porous cubic nanoshells for in vivo noninvasive SERS visualization of sub-millimeter microtumors. Nat Commun 13(1):5249

Lin X, Meng T, Chengcheng C et al (2021) Strongly phosphorescent and water-soluble gold(I)-silver(I)-cysteine nanoplatelets via versatile small biomolecule cysteine-assisted synthesis for intracellular hypochlorite detection. Biosens Bioelectron 193:113571

Lin X, Qinhui D, Yalin C et al (2024) Water-soluble photoluminescent Ag nanoclusters stabilized by amphiphilic copolymers as nanoprobe for hypochlorite detection. Chem 12:166

Liu Y, Li J, Chen M et al (2020) Palladium-based nanomaterials for cancer imaging and therapy. Theranostics 10(22):10057–10074

Liu W, Wang Y, Wang Y et al (2023) Black silver nanocubes@amino acid-encoded highly branched gold shells with efficient photothermal conversion for tumor therapy. ACS Appl Mater Interfaces 15(1):236–248

Liu H, Cunji G, Peijing X et al (2024) Biomimetic gold nanorods-manganese porphyrins with surface-enhanced Raman scattering effect for photoacoustic imaging-guided photothermal/photodynamic therapy. Small 20:e2401117

Loke YL, Adilet B, Pei-Wen W et al (2023) ROS-generating alginate-coated gold nanorods as biocompatible nanosonosensitisers for effective sonodynamic therapy of cancer. Ultrason Sonochem 96:106437

Luo Z, Xun Y, Yue Y et al (2012) From aggregation-induced emission of au(I)−thiolate complexes to ultrabright au(0)@au(I)−thiolate core−shell nanoclusters. J Am Chem Soc 134:16662–16670

Luo D, Andrew J, Xinning W et al (2020) Targeted radiosensitizers for MR-guided radiation therapy of prostate cancer. Nano Lett 20(10):7159–7167

Luo Q, Mengjian T, Fang L et al (2023) Multicolor biosensor for trypsin detection based on the regulation of the peroxidase activity of bovine serum albumin-coated gold nanoclusters and etching of gold nanobipyramids. Anal Chem 95:2390–2397

Lv Z, He S, Wang Y et al (2021) Noble metal nanomaterials for NIR-triggered Photothermal therapy in cancer. Adv Healthc Mater 10(6):e2001806

Ma Z, Yifan Z, Jin Z et al (2020) Ultrasmall peptide-coated platinum nanoparticles for precise NIR-II photothermal therapy by mitochondrial targeting. ACS Appl Mater Interfaces 12(35):39434–39443

Murphy TK (2004) Room temperature high yield synthesis-of multiple shapes of gold nanoparticles in aqueous-solution. J Am Chem Soc 126:8648–8649

Nardine SA, Brennan MR, Wilson WL, Murphy CJ (2014) Distance and plasmon wavelength dependent fluorescence of molecules bound to silica coated gold. ACS Nano 8:8392–8406

Nasrollahzadeh M, Sajjadi M, Dadashi J et al (2020) Pd-based nanoparticles: plant-assisted biosynthesis, characterization, mechanism, stability, catalytic and antimicrobial activities. Adv Colloid Interf Sci 276:102103

Nirmal GFL, Liu Y, Leong DT, Xie J (2016) Highly luminescent thiolated gold nanoclusters impregnated in nanogel. Chem Mater 28:4009–4016

Niu W, Chua YA, Zhang W et al (2015) Highly symmetric gold nanostars: crystallographic control and surface-enhanced Raman scattering property. J Am Chem Soc 137(33):10460–10463

Niu Q, Liling F, Yunxiang Z et al (2023) Sensitive and specific detection of carcinoembryonic antigens using toroidal metamaterial biosensors. Anal Chem 95:1123–1131

Odularu AT (2018) Metal nanoparticles: thermal decomposition, biomedicinal applications to cancer treatment, and future perspectives. Bioinorg Chem Appl 2018:9354708

Paiva-Santos AC, Herdade AM, Guerra C et al (2021) Plant-mediated green synthesis of metal-based nanoparticles for dermopharmaceutical and cosmetic applications. Int J Pharm 597:120311

Pothipor C, Jakmunee J, Bamrungsap S et al (2021) An electrochemical biosensor for simultaneous detection of breast cancer clinically related microRNAs based on a gold nanoparticles/graphene quantum dots/graphene oxide film. Analyst 146(12):4000–4009

Pradeep M, Dariusz K, Piotr K et al (2021) Uncovering the phytochemical basis and the mechanism of plant extract-mediated eco-friendly synthesis of silver nanoparticles using ultra-performance liquid chromatography coupled with a photodiode array and high-resolution mass spectrometry. ACS Sustain Chem Eng 10(1):562–571

Pu Y, Pons T (2023) Gold nanorod/titanium dioxide hybrid nanoparticles for plasmon-enhanced near-infrared photoproduction of hydroxyl radicals and photodynamic therapy. ACS Appl Mater Interfaces 15(43):49943–49952

Qiang Z, Li W, Moran C, Zeng J, Chen J, Wen L-P, Xia Y (2010) Seed mediated synthesis of ag nanocubes with controllable edge lengths in the range of 30-200 nm and comparision of theri optical properties. J Am Chem Soc 132:11372–11278

Qiao Z, Zhang J, Hai X et al (2021) Recent advances in templated synthesis of metal nanoclusters and their applications in biosensing, bioimaging and theranostics. Biosens Bioelectron 176:112898

Quan Z, Fei X, Haoyuan L et al (2022) A bioinspired ratiometric fluorescence probe based on cellulose nanocrystal-stabilized gold nanoclusters for live-cell and zebrafish imaging of highly reactive oxygen species. Chem Eng J 431:133954
Riedel R, Mahr N, Yao C et al (2020) Synthesis of gold-silica core-shell nanoparticles by pulsed laser ablation in liquid and their physico-chemical properties towards photothermal cancer therapy. Nanoscale 12(5):3007–3018
Santhakumar H, Nair RV, Maldepalli GD et al (2023) Engineering of tripeptide-stabilized gold nanoclusters with inherent photosensitizing property for bioimaging and photodynamic therapy. ACS Sustain Chem Eng 11(6):2102–2114
Sau TK, Rogach AL (2010) Nonspherical noble metal nanoparticles: colloid-chemical synthesis and morphology control. Adv Mater 22(16):1781–1804
Scaramuzza S, De Faria Clara MG, Vito C et al (2023) A laser synthesis route to boron-doped gold nanoparticles designed for X-ray radiotherapy and boron neutron capture therapy assisted by CT imaging. Adv Funct Mater 33(42):2303366
Shen J, Johannes K, Kai X et al (2021) Cancer cell membrane camouflaged iridium complexes functionalized black-titanium nanoparticles for hierarchical-targeted synergistic NIR-II photothermal and sonodynamic therapy. Biomaterials 275:120979
Singh P, Prakash H, Khushal S et al (2023) Palladium nanocapsules for photothermal therapy in the near-infrared II biological window. ACS Appl Mater Interfaces 15(33):39081–39098
Stamplecoskie KG, Kamat PV (2014) Size-dependent excited state behavior of glutathione-capped gold clusters and their light-harvesting capacity. J Am Chem Soc 136(31):11093–11099
Sun J, Jinping W, Wei H et al (2021) Camouflaged gold Nanodendrites enable synergistic photodynamic therapy and NIR biowindow II Photothermal therapy and multimodal imaging. ACS Appl Mater Interfaces 13(9):10778–10795
Tang Q, Fanyu Z, Licheng L et al (2023) Ultrasound-induced gold nanoparticle united with acoustic reprogramming of macrophages for enhanced cancer therapy. ACS Appl Mater Interfaces 15(44):50926–50939
Tang D, Minhui C, Bin W et al (2024) Near infrared-fluorescent dinuclear iridium(III) nanoparticles for immunogenic sonodynamic therapy. Adv Mater 36(38):e2406815
Wang Y, Counihan MJ, Lin JW et al (2020) Quantitative analysis of DNA-mediated formation of metal nanocrystals. J Am Chem Soc 142(48):20368–20379
Wang M, Yuqin L, Fangling L et al (2022) Iridium tungstate nanozyme-mediated hypoxic regulation and anti-inflammation for duplex imaging guided photothermal therapy of metastatic breast tumors. ACS Appl Mater Interfaces 14(51):56471–56482
Wu Z, Jin R (2010) On the ligand's role in the fluorescence of gold nanoclusters. Nano Lett 10(7):2568–2573
Wu HL, Kuo CH, Huang MH (2010) Seed-mediated synthesis of gold nanocrystals with systematic shape evolution from cubic to trisoctahedral and rhombic dodecahedral structures. Langmuir 26(14):12307–12313
Wu F, Yongjia L, Yan W et al (2020) Chlorin e6 and polydopamine modified gold nanoflowers for combined photothermal and photodynamic therapy. J Mater Chem B 8(10):2128–2138
Wu C, Xuancheng D, Bingqing J et al (2021) A transformable gold nanocluster aggregate-based synergistic strategy for potentiated radiation/gene cancer therapy. J Mater Chem B 9(9):2314–2322
Wu T, Chen K, Lai W et al (2022a) Bovine serum albumin-gold nanoclusters protein corona stabilized polystyrene nanoparticles as dual-color fluorescent nanoprobes for breast cancer detection. Biosens Bioelectron 215:114575
Wu Y, Li S, Chen Y et al (2022b) Recent advances in noble metal complex based photodynamic therapy. Chem Sci 13(18):5085–5106
Xiong Y, Rao Y, Hu J et al (2023) Nanoparticle-based photothermal therapy for breast cancer noninvasive treatment. Adv Mater 10:e2305140

Xu Y, Wang X, Ding C et al (2021) Ratiometric antifouling electrochemical biosensors based on multifunctional peptides and MXene loaded with au nanoparticles and methylene blue. ACS Appl Mater Interfaces 13(17):20388–20396

Yan T, Su M, Wang Z et al (2023) Second near-infrared plasmonic nanomaterials for photoacoustic imaging and photothermal therapy. Small 19(30):e2300539

Yang H, He H, Tong Z et al (2020) The impact of size and surface ligand of gold nanorods on liver cancer accumulation and photothermal therapy in the second near-infrared window. J Colloid Interface Sci 565:186–196

Yang K, Li J, Lamy De La Chapelle M et al (2021) A terahertz metamaterial biosensor for sensitive detection of microRNAs based on gold-nanoparticles and strand displacement amplification. Biosens Bioelectron 175:112874

Yang M, Li H, Li X et al (2022) Catalytic hairpin self-assembly regulated chameleon silver nanoclusters for the ratiometric detection of CircRNA. Biosens Bioelectron 209:114258

Ye X, Zheng C, Chen J et al (2013) Using binary surfactant mixtures to simultaneously improve the dimensional tunability and monodispersity in the seeded growth of gold nanorods. Nano Lett 13(2):765–771

Yin B, Hei HWK, Xinyue X et al (2023) A multilayered mesoporous gold nanoarchitecture for ultraeffective near-infrared light-controlled chemo/photothermal therapy for cancer guided by SERS imaging. Small 19(6):e2206762

Yu S, Dohyub J, Hong Y et al (2021) Plasmon-triggered upconversion emissions and hot carrier injection for combinatorial photothermal and photodynamic cancer therapy. ACS Appl Mater Interfaces 13(49):58422–58433

Yu T, Su S, Hu J et al (2022) A new strategy for microbial taxonomic identification through micro-biosynthetic gold nanoparticles and machine learning. Adv Mater 34(11):e2109365

Zhang C, Amin Z, Wenxiu H et al (2018) Mimicking pathogenic invasion with the complexes of Au22(SG)18-engineered assemblies and folic acid. ACS Nano 12:4408–4418

Zhang R, Linlin W, Xiaofei W et al (2020a) Acid-induced in vivo assembly of gold nanoparticles for enhanced photoacoustic imaging-guided Photothermal therapy of tumors. Adv Healthc Mater 9(14):e2000394

Zhang H, Wang Z, Wang F et al (2020b) In situ formation of gold nanoparticles decorated Ti(3) C(2) MXenes nanoprobe for highly sensitive electrogenerated chemiluminescence detection of exosomes and their surface proteins. Anal Chem 92(7):5546–5553

Zhang L, Yang G, Sijia S et al (2020c) pH-responsive metal–organic framework encapsulated gold nanoclusters with modulated release to enhance photodynamic therapy/chemotherapy in breast cancer. J Mater Chem B 8(8):1739–1747

Zhang H, Minhui C, Dongsheng T et al (2024) Localization of cancer cells for subsequent robust photodynamic therapy by ROS responsive polymeric nanoparticles with anti-metastasis complexes NAMI-A. Adv Mater 36(14):e2310298

Zhao P, Na L, Didier A (2013) State of the art in gold nanoparticle synthesis. Coord Chem Rev 257(3–4):638–665

Zhao J, Xue S, Ji R et al (2021) Localized surface plasmon resonance for enhanced electrocatalysis. Chem Soc Rev 50(21):12070–12097

Zhao Y, Jingwen X, Yuanyuan Z et al (2023) Biomimetic redox-responsive prodrug micelles with diselenide linkage for platinum nanozymes augmented sonodynamic/chemo combined therapy of colon cancer. Chem Eng J 472:120979

Zheng J, Cheng X, Zhang H et al (2021) Gold nanorods: the most versatile plasmonic nanoparticles. Chem Rev 121(21):13342–13453

Zhou X, Pu Q, Yu H et al (2021) An electrochemical biosensor based on hemin/G-quadruplex DNAzyme and PdRu/Pt heterostructures as signal amplifier for circulating tumor cells detection. J Colloid Interface Sci 599:752–761

Zhu J, Ai O, Zhuanglin S et al (2022) Sonodynamic cancer therapy by novel iridium-gold nanoassemblies. Chin Chem Lett 33(4):1907–1912

Zoughi S, Faridbod F, Moradi S (2024) Rapid enzyme-free detection of miRNA-21 in human ovarian cancerous cells using a fluorescent nanobiosensor designed based on hairpin DNA-templated silver nanoclusters. Anal Chim Acta 1320:342968

Chapter 2
The Application of Upconversion Nanomaterials in Tumor Diagnosis and Treatment

Changmin Hou and Wanqing Teng

Abstract Cancer often starts quietly and progresses very rapidly, which has led to a persistent search for more effective cancer treatment strategies. The means used in the clinical treatment of cancer mainly cover surgical resection, radiation therapy, and chemotherapy, and so on, and sometimes a combination of the above therapies will be used to eliminate cancer cells, but the overall therapeutic effect still has great room for improvement. With the integration of materials science, biology, chemistry, and medicine, the cancer nanodiagnosis and treatment system came into being. The system achieves imaging diagnosis of cancer with the help of integrated multifunctional structural combination design, can accurately transport drugs to the targeted site, and can achieve on-demand treatment according to stimulus response, which has become the general trend of future pharmaceutical development. Upconversion nanomaterials (UCNPs) have unique physical and chemical properties, and when they are combined with silica, polymers, and other substances to build a nanodiagnosis and treatment system, they have many powerful functions. This review will focus on the recent development of nanotherapeutic nanoplatforms based on upconversion nanoparticles, covering several important aspects of chemotherapy, photodynamic therapy, and photothermal therapy. We expect that this review will provide reference value for the development of accurate and efficient nanotherapies for cancer.

Keywords Upconversion nanomaterials · Imaging diagnosisNanotherapeutic · Nanomedicine · Combination therapy

---

C. Hou (✉) · W. Teng
Shandong Provincial Key Laboratory of Tumor Imaging Equipment Development and Tumor Diagnosis and Treatment Integration Technology, School of Medicine, Linyi University, Linyi, Shandong, People's Republic of China

X. Song et al. (eds.), *Functional Inorganic Materials for Tumor Diagnostics and Therapy*, https://doi.org/10.1007/978-981-95-0528-9_2

2.1 Construction of Upconversion Nanodiagnostic Reagents

Nano-anti-tumor system can achieve targeted drug delivery, imaging diagnosis, and on-demand treatment in response to stimulation through the combination of multi-functional and integrated structure, which is the inevitable trend of the development of nanomedicine in the future. UCNPs-based therapeutic and diagnostic nanosystems integrate multifunctional nanomaterials and UCNPs into a single nanoplatform to achieve multi-functions such as diagnosis, treatment, tracking biodistribution, and monitoring treatment progress. Generally, therapeutic diagnostic agents are composed of four components, namely imaging agents, therapeutic agents [anti-cancer chemicals, photothermal agents, photosensitizer PS, small interfering RNA (siRNA), and so on], carriers (polymers, silica, graphene, and so on), and surface modifiers (PEG, target molecules, stimulus response molecular valves, and so on). This part will focus on the combination method of UCNPs and carriers.

2.1.1 Silicon Dioxide

Silica materials have the advantages of good biocompatibility, large specific surface area, easy connection of functional groups, and tunable mesoporous channels due to their abundant silicon hydroxyl groups on the surface. Therefore, SiO_2-coated upconversion nanocomposites (UCNPs@SiO_2) have shown great application prospects in the field of nanomedicine and have attracted more and more attention. Silica coating includes the following two types: one is solid silica coating, which can improve the water dispersion, biocompatibility, and functional modification of UCNPs; the other is mesoporous silica coating, in addition to having the advantages of solid silica coating, its mesoporous pores can be loaded with therapeutic reagents and delivered to the lesion site.

At present, there are two main methods to achieve silica-coated UCNPs. The first is the classical Stober method, which is based on the sol-gel chemical coating of hydrophilic UCNPs, and then the hydrolysis and condensation reaction of silicon source to form a three-dimensional network structure to achieve the coating. The second method is the reverse micellar microemulsion method, which usually uses surfactant IgCO-520 to form a microreactor in the organic phase. After ligand exchange with oleate on the surface of UCNPs, the UCNPs are transferred to the center of the microreactor, and the coating is completed by hydrolysis and polymerization of silicon source under alkali catalysis. While exploring more coating methods, researchers used the above two coating methods to design a variety of flexible structure of silica coated UCNPs nanodiagnosis and treatment systems.

2.1.1.1 Mesoporous Silica Coated UCNPs Nanodiagnosis and Treatment System (UCNPs@mSiO_2)

Liu et al. (2012) successfully modified mesoporous silica layer on hydrophobic UCNPs surface using cationic surfactant hexadecyltrimethylammonium bromide (CTAB) as phase transfer agent and organic template agent. The experimental results show that appropriate sonication during the synthesis process can effectively avoid the aggregation of UCNPs and increase the mono-dispersity of UCNPs@mSiO_2 nanomaterials. Mesoporous silica has adjustable particle size, variable mesoporous size, and easily functionalized surface structure, so it can be assembled with a variety of functional molecules to design stimulus response valves to construct intelligent controlled drug delivery systems. At present, the commonly used internal stimuli are hypoxia, slightly acidic environment, redox potential, and overexpression of specific enzymes. Commonly used external stimuli include light, ultrasound, magnetic field, temperature, and so on.

2.1.1.2 Core–Shell Silica-Coated UCNPs Nanodiagnosis and Treatment System (UCNPs@dSiO_2@mSiO_2)

In the early work, Song et al. (2020) successfully prepared $NaYF_4$:Yb/Er@dSiO_2@mSiO_2 nanoparticles by calcination for the first time, and loaded the photosensitizer zinc phthalocyanine (ZnPc) in the mesoporous channel, which was used to generate singlet oxygen $^1O_2$ under 980-nm near-infrared light excitation to kill tumor cells. The TEM characterization results show that the synthesized $NaYF_4$:Yb/Er@SiO_2@mSiO_2 nanoparticles have uniform morphology, monodisperse state, and worm-like mesoporous structure. The research group (Dong et al. 2011) used the core–shell nanoparticles loaded with two photosensitizers to carry out two LRET processes with multicolor upconversion nanoparticles to realize binary photodynamic therapy and improve the efficacy of photodynamic therapy.

Although the core–shell structure of UCNPs/mSiO_2 prepared by calcination method can obtain a good mesoporous structure, it needs to remove surfactants at high temperature to form mesoporous structure, which is easy to cause agglomeration of nanoparticles and weakening of biocompatibility. Therefore, the organic template method was developed to solve the problems of the calcination method. The organic template method uses a surfactant, usually CTAB or hexadecyl trimethyl ammonium chloride (CTAC), to transfer hydrophobic UCNPs from the organic phase to the aqueous phase. In this process, CTAB or CTAC can be used as a template molecule for phase transfer and mesoporous formation, and then a silicon shell is coated on the surface of UCNP containing the template by Stober method. Finally, the template molecules were removed by ion exchange to form mesopores. In 2012, Cichos and Karbowiak (2014) used CATB as a template to controllably coat the surface of hydrophobic NaGdF4 nanoparticles with a thin mesoporous silicon shell. Liu et al. (2012) systematically studied a series of factors affecting the coating such as the amount of CTAB used, reaction temperature, and ultrasonic treatment in the same year. By precisely controlling these influencing factors, $NaYF_4$: Tm/Yb/Gd@mSiO_2 nanoparticles with uniform size, monodisperse, impurity free, and controllable particle size can be prepared for dual-mode fluorescence imaging and magnetic resonance imaging. Xu et al. (2011) synthesized core–shell Gd_2O_3: Er@SiO_2@mSiO_2 nanoparticles through a two-step sol-gel process by modified Stober method. They first coated a layer of SiO_2 shell on the surface of UCNPs to obtain Gd_2O_3: Er@SiO_2. Gd_2O_3: Er@SiO_2@CTAB was formed by electrostatic assembly of the negatively charged silicon hydroxyl group on the surface of Gd_2O_3: Er@SiO_2 and the positively charged CTAB template, and then the silicon shell was formed by condensation of the silicon source. After removing the template, Gd_2O_3:Er@mSiO_2 with mesoporous structure was obtained. Sun et al. (2019) prepared a series of core–shell structure nanoparticles in this way, such as Fe_3O_4@SiO_2@mSiO_2, $NaYF_4$:Yb/Er@SiO_2@mSiO_2, and $NaYF_4$:Yb/Er@ $NaGdF_4$:Yb/Er@mSiO_2 are used for drug delivery and imaging.

2.1.1.3 The Yolk–Eggshell Structure of UCNPs Nanodiagnosis and Treatment System (UCNPs@Cavity@mSiO_2)

The yolk–eggshell structure is a cavity between UCNPs and mSiO_2, forming a yolk–eggshell structure, which can have a higher loading capacity than the mSiO_2 shell (Li et al. 2013a). At present, the yolk–eggshell structure UCNPs@cavity @ mSiO_2 is mainly prepared by the surface protection etching method. The main principle of this method is to coat the surface of the nanoparticles with a surface protectant, and then use an etching agent to etch the surface, which protects the surface from etching and thus only erodes the inside of the nanoparticles to form a yolk–eggshell structure. Commonly used surface protectants are polyvinylpyrrolidone, polyethylenimine, and poly dimethyl diallyl ammonium chloride, and commonly used etching agents include hot water, hydrofluoric acid, sodium oxide, and sodium carbonate. Cheng research group (Cheng et al. 2022) successfully prepared nanoparticles with yolk and eggshell structure for the first time using the surface protection etching method. Molkenova et al. (2024) proposed a general method for preparing yolk–eggshell structure nanoparticles. Firstly, the organic template method was used to prepare the core–shell structure Fe_3O_4@mSiO_2 nanoparticles, and then polyethylene imine was used as a surface protector to cover the surface of Fe_3O_4@ mSiO_2, and then hot water was used as an etching agent at 50 °C. The yolk–shell structure can be obtained. By modifying the surface of nanoparticles with fluorescent targeting functional molecules and loading drugs in the cavity, targeted fluorescence imaging and magnetic field-guided drug delivery can be realized. Using this method, UCNPs/mSiO_2 and nanorods/mesoporous SiO_2 nanoparticles with yolk–eggshell structure can be prepared. Fan et al. (2013) developed a novel method for the preparation of UCNPs carriers with yolk–eggshell structures. They coated the surface of UCNPs with two successive layers of SiO_2, and then coated the surface protector polyvinylpyrrolidone PVP on the surface of UCNP@d1-SiO_2@D2-SiO_2, using hot water etching to obtain yolk–eggshell structured nanoparticles. In 2014,

Liu et al. (2021a) improved this method. Firstly, UCNPs were coated with SiO_2 by reverse micelliform microemulsion method, and then $mSiO_2$ was coated with organic template method to obtain UCNPs@SiO_2@$mSiO_2$. Finally, PVP was used as surface protective agent, and hot water was used as etching agent to obtain the yolk–eggshell structure.

2.1.2 UCNPs@Polymer Therapeutic Nanoagents

Diagnostic nanoagents based on UCNPs must have certain aqueous phase stability, good biocompatibility, and easy functionalization. However, usually synthesized UCNPs are somewhat hydrophobic. Organic polymers can make hydrophobic UCNPs have certain aqueous phase dispersibility, provide active chemical groups to facilitate functionalization (Dai et al. 2016), improve the circulation time of nanoparticles in the blood, and reduce nonspecific uptake by cells in the reticuloendothelial system (Cheng and Lin 2015). At present, various polymers have been widely used to modify UCNPs such as polyacrylic acid (PAA) (Naccache et al. 2009; Pegaz et al. 2006), polyethylene imine (PEI) (Chen et al. 2011, 2014), polyallylamine (PAAm) (Wang et al. 2011a; Xia et al. 2014), polyethylene glycol (PEG)-grafted amphiphilic polymers (Wang et al. 2011b), Peg-phospholipid (Sun et al. 2023; Xiang et al. 2023; Zhan et al. 2011), polydopamine (PDA) (Li et al. 2013a), TWEEN (Ren et al. 2012; Tian et al. 2014), and many more.

According to the different properties of polymers, many strategies have been used to construct therapeutic nanomaterials that combine upconversion nanoparticles (UCNPs) with polymers. Among them, the ligand exchange method is used to coat the surface of the hydrophobic upconversion nanoparticle with a hydrophilic polymer, using a polydentate ligand. In 2013, Li et al. (2013b) conducted a study using ligand exchange and polyethylenimide (PEI) as a surface modifier. In this way, the UCNPs can not only be transferred to the aqueous phase, but also provide a free functional group, amino group, which can be used to link trans-platinum (IV) prodrugs. The upconversion nanoparticle nanodiagnostic system thus constructed has the unique advantage of being able to achieve near-infrared light-induced platinum prodrug release under the guidance of three-mode imaging, namely upconversion luminescence (UCL), computed tomography (CT), and magnetic resonance imaging (MRI).

In addition, there is a widely used amphiphilic polymer attachment strategy. The strategy was based on the van der Waals interaction between the hydrophobic chain segment of the amphiphilic polymer and the hydrophobic ligand on the UCNPs, so that the amphiphilic polymer was adsorbed on the surface of the UCNPs. And the hydrophilic chains exposed to amphiphilic polymers allow the UCNPs to achieve good dispersion in water. It was worth mentioning that Liu's research team made an innovative attempt by synthesizing an amphiphilic polymer called C18PMH-polyethylene glycol (PEG) and then coating it on the surface of UCNPs modified with oleic acid. Through this operation, a pegylated UCNPs with a "hydrophobic pocket" and a hydrophilic surface was obtained. It had important application value and can be used to physically load the anti-cancer drug doxorubicin (DOX), and the release of doxorubicin can be controlled by adjusting the pH value. This strategy is both simple and flexible and can be used to load other anti-cancer drugs. Compared with the ligand exchange method, the amphiphilic polymer attachment strategy method can prevent water molecules from entering the luminescence center and better maintain the characteristics of UCNPs.

Layer-by-layer (LBL) technology based on electrostatic attraction is a method capable of implanting UCNPs with excellent water solubility and precise control of coating Layer thickness. Liu et al. (2012) used LBL technology to construct a nano-agent for the diagnosis and treatment of UCNPs based on pH-sensitive photodynamic therapy. Mn^{2+} doped UCNPs can emit red fluorescence under the excitation of near-infrared light, which can be used for UCL fluorescence imaging and photodynamic therapy by activating photosensitior chlorin e6 (Ce6). Polymer-functionalized Ce6 (PAH-Ce6-SA) was deposited on the surface of UCNPs through electrostatic adsorption, which were functionalized with Poly(allylamine hydrochloride) (PAH). The load of photosensitizer can be optimized by adjusting the PAH/ PAH-CE6-SA layer number. Ph-sensitive photodynamic therapy was achieved by using a pH-sensitive polymer (DMMA-PAH-PEG). This flexible design strategy and ease of preparation have stimulated great interest in the development of new nanoplatforms for therapeutic diagnostics.

2.1.3 UCNPs@Noble Metal (Au or Ag) Therapeutic Nanoparticles

Localized surface plasmon resonance (LSPR) can improve the luminescence efficiency of UCNPs (Ge et al. 2013; Schietinger et al. 2010; Xu et al. 2017). UCNPs@ noble metal (Au or Ag) nanocomposites have been applied in cancer therapy (Cheng et al. 2012; Li et al. 2017). To construct UCNPs@noble metal nanocomposites, the surface of UCNPs is usually modified to have good hydrophilicity and carry positive charges or ligands to attract negatively charged metal nanoparticles. Song et al. (2020) designed the core–shell structure $NaYF_4$: Yb, Er@Ag nanocomposites for UCL fluorescence imaging and photothermal therapy. In their experiments, thioglycolic acid (TGA) was used to transfer UCNPs into the aqueous phase and attract Ag^{+} through the thiol group. By adjusting the thickness of the silver layer, the wavelength of the LSPR is adjusted to 980 nm. The Au shell has a better NIR absorption compared to the Ag shell. Xiao's group coated Au shells on the surface of UCNPs (Xiao et al. 2013), which showed high tumor enrichment and good photothermal therapy effect.

2.1.4 Synthesis of Therapeutic Diagnostic Nanoagents Based on Hybrid Structure UCNPs

Silica has various active functional groups (such as –COOH, $-NH_2$, –SH, and many others), so UCNPs can use silica to interact with active substances such as biomolecules to obtain functional nanoparticle combinations. For example, targeting groups such as folic acid can be flexibly modified on the surface of amino-functionalized UCNPs@SiO_2 nanocomposites to achieve targeted diagnosis and treatment (Li et al. 2010b). Inorganic nanoparticles such as CuS can be loaded on the surface of UCNPs@ SiO_2-NH_2 nanoparticles for enhancing radiotherapy and photothermal synergistic therapy (Liu et al. 2012). Assisted by SiO_2 shells, Au nanoparticles are also incorporated with UCNPs for computed tomography imaging, photothermal therapy, gene therapy, and biosensing (Dong et al. 2020; Fang et al. 2023; Song et al. 2017; Zhong et al. 2017). Zhang et al. (2021a) constructed a novel ZN-functionalized UCNPs@SiO_2 nanocomposite by covalent grafting, which can be used for multimodal bioimaging and on-demand drug release with specific pH stimulation.

2.1.5 Synthesis of Therapeutic Diagnostic Nanoagents for DNA-Based UCNPs

Composite nanomaterials combining DNA and UCNPs have shown significant development trends in multiple fields. In this composite system, DNA-functionalized UCNPs exhibited multifaceted functional characteristics. They can not only serve as signal markers and carriers and play important roles, but also achieve a series of complex biological functions through pre-precisely designed DNA sequences, covering aspects such as target recognition, imaging, therapy, catalytic reactions, and sensing detection. This situation is similar to protein engineering techniques. Through meticulous and professional design of the structure and sequence of DNA, specific and unique functional properties can be imparted to nanomaterials. It was worth mentioning that long-chain DNA with multiple functional regions has unique advantages in drug combination and tumor targeting. It was capable of real-time drug release monitoring and provides strong support for pharmacokinetic studies, thereby expanding broad space and diverse application prospects for the in-depth application of composite nanomaterials in numerous cutting-edge fields such as biomedicine, and bringing new opportunities and challenges to relevant research and applications.

In 2021, Zhang et al. (2019a) reported an upconversion luminescence driven DNA-azobenzene nanopump for rapid and efficient drug release by assembling azobenzene functionalized DNA chains on UCNPs, it was clear that UCNPs was successfully synthesized. Doxorubicin (DOX) was inserted into the DNA double helix. Under NIR irradiation, the ultraviolet and visible light emitted by UCNPs caused continuous isomerization of azo compounds. This triggers the hybridization and dehybridization of DNA molecules, thereby enabling controllable release of DOX. The results showed that within a short time, the release amount of DOX is as high as 86.7%. Perinuclear aggregation and enhanced anti-cancer therapy of DOX targeting cancer cell nuclei were achieved by systemically assembling HIV-1 TAT peptides and hyaluronic acid. Such efficient drug delivery nanopumps could contribute to the development of chemotherapy.

In order to activate the DNA nanomachines in cancer cells within the complex microenvironment of solid tumors, the Ju group designed a transmembrane DNA logic computing strategy (Fig. 2.1) (Zhang et al. 2021b). Among them, the DNA nanomachine is based on upconversion nanomaterials and modified with specially designed DNA strands. In addition, LA-apt was a DNA strand that can anchor to overexpressed receptors on the cancer cell membrane and serves as input 1. The intracellular miRNA-21 acts as input 2. The combination of these two input signals greatly enhances the specificity of the DNA nanomachine. After specific hybridization with input 1 on the cell membrane, the miRNA-21 recognition region of the DNA nanomachine is exposed, and at the same time, it enters the interior of cancer cells. The resulting intermediate further undergoes cascade hybridization with input 2 inside the cell, thereby activating the photosensitizer Rose Bengal to generate reactive oxygen species. Photodynamic therapy is carried out at precise locations. This dual logic gate control can target treatment sites with high precision to a large extent.

Chen et al. (2024) designed a Ru(II)-polypyridyl complex (RuC)-loaded upconversion nanoplatform (UCSNs-R) was rationally established for efficient therapy of gliomas by the elaborate manipulation of DNA separation and $^1O_2$ generation under

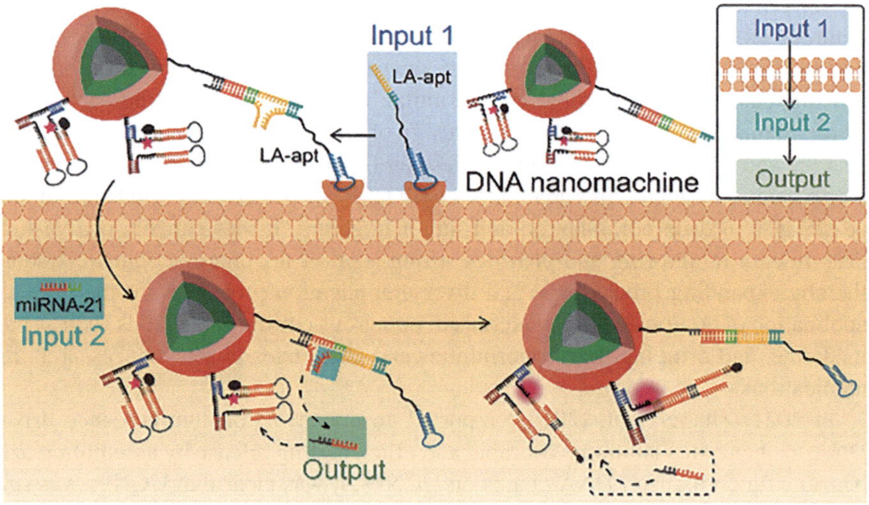

Fig. 2.1 Structural diagram of the transmembrane DNA computation strategy (Zhang et al. 2021b)

NIR laser irradiation (Fig. 2.2). The UCSNs can not only implement high loading of DPPZ-containing RuC, but also effectively convert 980 nm laser to UV light, thus facilitating abundant $^1O_2$ production. The DPPZ ligand of RuC can intercalate into DNA through the minor groove, further facilitating the photocleavage reaction with DNA. Furthermore, the released DPPZ-containing RuC, with DNA "light-switch" effect and high DNA binding affinity, can manipulate DNA phase separation in living cells, leading to DNA damage and ultimately preventing tumor cell proliferation. Additionally, Gd doping enables UCSNs-R to monitor the therapeutic process by magnetic resonance imaging (MRI). Assessments at the cellular and animal levels revealed that UCSNs-R can significantly inhibit tumor proliferation under NIR laser exposure by promoting the rapid production of $^1O_2$ and enhancing DNA phase separation. Moreover, UCSNs-R can modulate DNA phase separation by evidently inhibiting DNA replication under NIR laser irradiation, as demonstrated by RNA-sequencing. Overall, their work presented here the first nanomodulator of DNA phase separation in living cells by integration of the RuC based anti-cancer complexes and the UCSN, thereby providing a strategy to design a DNA phase separation nanoinducer for cancer treatment.

In 2024, Sun et al. designed and fabricated a hierarchical nanoprobe assembled by DNA programming in order to achieve high-sensitivity imaging of miRNAs and precise deep cancer treatment under image guidance. This nanoprobe was based on upconversion nanoparticles and used DNA strands as the supporting network, with bright-emitting quantum dots (QD) and emissive quencher gold nanoparticles (GNP) programmatically assembled on the upconversion nanoparticles. The DNA-modified GNP can precisely identify intracellular miRNAs, and at the same time, the catalytic decomposition of GNP will turn on and amplify the fluorescence signal, thereby achieving sensitive imaging of deep cancers. In addition, the near-infrared excitation energy collected by UCNPs was transferred to the surface of the bonded QD, thereby facilitating the photosensitization of the aggregation-induced emission (AIE) photosensitizer composed of dithiols on the QD surface for photodynamic treatment. The constructed hierarchical nanoprobe can not only accurately

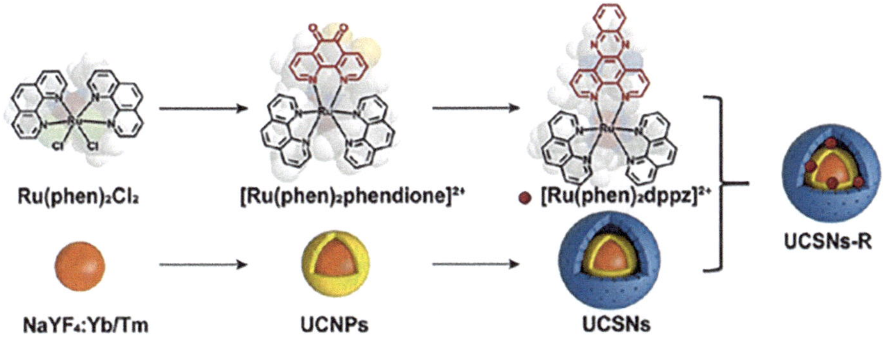

Fig. 2.2 Schematic illustration of the fabricated procedure and synergistic cancer therapy of UCSNs-R. Fabrication of the DNA phaseseparation nanomodulator (Chen et al. 2024)

distinguish normal cells from cancer cells but also provides a novel idea for treating deep cancers. The results show that the hierarchical nanoprobe successfully inhibits the growth of deep tumors and achieves highly sensitive imaging.

2.2 Application of UCNPs in Tumor Diagnosis and Treatment

2.2.1 Chemotherapy

Chemotherapy indeed presents certain distinctive benefits in cancer treatment. However, it is marred by a lack of precise targeting toward the affected area, thereby leading to significant side effects. Following the oral or intravenous administration of anti-cancer medications, merely a trifling quantity of the drugs manages to reach the tumor tissue, resulting in a rather low drug utilization rate and necessitating frequent dosing. On the one hand, it promotes the development of drug resistance in tumor cells. On the other hand, owing to the lack of selectivity, it kills normal cells, giving rise to severe systemic toxic side effects (Li et al. 2010a; Wang et al. 2014a). In response to the diverse shortcomings of chemotherapy, researchers have devised a nanodrug sustained release system. This system loads the drug into the nanocarrier, safeguarding the drug's activity from premature depletion. The nanocarrier passes through the intercellular space and is absorbed by the cell tissue, and the drug is slowly released. Subsequently, researchers designed a controlled drug release system. By selectively modifying the functional groups on the surface of the nanocarrier, a nanovalve was installed. The "open" or "closed" state of the nanovalve could be regulated by external stimuli, ensuring that the drug was released within the tumor tissue. This approach not only averts drug leakage in normal tissues but also enables controlled drug release, thereby realizing controlled chemotherapy (Zhang et al. 2019a; Li et al. 2010b; Wang et al. 2014b). Rare earth upconversion Luminescent materials play an important role in the design and construction of stimulus-release drug systems. This part will focus on various tumor microenvironment-stimulated drug delivery and light-controlled drug delivery systems.

In 2021, Wahab et al. (2023) designed a NIR-controlled nanoscale proton supplier, which can rapidly release H^+ under 980-nm laser irradiation to reduce the pH value at the tumor site, thereby effectively inhibiting cell invasion and migration and improving anti-cancer effect. Increasing hydrogen ion concentration decreases cofilin activity, impeds F-actin polymerization, and further affects the viability of lamellipodia. In addition, increased intracellular H+ content can enhance ·OH and effectively inhibit tumor growth. Therefore, this nanomedicine can not only effectively inhibit tumor cell movement to reduce invasion and migration, but also play an effective role in the field of nanomedicine therapy.

Hu et al. (2019) successfully prepared a pH-sensitive drug delivery nanoplatform, which had unique advantages and integrates the function of real-time monitoring of drug release in living cells (Fig. 2.3). This nanomaterial used UCNPs as the core, and then coated its surface with mesoporous silica. Subsequently, a metal-phenolic network of tannic acid and Cu^{2+} ions was modified. Specifically, the mesoporous structure of silica can encapsulate a large amount of DOX. Moreover, the metal-phenolic network structure on the surface had degradation properties and can trigger release behavior in an acidic environment. Meanwhile, when DOX was encapsulated in the mesopores, a fluorescence resonance energy transfer (LRET) phenomenon occurred with UCNPs, resulting in quenching. When the pH condition triggered the release of DOX, the fluorescence resonance energy transfer process will gradually terminate, and correspondingly, the upconversion luminescence intensity will increase accordingly, thereby realizing the real-time monitoring of the doxorubicin release process. In view of this, this nanoplatform will play an important role in promoting the development of an efficient drug delivery system (DDS) for cancer treatment and provides a new idea and powerful tool for the research and development of drug delivery systems in the field of cancer treatment.

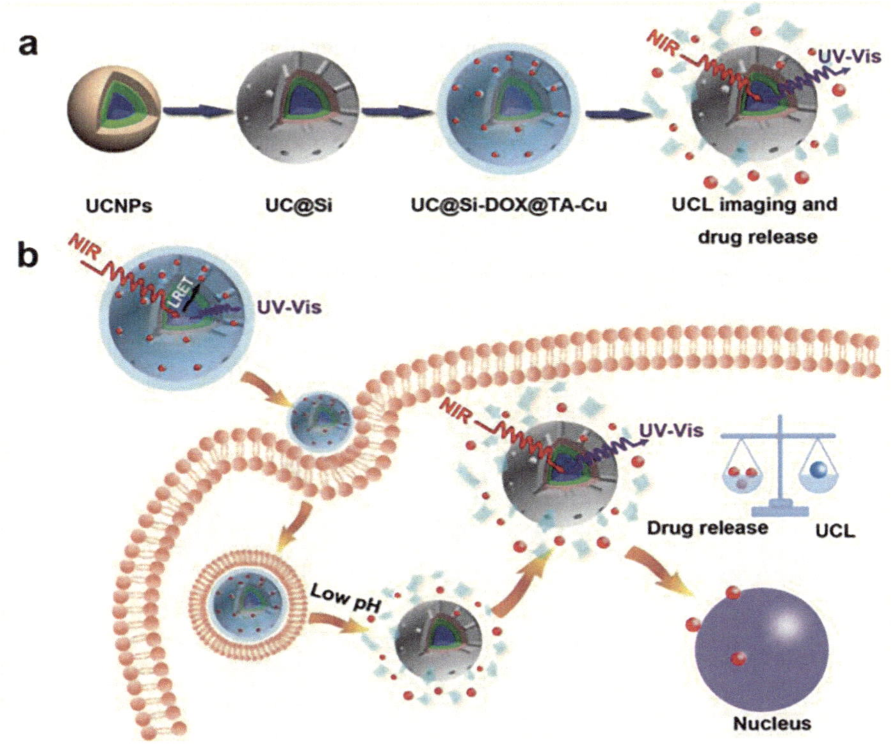

Fig. 2.3 (**a**) Schematic showing the synthesis of nanoplatform. (**b**) Schematic of pH-sensitive drug delivery nanoplatform (Hu et al. 2019)

Yang et al. (2022b) formed a mesoporous shell assembled by ultra-small tin dioxide (SnO_2) on the surface of UCNPs by hydrothermal deposition method, used mesoporous channels to encapsulate anti-tumor drug adriamycin hydrochloride (DOX), and further modified bovine serum albumin (BSA) to provide biocompatibility of the material. An integrated multifunctional nanocapsules (UCNPs@SnO_2-DOX/BSA) was formed. Controlled drug release combined with optical therapy can effectively enhance tumor cell death and has multimodal imaging capabilities, which is expected to stimulate powerful imaging guidance and real-time monitoring of anti-tumor synergistic treatment effects (Fig. 2.4).

Liu et al. (2024) used UCNPs as carriers to modify a multi-targeted octahedral platinum (IV) prodrug in order to improve the therapeutic effect of cisplatin. The multi-targeted octahedral platinum (IV) prodrug with octadecyl aliphatic chains and phenylbutyric acid (PHB) was successfully loaded by utilizing hydrophobic interactions. In addition, in order to enhance the biocompatibility and tumor specificity of the nanomaterial, it was combined with distearoylphosphatidylethanolamine-polyethylene glycol 2000 (DSPE-PEG2000) and arginine-glycine-aspartic acid (RGD) peptides. This nanoparticle (UCNP/Pt(IV)-RGD) can accurately identify tumor cells and perform in situ triggering, releasing a large amount of platinum (II) and PHB (Fig. 2.5). Eventually, the tumor cells are successfully killed. The experimental results show that this nanoparticle has a good anti-tumor effect and provides new ideas for future combination therapies for cancer treatment.

In order to solve the shortcomings of small molecular compounds such as poor water solubility, no cell tissue targeting, and easy toxic side effects caused by excessive use, Yang et al. (2022a) research group from South China Normal University

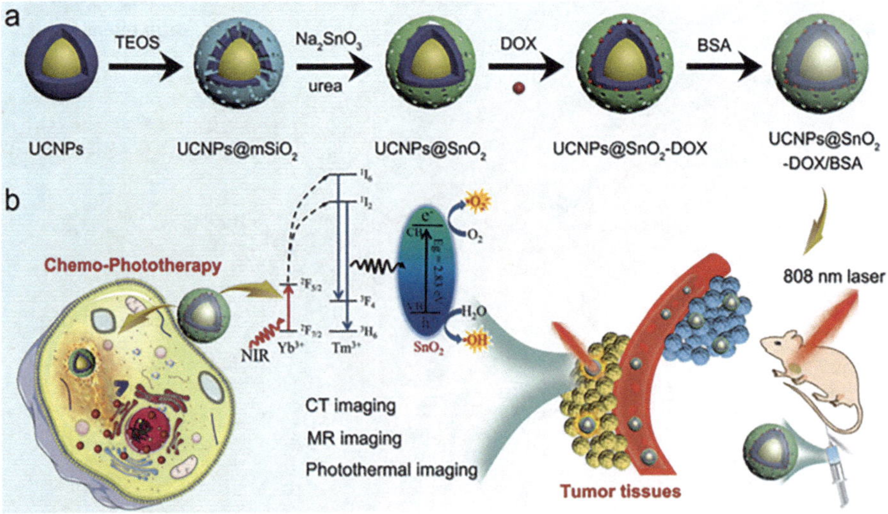

Fig. 2.4 Preparation and treatment mechanism of UCNPs@SnO_2-DOX/BSA nanocapsules. (**a**) Fabrication of UCNPs@SnO_2-DOX/BSA nanocapsules. (**b**) Schematic representation of nanocapsules for multimodal imaging-guided stimuli-responsive chemophototherapy (Yang et al. 2022c)

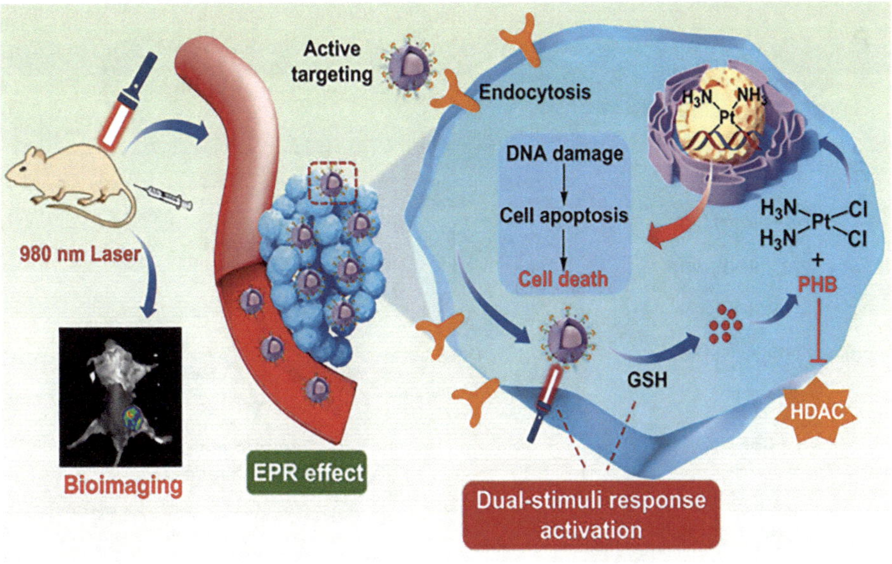

Fig. 2.5 The mechanism of UCNP/Pt(IV)-RGD for tumor therapy (Liu et al. 2024)

constructed a multifunctional UCNP: On the one hand, as a drug carrier, it can effectively load small-molecule drug KGN and target MSC for delivery, and at the same time, it can release KGN efficiently under near-infrared light irradiation to specifically induce MSC chondro differentiation, so as to realize "light-induced" MSC chondro differentiation in vitro and in vivo. On the other hand, with its excellent optical properties, UCNP after drug release can be used as a fluorescent probe to track MSC in vivo for a long time, providing a certain scientific basis for the therapeutic effect of MSC (Fig. 2.6).

2.2.2 Photodynamic Therapy

Photodynamic therapy (PDT) employs photosensitizer (PS) to absorb excitation light of a specific wavelength (Chan et al. 2017; Song et al. 2024). In this process, electrons shift from the ground state to the excited state. Subsequently, a portion of the electrons in the excited state return to the ground state, thereby emitting fluorescence via radiation. The remaining electrons progress to the triplet state and exchange energy with oxygen molecules or water molecules in the ambient environment, generating biotoxic reactive oxygen species (ROSs) and fulfilling the objective of eradicating tumor cells (Jiang et al. 2024; Wang et al. 2024). There are three crucial elements within the photodynamic process: light of a specific wavelength, photosensitizer, and oxygen or water molecules in the vicinity of the tissue. In comparison to clinical chemotherapy and radiotherapy, photodynamic therapy exhibits

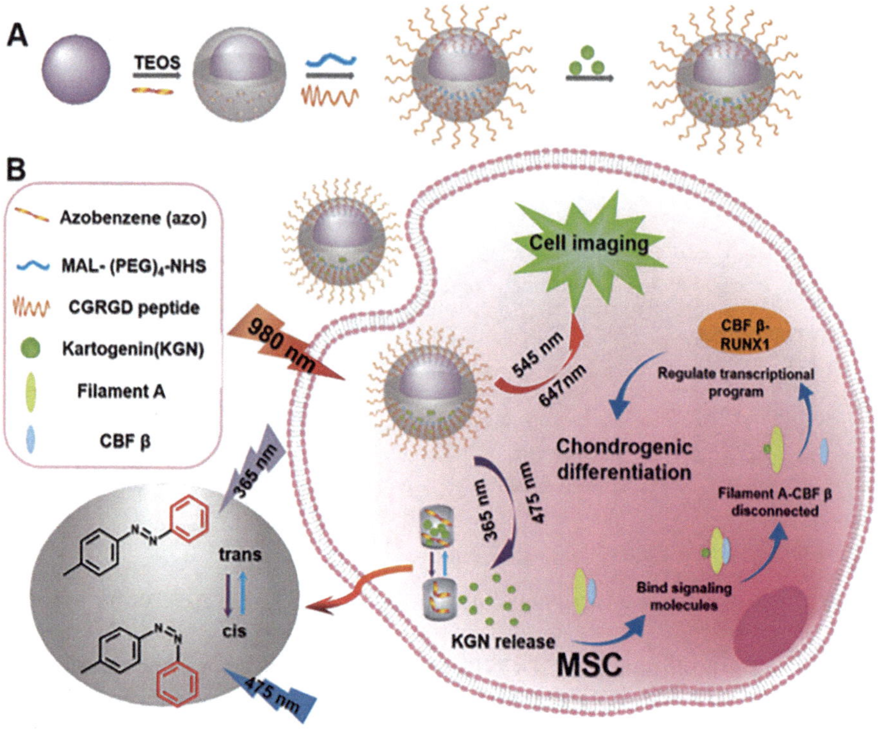

Fig. 2.6 (**a**) Schematic illustration of the synthesis of the multifunctional UCNP nanoplatform, and the photocontrolled chondrogenic differentiation, and (**b**) long-tern tracking of MSCs in vitro and in vivo by the UCNP nanoplatform through NIR light (Yang et al. 2022b)

several distinctive advantages, including high selectivity, favorable cost-effectiveness, a high cure rate, and less damage to body tissues following surgery (Nsubuga et al. 2024; Zhang et al. 2020). It has been successfully applied to the treatment of superficial tumors or organs such as skin, head and neck, esophagus, pancreas, and bladder. At present, PDT is usually based on UV-visible light irradiation. However, due to the light absorption and scattering of biological tissues, UV-visible light has poor tissue penetration ability, which limits the ability to treat tumors. UCNPs as carriers of photosensitizers provide a new strategy to solve this problem. UCNPs NIR excitation light has a deeper tissue penetration ability, which can effectively stimulate PSs and produce strong activity of ROSs. At the same time, it can improve the problems of PSs in vivo delivery difficulty and low targeting and achieve the purpose of directly killing cancer cells or destroying blood vessels at the tumor site and activating the host immune system (Jiang et al. 2024; Wang et al. 2024; Zhang et al. 2020).

In 2021, Chen et al. (2024) loaded blebbistatin onto the upconversion nanoparticles coated with mesoporous silica. The loading of blebbistatin not only enables efficient conversion from near-infrared light to blue light. In addition, blebbistatin was capable of direct light-induced release of oxygen-independent hydroxyl radicals (·OH). Therefore, this design strategy shows a near-infrared light-triggered photodynamic and sonodynamic combined therapy (PACT) treatment idea for generating ·OH without oxygen. The experimental results show that UCSNs-B has a highly efficient therapeutic effect and provides new ideas for developing efficient nanomedicines for cancer treatment.

In 2023, Tan et al. (2024) conducted a key research. They developed an innovative near-infrared responsive nanoplatform, which was specifically designed for efficient antibacterial photodynamic therapy (PDT). The researchers ingeniously integrated lanthanide-doped UCNPs, platinum-modified porphyrin-metal organic frameworks (PMOFs), and near-infrared dye IR808 into a core–shell structure. In this nanoplatform, the PMOF shell played a dual role: on the one hand, it acted as a photosensitizer, capable of absorbing the upconversion luminescence (UCL) from UCNPs and using this energy to generate singlet oxygen ($^1O_2$), which is a key active component of PDT; on the other hand, it also serves as a porous carrier for loading the IR808 dye. In addition, it was found that the heavy atom effect of Pt(II) plays a crucial role in increasing the intersystem crossing efficiency of IR808 and the porphyrin ligand Pt-TCPP, which not only enhances the triple sensitization effect of the dye on UCL but also boosts the $^1O_2$ production by the UCNP photosensitizer. The experimental results showed that this newly developed nanoplatform exhibited an extremely high $^1O_2$ production capability and showed remarkable bactericidal efficacy in tests against methicillin-resistant Staphylococcus aureus (MRSA), achieving a killing effect of 5.27 log10 (>99.999%) of bacteria were killed. This achievement highlights the great potential of this nanoplatform as a new and highly efficient antibacterial treatment modality.

Chen et al. (2018) proposed a deep PDT technique based on UCNPs enriched with photosensitizers and synthesized PPIX-PEI-UCNP@FA nanoparticles for deep PDT. The composite material consisted of PPIX molecules, UCNPs doped with ytterbium (Yb) and erbium (Er), PEI, and folic acid (FA). Moreover, an amino rich PEI with high branching degree was used to connect the surface bifunctionalized UCNPs to construct a spatial nanosphere structure with UCNPs as the core, which had a high specific surface area and achieved a PPIX loading rate of up to 0.69 wt.%. At the same time, amino groups were used to effectively reduce the toxicity of PEI. It can achieve a deep tumor killing effect of more than 10 mm.

Shi et al. (2024) developed a nanomaterial with a luminescence mechanism completely different from that of traditional upconversion luminescent nanoparticles. They prepared lanthanide ion-doped upconversion luminescent bismuth carbonate ($Bi_2O_2CO_3$) nanosheets (NS) with an ultra-thin two-dimensional (2D) structure. The bismuth (Bi), ytterbium (Yb), and erbium (Er) elements contained in the NS can not only enhance the radiosensitizing effect, but also the lower content of Er ions will emit brighter red light. The NS has both radiotherapy and photodynamic therapy effects (Fig. 2.7). Compared with traditional single treatment methods, this combined treatment greatly improves the treatment effect. The experimental results also show that it has a strong therapeutic ability for tumor cells. At the same time, it

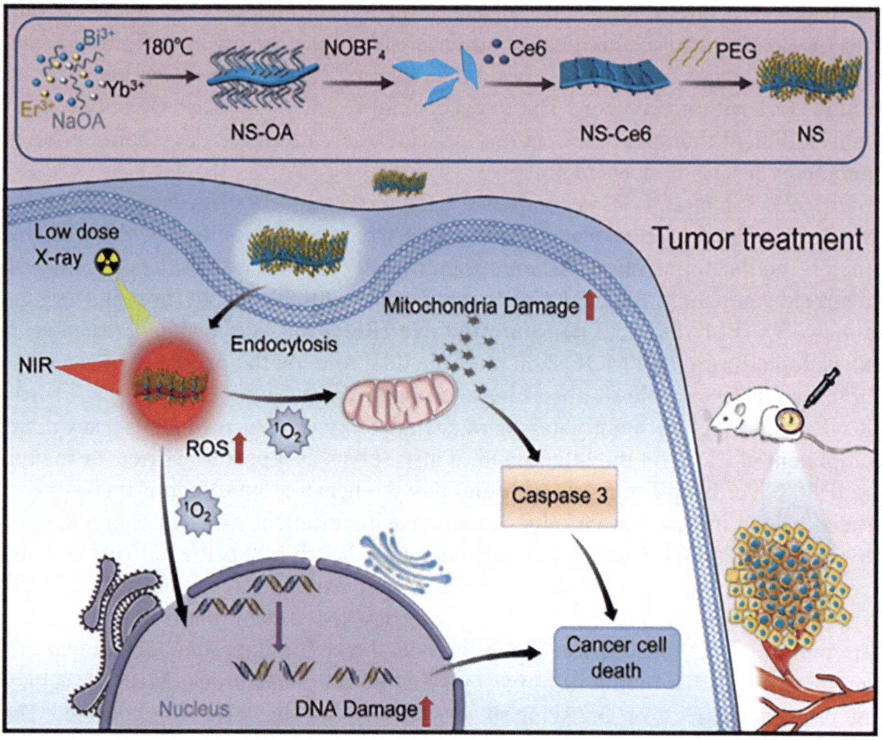

Fig. 2.7 Schematic illustration of the fabrication and working mechanism of nanosheets as radiosensitizer and photodynamic therapy platform for tumor synergistic therapy (Shi et al. 2024)

largely avoids radiation damage caused by radiation and thermal damage caused by photodynamic therapy.

2.2.3 *Photothermal Therapy*

Photothermal therapy (PTT) is a new tumor treatment method, which mainly uses materials with high photothermal conversion efficiency as photothermal agents to absorb light energy and convert it into heat under NIR light irradiation, and can selectively heat tumor cells or tissues (Li et al. 2023; Xu et al. 2021). Due to the abnormal morphology of blood vessels in tumor tissue, they are fragile and easy to rupture. When attacked by high temperature, they are easy to rupture, resulting in the reduction of local blood flow, which leads to local hypoxia and heat dissipation difficulties, thus killing tumor cells. Compared with traditional treatment methods, PTT has the advantages of deep-tissue penetration, less damage to normal tissues, repeatable treatment, and relatively few toxic and side effects (Atkins et al. 2021;

Zhang et al. 2019b). The effectiveness of PTT mainly depends on the photothermal conversion efficiency of the photothermal agent. An ideal metal nanomaterial for photothermal therapy should have the advantages of strong and adjustable surface plasmon resonance absorption effect, easy transport, low toxicity, and easy binding to tumor cells, such as gold and silver nanoparticles. CuS (Chen et al. 2018), Cu_9S_5 (Zhao et al. 2019), and LaB_6 (Song et al. 2020) nanoparticles with fixed NIR absorption bands are also a novel class of photothermal agents that can be used in PTT (Zhang et al. 2018).

Liu et al. (2021b) developed the novel near-infrared (NIR) photoresponsive hybrid hydrogel (UCNP-Au-Alg) by upconverting the complexation of lanbium-Au hybrid nanoparticles (NP) and alginate. UCNP-Au-Alg hydrogel was rapidly heated under 808-nm NIR irradiation to eradicate subcutaneous solid tumors in a mouse model. The immobilization of the damaged bone through the interaction between Ca^{2+} and hydrogel enables the hydrogel to perform the function of supporting and immobilizing the damaged bone, thus showing its potential in the repair and treatment of bone-related complications.

Zhou et al. (2018) developed a multifunctional nanoparticle with tumor targeting, which combined tumor starvation and low-temperature PTT to achieve better therapeutic effects (Fig. 2.8). Glucose oxidase (GOx) was loaded into the nanosystem using porous and hollow Prussian blue nanoparticles (PHPBNs) and then hyaluronic acid (HA) was modified on its surface by disulfide bonds with reducing properties. The combination of tumor starvation therapy and low-temperature photothermal therapy can effectively improve the therapeutic effect, which has a good

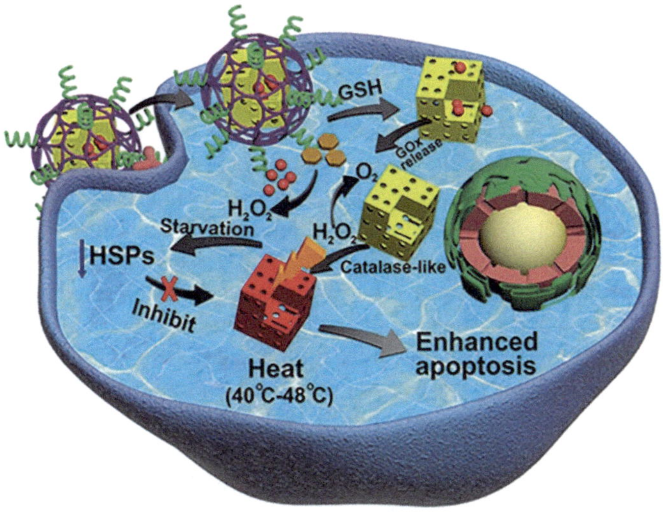

Fig. 2.8 Schematic presentation of the multifunctional porous hollow Prussian Blue nanoparticles (PHPBNs) for combined tumor starvation and enhanced low-temperature photothermal therapy against hypoxic tumors (Zhou et al. 2018)

practical significance for the establishment of combined therapy system in clinical treatment.

Wang et al. (2016) synthesized a porous $NaLuF_4$. This material was doped with Yb^{3+}, Er^{3+}, and Gd^{3+} and conjugated with cypate, and its potential in upconversion luminescence/magnetic resonance dual-modality molecular imaging was tested at the same time. This technique can be used to guide tumor therapy. When the small interfering RNA gene against heat shock protein 70 (UCNP-cy-siRNA) is loaded onto the upconversion nanoparticles conjugated with cypate (UCNP-cy), cell damage will be enhanced. However, thanks to the synergistic effect of gene silencing and photothermal therapy, UCNP-cy-siRNA exhibited a significant anti-tumor effect in vivo, and the drug dose used was low, and the side effects were minimal. This discovery provided a clear direction for the development of the next-generation multifunctional nanoplatforms, which can be used for synergistic tumor therapy under the guidance of multimodal imaging, thereby improving the accuracy and effectiveness of treatment.

In 2023, Wang et al. designed an innovative brain-targeting near-infrared theranostic system, which has two fluorescence resonance energy transfer (FRET) pathways and excellent spectral matching performance, aiming to maximize energy utilization for the synergistic photodynamic and photothermal therapy of gliomas (Fig. 2.9). This system is constructed based on multiple optimization adjustments, using thulium (Tm)-doped upconversion nanoparticles (UCNPs), zinc tetraphenylporphyrin (ZnTPP), and copper sulfide (CuS) nanoparticles as the main building materials. The research team first precisely adjusted the doping ratio of Tm, which significantly increased the relative emission intensity of UCNPs at 475 nm by a factor of 11.5. Meanwhile, the J-aggregates formed by ZnTPP greatly enhanced its absorption capacity at 475 nm, with an increase of 163.5 times. This combination not only optimizes the FRET efficiency between UCNPs and ZnTPP but also provides a solid foundation for efficient near-infrared photodynamic therapy. In addition, the system cleverly utilizes the emission light at 800 nm to enhance the photothermal effect of CuS nanoparticles through the second FRET pathway, thereby implementing photothermal therapy. After being modified by a brain-targeting peptide, the system shows powerful therapeutic efficacy and can effectively trigger the synergistic phototherapy ablation of glioma cells. When the system successfully crosses the blood–brain barrier and precisely targets the glioma, it can significantly prolong the survival time of orthotopic glioma mice. The successful application of this advanced spectral modulation technology and dual-site FRET strategy opens up new avenues for future light energy utilization in the treatment of brain diseases and inspires more innovative thinking about the application of upconversion nanoparticles in the medical field.

Li et al. (2018) fabricated a multifunctional nanocomposite, UCNP@NGO, for upconversion fluorescence imaging and photothermal therapy. This was achieved by complexing the PTT agent NGO with the upconversion luminescence probe UCNP through carboxyl groups. The resultant theranostic nanoplatform demonstrated a high upconversion luminescence intensity, suitable for imaging purposes, and an enhanced therapeutic efficacy due to the PTT effect. This innovative

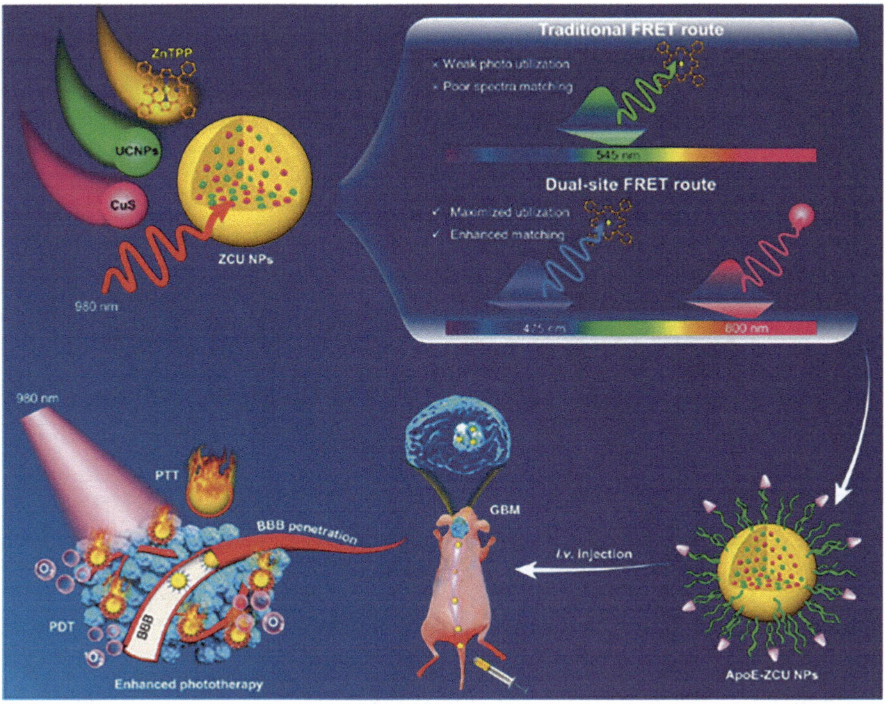

Fig. 2.9 Illustration for the advantage comparison between traditional FRET and dual-site FRET routes of ApoE-ZCU NPs, which powerfully support the brain-target phototherapy of orthotopic glioma (Wang et al. 2023)

nanocomposite integrated upconversion luminescence (UCL) imaging and PTT into a unified nanoplatform, serving as a theranostic probe for combined UCL imaging and PTT in cancer treatment (Fig. 2.10). As both the imaging and PTT modalities within this nanoplatform are light-triggered, only the tumor area was targeted, conferring significant advantages over traditional cancer therapies in terms of increased cancer cell killing specificity and reduced side effects. Furthermore, a synergistic effect between the combined UCL imaging diagnosis and photothermal therapy was validated, with the anticipation of diminishing dosage-limiting toxicity and tissue damage caused by overheating, thereby enhancing the overall therapeutic efficiency.

Wang et al. (2020) put forward a theranostic platform that combined CCM-based biomimetic UCNPs and AuNPs with an innovative simultaneous dual-modal imaging system, as shown in Fig. 2.11. In this platform, AuNPs and UCNPs were enclosed by CCM, resulting in the formation of CC-AuNPs and CC-UCNPs, respectively. Such encapsulation promoted immune evasion and allowed for highly specific targeting of homologous cancers, as illustrated in Fig. 2.15a. A dual-modal imaging system, based on a novel detector named OptX that had the ability to detect both high-energy X-rays and optical wavelength photons, was designed and put into practice. This system enabled both upconversion fluorescent imaging and X-ray

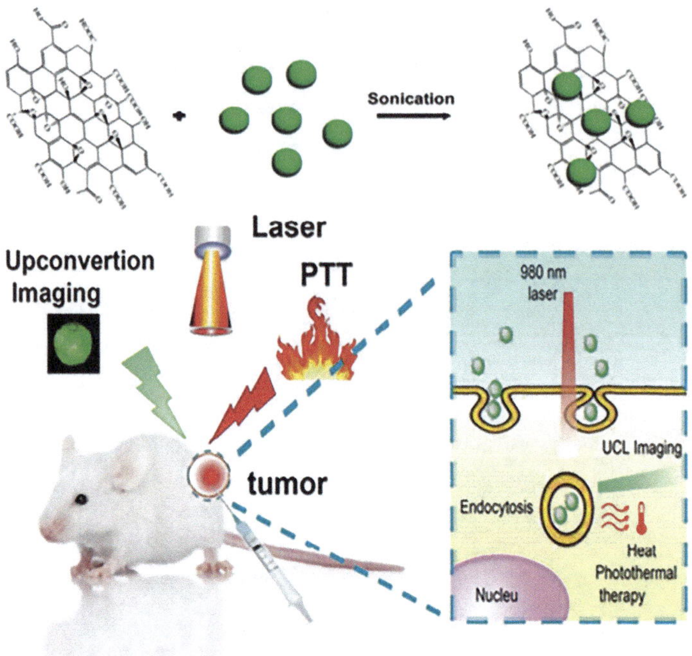

Fig. 2.10 Illustration of complexation of UCNP@NGO as a theranostic agent for near-IR imaging and photothemal therapy (Li et al. 2018)

imaging using only one detector, removing the need for additional markers or algorithms during the registration and fusion of the two imaging modalities. It has been confirmed that CC-UCNPs could evade the reticuloendothelial system (RES) and accurately target tumor sites. Together with the X-ray imaging provided by the established system, CC-UCNPs offered valuable information for guiding and evaluating PTT therapy, as depicted in Fig. 2.15b. It is worth noting that CC-AuNPs, when used as a PTT reagent, showed better performance than conventionally functionalized AuNPs. This development provides an outstanding methodology and platform for research on imaging-guided PTT for cancer treatment.

Song et al. (2020) found that gold and silver nanomaterials have surface plasmon resonance (SPR) properties, so they prepared $NaYF_4$:Yb/Er@Ag core–shell nanocomposites. This material was capable of performing effective photothermal therapy (PTT) under 980-nm laser excitation, but this material can only perform a single type of cancer treatment. Yang et al. constructed TiO_2@$Y_2Ti_2O_7$@YOF:Yb/Tm nanocomposites with a multi-layer core–shell structure. Under the irradiation of a 980-nm laser, this material can not only perform PTT but also PDT, showing an efficient combined therapeutic effect. Specifically, as shown in, due to the formation of the core–shell composite structure, the distance between the valence band and the conduction band is reduced, and thus the $Y_2Ti_2O_7$ photocatalyst generates a large amount of reactive oxygen species (ROS). Meanwhile, a large amount of thermal

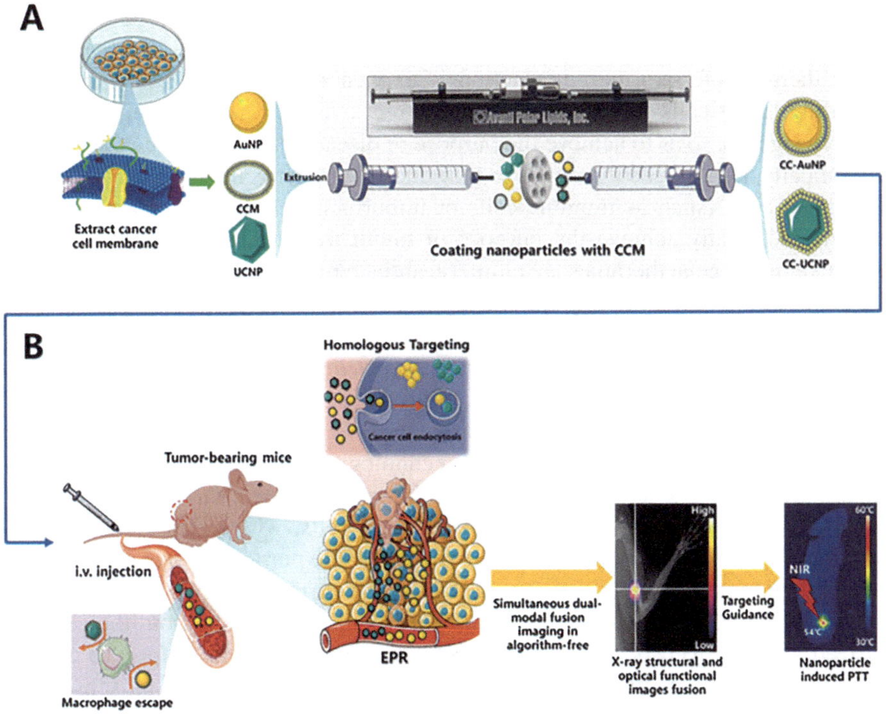

Fig. 2.11 Schematic of the synthesis and application of cancer cell membrane-coated gold nanoparticles (CC-AuNPs) and cancer cell membrane-coated upconversion nanoparticles (CC-UCNPs) combined with a novel simultaneous dual-modal imaging platform for guided photothermal therapy (PTT) for cancer. The cancer cell membrane-coated UCNPs were used for highly specific tumor imaging and provided information for the PPT of cancer, which is based on the cancer cell membrane-coated AuNPs. (**a**) Preparation procedure of CC-NPs. (**b**) Schematic of immune escape and homologous targeting CC-NPs for dual-modal imaging-guided PTT (Wang et al. 2020)

energy is generated during the non-radiative process and the recombination of electron–hole pairs. Eventually, a large amount of ROS and thermal energy work together to cause cancer cell apoptosis. This material can reduce the damage of the laser to skin tissue to a certain extent.

In recent years, PTT combined with UCNPs has made significant advancements in the field of cancer treatment. UCNPs were capable of converting low-energy NIR light into high-energy visible or ultraviolet light, which enables their activation in deep tissues, overcoming the limited light penetration depth issue in traditional phototherapy. By integrating UCNPs with photothermal materials (such as gold nanoshells, graphene, and so on), researchers have developed various efficient PTT platforms. With technological advancements and increased interdisciplinary collaboration, we have reason to believe that this novel therapeutic approach will play an increasingly important role in future cancer treatments.

2.2.4 Gene Therapy

Gene therapy refers to the introduction of exogenous target genes into the corresponding target cells through vectors to correct over-activated or compensate for defective genes, so as to achieve the purpose of disease treatment (Liu et al. 2023b). Specifically, tumor gene therapy is to introduce the target gene into the corresponding target cells (such as immune cells or tumor cells) through the corresponding vector, and finally achieve the purpose of tumor treatment. At present, the most effective tumor gene therapies are chimeric antigen receptor T cell therapy (CAR-T) based on immune cell therapy and oncolytic virus (OVs)-based tumor gene therapy (Liu et al. 2023a).

The main measures of gene therapy are: (1) *Gene correction*: correct the abnormal bases in the disease-causing genes, while the normal part is retained. (2) *Gene replacement*: refers to the use of normal genes to replace disease-causing genes in situ through homologous recombination technology, so that the DNA in the cell can completely return to a normal state. (3) *Gene Supplementation*: This approach involves introducing normal genes into somatic cells, where they integrate non-specifically into the genome and begin to express. The purpose is to compensate for the impaired function of defective genes or to enhance the activity of existing genes. Importantly, this method does not remove the pathogenic genes from the cells but instead aims to override their negative effects. (4) *Gene Inactivation*: To achieve gene inactivation, specific types of antisense nucleic acids, such as antisense RNA and antisense DNA, along with ribozymes, are delivered into cells. These molecules target and interfere with the expression of problematic genes at both the transcriptional and translational stages, effectively blocking their abnormal activity and achieving therapeutic goals. (5) *Suicide gene therapy*: a suicide gene, derived from certain viruses or bacteria, encodes an enzyme that can convert a non-toxic or minimally toxic prodrug into a cytotoxic agent within the cell. As a result, the cell expressing the suicide gene will undergo self-destruction upon exposure to the appropriate prodrug. This strategy is particularly useful for targeting and eliminating undesirable cells, such as cancer cells. (6) *Immunotherapy*: immunogene therapy is to introduce the corresponding and epitope genes that produce antiviral or tumor immunity into the body's cells to achieve the purpose of treatment, such as the introduction and expression of cytokine genes. (7) *Drug-resistant therapy*: Drug-resistant gene therapy is to introduce anti-drug toxicity genes into human cells in order to improve the body's ability to tolerate chemotherapy drugs during tumor treatment, so that the body can tolerate a higher dose of chemotherapy (Zhang et al. 2022; Zhao et al. 2023).

To enhance the endosomal escape of the delivered gene, Zhang et al. (2021a) employed photochemical internalization (PCI) to combine UCNPs with sodium tetraphenylporphyrin-sulfonate (TPPS2a, a photosensitized agent) and photomorpholine (a nucleic acid analogous for gene knockdown). It can jointly load the signal transduction and transcriptional activator 3 to improve the therapeutic effect. When excited by NIR, UCNPs triggered TPPS2a to produce RO, which caused

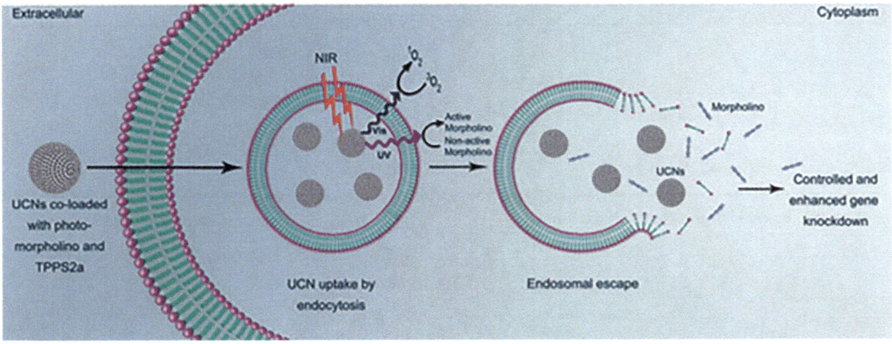

Fig. 2.12 Gene knockdown with UCNP-based photoactive delivery platform (Zhang et al. 2021b)

damage to the local endosomal vesicle wall, prompting particle release. In addition, the ultraviolet light emitted by the upconversion nanoparticles can cut the sense chain of photomorpholine and release the antisense morpholine, which can further reduce the expression of the signal transduction and transcriptional activator 3 (STAT3) gene (Fig. 2.12). This design idea provides clinical application potential for gene therapy of cancer.

Wang et al. (2014b) loaded small interfering RNA (siRNA) targeting the Plk1 oncogene with therapeutic effects onto UCNPs. The nanoparticles can target cancer cells accurately. Under NIR irradiation, they will deliver PDT and siRNA to cancer cells, thereby causing cancer cell apoptosis. This is a simple and effective treatment method. Through a simple polymer coating method, it combines PDT with gene therapy to kill cancer cells simply and efficiently.

Cheng et al. (2022) devised a NIR light-activated delivery platform for DNAzymes to enhance gene silencing efficacy. This nanoplatform integrated therapeutic DNAzymes, photosensitizers (Ps), and UCNPs, with the UCNPs capable of converting NIR light into visible light. Due to the transfer of energy from UCNPs to Ps, this system allowed near-infrared light activation to produce cytotoxic reactive oxygen species, which facilitates endosomal escape of DNA zyme, thereby improving gene silencing efficacy.

Liu's group designed the first CRISPR-cas13d-based nanoprodrug for enhancing cancer therapy by using CRISPR technology through orthogonal photo modulation of gene editing and prodrug release, based on Pt (IV) prodrugs and orthogonal emission UCNPs (Liu et al. 2023a) (Fig. 2.13). The green light-activated Pt (IV) prodrug and visible UV-activated Cas13d gene editing kit was encapsulated in UCNPs to construct the nanoprodrug (URL). The results showed that URL maintained good orthogonal emission behavior under excitation at 808 nm and 980 nm, which could perform wavelength-selective photoactivation of Cas13d and prodrug, down-regulate drug resistance-related genes, and induce chemical photodynamic therapy. In addition, the advantages of optoelectronic modulation of URLs in overcoming drug resistance were highlighted by integrating URLs with Boolean logic gates that programmatically modulate multiple cell behaviors. Importantly, in vivo studies

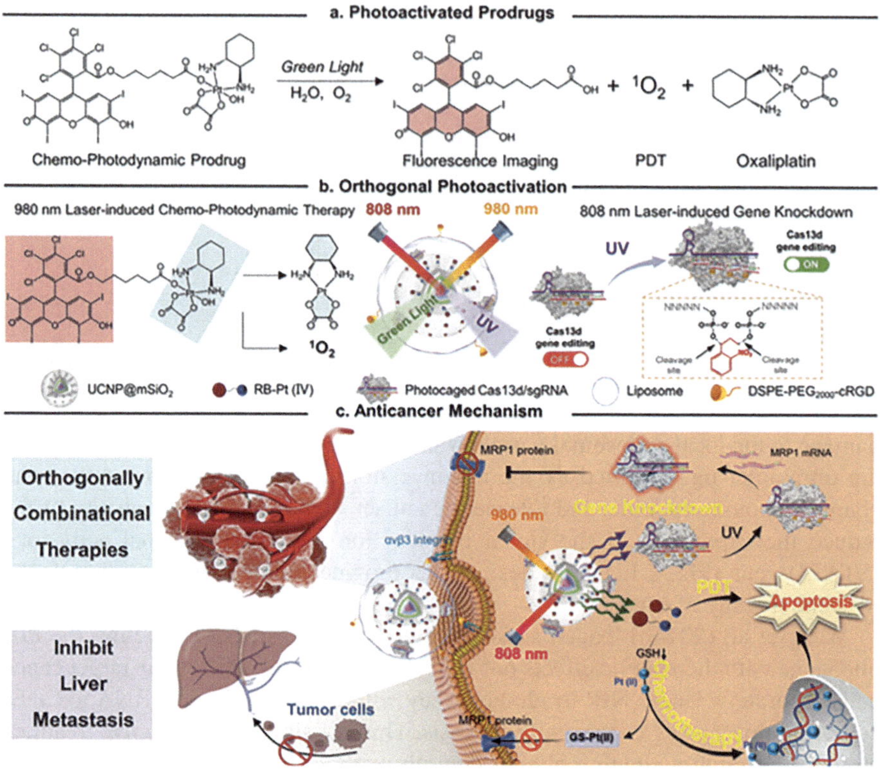

Fig. 2.13 Schematic diagram of the design of orthogonally activated CRISPR. (**a**) It showed the chemical composition of the chemo-photodynamic prodrug as well as the principle of its photoactivation. (**b**) The orthogonal photoactivation process included chemo-photodynamic therapy triggered by the oxaliplatin prodrug and gene editing mediated by Cas13d. (**c**) The nanoprodrug can be activated on demand, featuring sequential MRP1 knockdown and effective glutathione-mediated release of oxaliplatin(II). This helps reverse drug resistance and thus brings about improved combination therapy in vivo (Liu et al. 2023b)

have shown that URLs can promote Pt (IV) prodrug activation and ROS generation, and induce targeted drug accumulation on a large scale through cas13d-mediated resistance attenuation. It provides the ultimate chemophotodynamic therapy performance, effectively eradicating primary tumors and preventing further liver metastases.

Song et al. (2024) developed a multifunctional DNA/UCNPs complex that can control the co-delivery of CRISPR-cas9, hemoglobin, and protoporphyrin (PP) for synergistic PDT. Ultra-long single-stranded DNA (ssDNA) was prepared by rolling circle amplification (RCA). The ssDNA contains a single guide RNA (sgRNA) recognition sequence loaded with Cas9 ribonucleoprotein (RNP), a g-quadruplet sequence loaded with hemin and PP, and a linker sequence bound to UCNP. Cas9 RNP cleaves antioxidant regulatory nuclear factor e2-related factor 2 (Nrf_2) to improve the sensitivity of cancer cells to ROS and enhance the synergistic PDT effect. g-Quadruple/hemoglobin DNAzyme mimics horseradish peroxidase (HRP) to catalyze the conversion of endogenous H_2O_2 to O_2 to overcome tumor hypoxia. The introduced UCNPs convert near-infrared radiation penetrating deep tissues into light with shorter wavelengths and excitation PP converted abundant O_2 into $^1O_2$. The combination of gene editing and PDT enhanced apoptosis by increasing the accumulation of $^1O_2$ in cancer cells, and this synergistic PDT has shown significant therapeutic effects in mouse models of breast cancer.

In summary, the integration of UCNPs with gene therapy holds significant potential in cancer treatment. By leveraging the unique properties of UCNPs, future research and development efforts can pave the way for more effective, targeted, and personalized cancer treatments. As the field continues to evolve, the potential for UCNP-based gene therapy to transform cancer care becomes increasingly evident.

2.2.5 *Immunotherapy*

Tumor immunotherapy is a therapeutic strategy that utilizes immunological principles and methods to enhance the immunogenicity of tumor cells and their sensitivity to effector cell killing, while simultaneously activating and enhancing the host's anti-tumor immune response. This process involves the introduction into the patient of immune cells and effector molecules that work in concert with the host's own immune system to work together to destroy tumor cells and inhibit tumor growth (Wang et al. 2019, 2022). Unlike traditional treatments, tumor immunotherapy primarily works by modulating and strengthening the immune system to trigger an immune response against tumors. This approach not only directly attacks the tumor, but also boosts the body's overall defense capabilities to recognize and destroy cancer cells more effectively.

The immune system plays a highly significant role in human health. It is crucial to understand how the immune system functions and how it regulates in order to protect the body and cope with diseases (Park et al. 2020). The combination of immune biotechnology and nanomaterials can provide more effective and safer immunotherapy strategies. In 2013, the immunotherapy method was named the top ten scientific breakthroughs of the year by Science (Wang et al. 2019; Wu et al. 2022). At present, there are three main methods of tumor immunotherapy (Zhu et al. 2023): (1) Tumor vaccine; (2) Adoptive immunotherapy; (3) Immune checkpoint blockade.

Du et al. (2024) prepared a nanoformulation with a core–shell structure by combining UCNPs with metal-organic framework materials (UCNP@MOF). The porous structure of the MOF material in this nanoformulation can load a large number of immunotherapeutic drugs, and under near-infrared (808 nm) irradiation, it can also perform Type I and Type II photodynamic therapies, achieving coordinated therapeutic linkage. The reduction of Cu^{2+} to Cu^{+} on the MOF shell node has been shown to promote the production of reactive oxygen species by depleting the overexpressed glutathione (GSH) compound, while hydrogen peroxide enriched in the tumor microenvironment has been found to greatly enhance the PDT effect. Importantly, MOF pore-loaded indoleamine 2,3-dioxygenase (IDO) inhibitors, as well as s-nitroso-n-acetylpenicillamine (SNAP), upregulate tumor-infiltrating lymphocytes in vivo and induce significant immunotherapeutic effects. This core–shell UCNP@MOF nanoformulation has demonstrated efficacy for both primary and distant tumors in xenograft tumor models, promising to treat metastatic solid tumors.

Wang et al. (2022) ingeniously devised a smart UCNPs@Cu-Cys-GOx (UCCG) nanosystem predicated on TME-responsive UCNPs. This nanosystem amalgamates natural enzymes and nanozymes with the aim of augmenting the in situ generation of reactive oxygen species (ROS) for the purpose of cancer starvation, chemokinetics, and immunotherapy, as depicted in. One of the most prominent merits of this material lies in its ability to remain dormant or inactive (switched off) within normal tissues. However, specifically within the tumor microenvironment (TME), it can be selectively activated (switched on) through a cascade of enzymatic reactions. This activation promotes ROS production by implementing strategies such as open source (self-supplying H_2O_2) and curtailing consumption [via the depletion of glutathione (GSH)]. Significantly, the augmented oxidative stress induced by the UCCG nanocomposites managed to reverse the immunosuppressive TME and stimulate anti-tumor immune responses. Concurrently, the combination of UCCG with the PD-L1 antibody-triggered starvation and chemokinetic synergistic therapy efficaciously curbed the growth and metastasis of primary tumors. Additionally, the UCNPs incorporated within the UCCG formulation exhibit an enhancement in upconversion luminescence, which can be harnessed for the real-time visualization of the augmented ROS generation.

Li et al. (2022) developed an NIR-triggered nitric oxide (NO)-releasing nano-gas reservoir (UNTPs) for reprogramming ITME in combination with ICB therapy and enhancing immunotherapy efficacy. This nitric oxide-emitting nano-gas reservoir consists of an UCNPs core stabilized by a biodegradable polymer donor, polyethylene glycol-polynitrate-pen-ethylene copolymer (mPEG-PNTC-PEI). Ultraviolet (UV) stimulation accelerates the release of NO from UNTPs nano-gas reservoirs, and UV stimulation is converted from the core of UCNPs under near-infrared irradiation. Transient NO burst directly kills tumors with strong response ability, significantly promotes immunogenic cell death (ICD) of tumor cells, thereby promoting the maturation of dendritic cells (DCs) and the infiltration of T cells. In addition, ITME was remodeled into an immune-stimulating environment with potent distant tumor suppression and long-term survival by ICB, through a large number of NO and αPD-1 antibodies, inhibits PD-L1 expression and induces tumor-associated macrophage (TAM) polarization to an anti-tumor M1 phenotype.

The treatment of hypoxic tumors is still at a critical point of breakthrough. To address this issue, Shao et al. (2020) developed upconversion nanoparticle @ porphyrinic metal-organic framework nanocomposites (UCSs) with a core–shell structure. UCSs used UCNPs as the core, and through simple surface engineering of them, nanocomposites with a core–shell structure can be synthesized efficiently and in large quantities. This special core–shell heterogeneous structure enables effective energy transfer, thereby generating a large amount of ROS under NIR light irradiation, which can kill cancer cells. In addition, the researchers also encapsulated the hypoxia-activated prodrug tirapazamine (TPZ) in the nanopores (TPZ/UCSs), thereby realizing hypoxia-activated chemotherapy. The successful preparation of TPZ/UCSs provides a unique solution for the treatment of hypoxic tumors. This kind of nanoparticle can achieve both in vivo and in vitro treatment of hypoxic tumors, thus enabling multimodal treatment. Moreover, when this nanoparticle platform is used in combination with the anti-programmed death ligand 1 (α-PD-L1) immune checkpoint inhibitor, it can promote the abscopal effect by triggering the specific infiltration of cytotoxic T cells into tumors, thereby effectively inhibiting the growth of distant tumors that have not been directly treated. Overall, this study shows a powerful multifunctional nanoplatform that combines near-infrared light-triggered photodynamic therapy, hypoxia-activated chemotherapy, and immunotherapy, aiming to overcome the challenges and limitations in existing tumor treatments. This innovative combination therapy not only enhances the effect of local tumor control but also activates the systemic immune response to combat distant metastatic lesions, providing new ideas and tools for cancer treatment (Fig. 2.14).

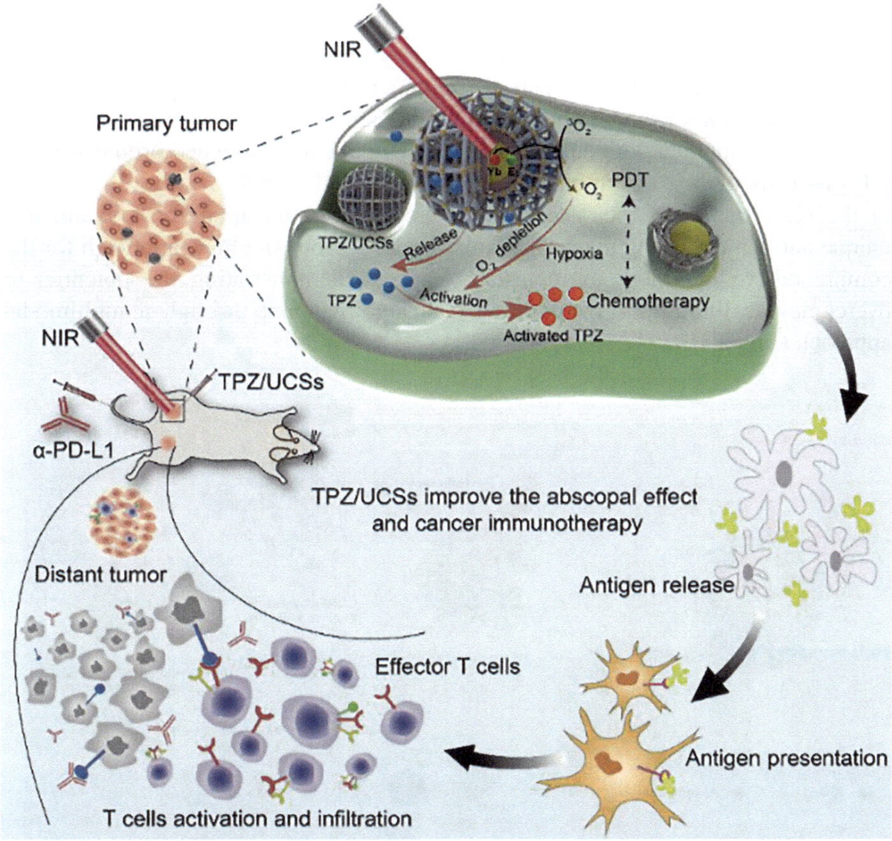

Fig. 2.14 Schematic illustration of the structure of TPZ/UCSs and their application for tumor treatment through the combination of PDT, hypoxia-activated chemotherapy, and immunotherapy (Shao et al. 2020)

In order to effectively treat metastatic cancers, Wang et al. (2019) designed an antigen-capturing nanoplatform based on UCNPs, aiming to promote the synergistic effect between phototherapy and immunotherapy. The nanoplatform was constructed through self-assembly technology, specifically by combining distearoylphosphatidylethanolamine-polyethylene glycol-maleimide (DSPE-PEG-maleimide) and indocyanine green (ICG) onto UCNPs, and further loading the photosensitizer rose bengal (RB). Under NIR laser excitation, ICG improves the photodynamic therapy efficiency of the UCNP/ICG/RB-maleimide complex and simultaneously enables selective photothermal therapy. It was particularly worth noting that due to the presence of the maleimide group, the platform can capture and retain in situ the tumor-derived protein antigens from tumor cells after phototherapy treatment. This characteristic facilitated the effective uptake and presentation of tumor antigens by antigen-presenting cells, thereby enhancing the immune response. The experimental results showed that the synergistic photothermal, photodynamic, and immunological effects generated by using the light-activated UCNP/ICG/RB-maleimide nanoplatform can induce a strong tumor-specific immune response. In the mouse model, intratumoral injection of this nanoplatform followed by non-invasive near-infrared laser irradiation successfully destroyed the primary tumors and inhibited the growth of untreated distant metastatic tumors. In particular, in the 4T1 mammary tumor model, this combination therapy performed excellently, especially when combined with anti-cytotoxic T lymphocyte-associated protein 4 (anti-CTLA-4), about 84% of tumor-bearing mice achieved long-term survival, and 34% of the mice developed persistent tumor-specific immune memory. Overall, this unique antigen-capturing nanoplatform provides a promising new approach for the comprehensive treatment of metastatic cancers, demonstrating the potential to overcome the limitations of traditional cancer treatment through a multimodal approach (Fig. 2.15).

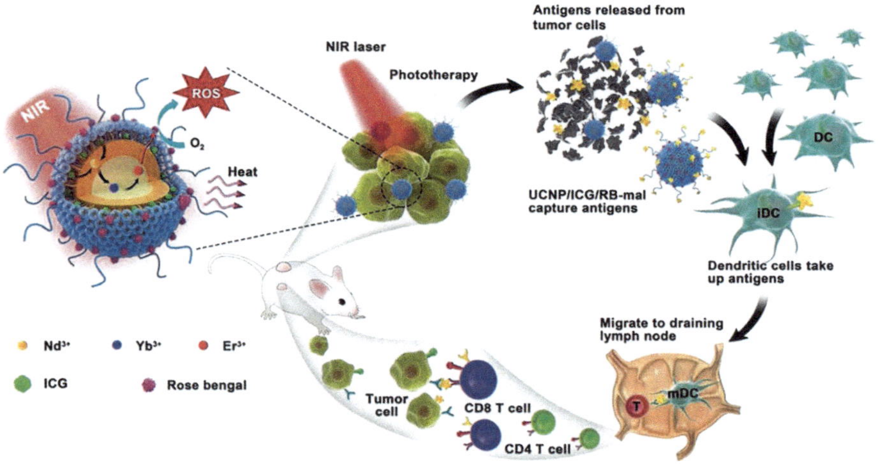

Fig. 2.15 Schematic illustration of both fabrication and mechanism of near-infrared (NIR)-triggered antigen-capturing nanoplatform for synergistic photoimmunotherapy (Wang et al. 2019)

According to research findings, there were some microorganisms with colonization ability that can survive in the tumor microenvironment and express special heterologous genes, thereby severely affecting the therapeutic effect and generating extratumoral toxicity. Zhu et al. (2023) combined *Escherichia coli* Nissle 1917 (EcN), which can sense blue light and release the encoded flagellin B (flaB), with UCNP materials for immunotherapy with high precision and controllable bacterial treatment. As shown in, under NIR irradiation, UCNPs will emit blue light, and the emitted blue light will activate EcN to release flaB. Then, flaB will specifically recognize the macrophage cell membrane and subsequently activate the pathway immune response. By combining optogenetic techniques with the natural properties of specific bacteria, researchers can develop precisely controlled treatment methods that both enhance the anti-tumor effect and reduce the impact on normal tissues. This innovative approach not only promotes the regression of local tumors but also stimulates systemic anti-tumor immune responses, opening up a new path for cancer treatment. These studies emphasize the great potential of the near-infrared nano-optogenetic platform in cancer immunotherapy, indicating that it may become an important part of future personalized medicine.

In this study, Choi and Kim (2023) developed an innovative nanocarrier system that combines immunotherapy, which aimed to attack tumors by stimulating the immune system, with photodynamic therapy (PDT) to selectively destroy only cancer cells. The researchers synthesized multi-shell structured upconversion nanoparticles (MSUCNs) with high photoluminescence (PL) intensity for the combined treatment of near-infrared light-induced photodynamic therapy and a specific immune checkpoint inhibitor. By optimizing the doping ratio of ytterbium ions (Yb^{3+}) and constructing a multi-shell structure, the synthesized MSUCNs exhibited a photoluminescence efficiency improvement of up to 260–380 times compared to the core particles and were capable of emitting light at multiple wavelengths. Subsequently, in order to achieve targeted delivery, photodynamic treatment, and immunomodulatory functions, the surface of MSUCNs was modified to carry folic acid (FA) as a tumor targeting ligand, chlorin e6 (Ce6) as a photosensitizer, and 1-methyl-tryptophan (1MT) as an indoleamine 2,3-dioxygenase (IDO) inhibitor. When these modified MSUCNs (F-MSUCN3-Ce6/1MT) bind to folate receptor-positive HeLa cancer cells, they show significant targeted cellular uptake ability. Under the irradiation of 808-nm near-infrared light, the F-MSUCN3-Ce6/1MT nanocarriers generate ROS, which leads to cancer cell apoptosis and activates $CD8^+$ T cells. By binding to immune checkpoint inhibitory proteins and blocking the IDO pathway, these nanocarriers enhance the body's anti-tumor immune response. Therefore, the F-MSUCN3-Ce6/1MT nanocarriers represent a potential synergistic anti-cancer treatment strategy that combines IDO inhibitor-based immunotherapy with enhanced near-infrared-triggered photodynamic therapy and may provide an efficient and low-side-effect new approach for cancer treatment.

Zhang et al. (2024) designed an innovative nanodrug system, which uses cancer cell membranes (CM) for camouflage, aiming to achieve multi-level targeting and enhance immune responses. This mitochondria-targeted nanodrug (UCNP@COF@FeCO/TPP@CM, abbreviated as UCFT@CM) integrates photodynamic therapy

(PDT), carbon monoxide gas therapy, and immunotherapy to achieve synergistic treatment effects. With the assistance of cancer cell membranes, the UCFT@CM nanodrug significantly enhances the homologous targeting ability. Once entering the cell through endocytosis, these nanodrugs will accumulate in the mitochondria of cancer cells and utilize the microenvironmental characteristics of mitochondria to control the release of carbon monoxide (CO) while increasing the efficiency of photodynamic therapy. This process leads to severe mitochondrial oxidative stress, thereby triggering a strong immune response. In addition, the new immune signal related to mitochondrial oxidative stress, mitochondrial transcription factor A (TFAM), is activated, which is closely associated with immunogenic cell death (ICD). The activation of TFAM not only intensifies the intracellular stress response but also promotes stronger immunogenic signal transmission, further enhancing the anti-tumor immune response. In conclusion, this sequential targeting strategy made full use of the unique properties of mitochondria, provides a new method for precise targeting of organelles, and can effectively enhance immune responses, opening up new avenues for cancer treatment. This approach demonstrates the potential of combining different treatment modalities to achieve more efficient and precise cancer treatment.

Qian et al. (2024) formed a core–shell structure by using UCNPs as the core and bridging a mesoporous silica layer with tetrasulfide bonds. In addition, Fe^{2+}, curcumin, and indoximod were modified to implement chemodynamic therapy/chemotherapy/immunotherapy. Driven by the tumor microenvironment (TME) specificity, this nanoplatfrom can not only activate immunogenic cell death but also trigger systemic anti-tumor immunity and alleviate immunosuppression. The experimental results showed that this nanoplatfrom can not only effectively inhibit the growth of primary tumors but also initiate immune surveillance to prevent the recurrence of tumor cells. This idea provided potential for multimodal treatment combined driving.

Wu et al. (2022) developed a two-stage synergistic treatment platform, utilizing upconversion nanoparticles (UCNPs) and engineered bacteria as therapeutic mediators to effectively treat primary tumors and distant metastatic tumors (see Fig. 2.16). The platform consists of two key steps: In the first stage, UCNPs doped with erbium ions (Er^{3+}) act as light switches to activate the photosensitizer zinc phthalocyanine (ZnPc) under near-infrared light irradiation, generating singlet oxygen ($^{1}O_2$) for targeted destruction of primary tumor cells. This process achieves precise treatment of primary tumors through highly efficient PDT. In the second stage, UCNPs doped with thulium ions (Tm^{3+}) emit blue-violet light upon excitation by near-infrared light, and this specific wavelength of light can activate the engineered bacteria, prompting them to generate and release interferon-γ (IFN-γ) into the intestinal environment. This not only enables direct combat against tumor cells in the intestine but also, through synergy with photodynamic therapy, modulates the immune pathways and activates the systemic immune system, thereby generating a combined immunotherapy effect to inhibit the growth of distant tumors. As an innovative programmed combination therapy, this platform demonstrates the potential to jointly activate the body's immune system during photodynamic and immunotherapy

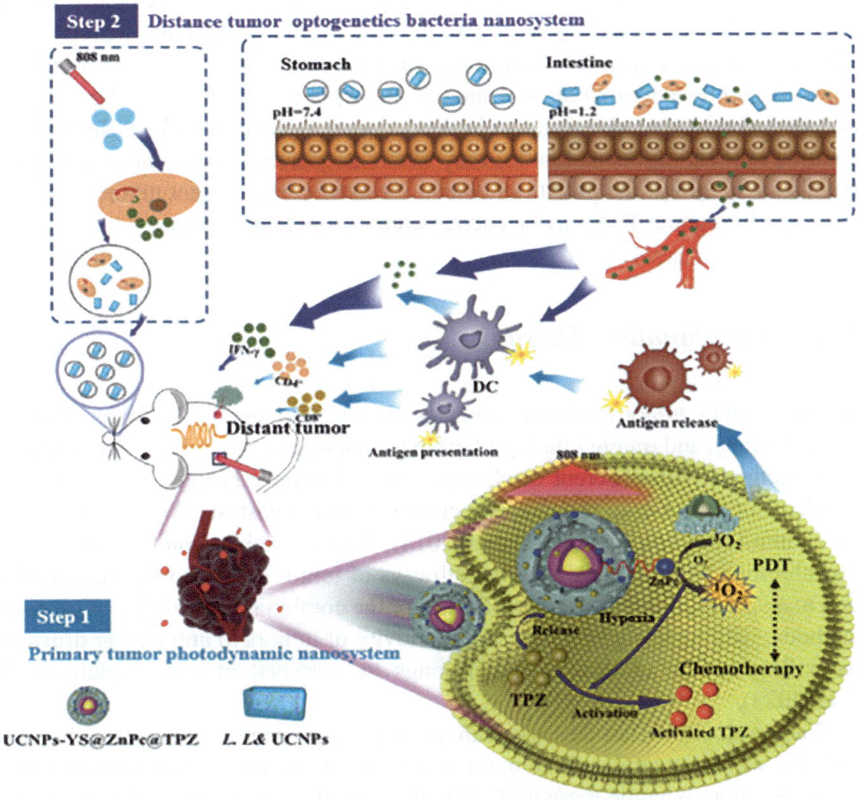

Fig. 2.16 Programmed synergistic platform for the treatment of the primary tumor and the distant tumor through the combination of the UCNP-based PDT nanosystem and the engineered bacteria-based immunotherapy system (Wu et al. 2022)

treatments and proves its ability to effectively inhibit tumor metastasis. This study provides important insights for the development of new cancer treatment strategies, especially for those tumors that are difficult to reach or have already metastasized.

To sum up, we summarized the application of tumor photoimmunotherapy based on UCNPs. As an emerging treatment strategy, photoimmunotherapy based on UCNPs has attracted widespread attention. This method combines the advantages of PDT and immunotherapy, aiming to activate nanoparticles through non-invasive light irradiation, thereby generating a series of biochemical reactions that can trigger immunogenic death of tumor cells. When upconversion nanoparticles are irradiated by near-infrared light of a specific wavelength, they can convert low-energy light into high-energy visible light, and this conversion can trigger a local photodynamic reaction at the tumor site. This process can not only directly damage tumor cells but also stimulate the body's immune system to identify and attack cancer cells. Studies have shown that this process helps to promote the maturation of dendritic cells, which are important antigen-presenting cells and are crucial for

therapy (PDT). Compared with single-mode therapy, this combination of Mir-21-activated phototherapy significantly improves the therapeutic efficacy of breast cancer, not only accurately targeting cancer cells, but also providing efficient imaging and therapeutic functions. This study demonstrates the great potential of UCNP@Au nanostructures as a multifunctional platform that combines precise biosensing with efficient photothermal and photodynamic therapy, providing new tools for early diagnosis and personalized treatment of cancer.

Cen et al. (2023) designed a robust tumor synergistic therapy nanoplatform that combines dual photosensitizer-loaded UCNPs and Ferronnic acid (FeTA) nanocomplexes. Dual-photosensitizer-loaded UCNPs induce PDT effects by producing cytotoxic ROS on demand under NIR light irradiation. FeTA can produce Fe^{2+} in the tumor microenvironment and react with H_2O_2 in the tumor microenvironment to induce chemodynamic therapy (CDT). The successful application of this platform demonstrates the great potential of combining multimodal therapies with intelligent activation mechanisms in cancer treatment.

Jin et al. (2020) prepared ROS-responsive micelle-coated multifunctional UCNPs for the synergistic effects of tumor-targeted and NIR-triggered PDT in combination with chemotherapy (Fig. 2.18). In addition, both PDT and chemotherapeutic agents can activate anti-tumor immune cell death and CD^{8+} and CD^{4+} T cell

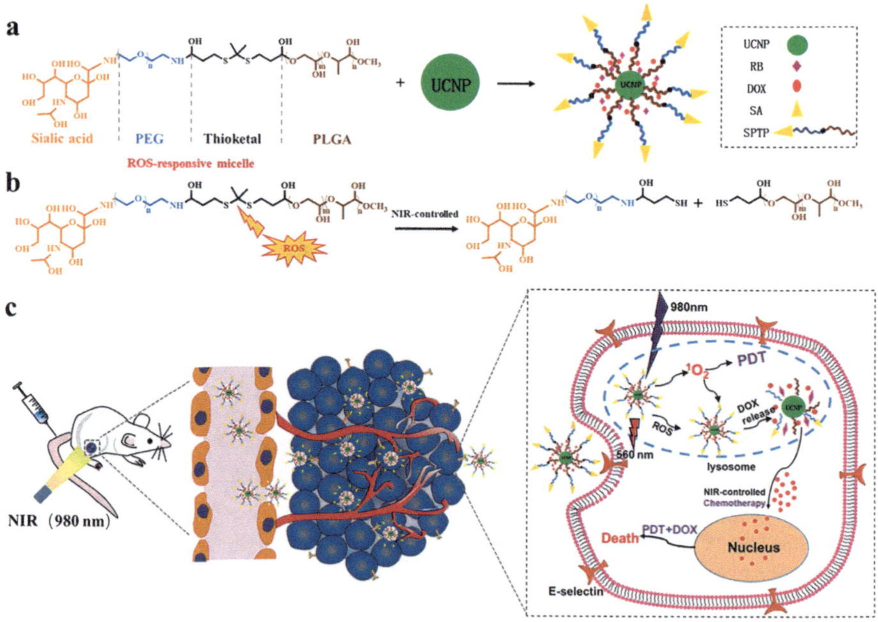

Fig. 2.18 Illustration of systemic anti-tumor immunity. The schematic diagrams include: (**a**) the rational design and preparation process of SPTP@UCNP - RB - DOX; (**b**) the reaction mechanism of the self - destructible polymer; (**c**) the principle of achieving anti - cancer applications through cascade - amplified chemo - PDT combined with systemic anti - tumor immunity in a TNBC murine model (Jin et al. 2020)

infiltration by inducing immunogenicity. Through experiments, multifunctional nanocarriers were injected intravenously under non-invasive near-infrared irradiation to effectively destroy orthotopic tumors and inhibit lung metastasis in metastatic triple-negative breast cancer models by cascade amplification of chemical photodynamic therapy and systemic anti-tumor immunity.

Zhang et al. (2023) constructed a unique nanoconjugate (UCNP-Ce6/AIEgen) for enhancing dual-pathway PDT. This nanoconjugate achieved efficient PDT by co-modifying UCNPs with chlorin e6 (Ce6) and a luminogen with aggregation-induced emission (AIE) characteristics (AIEgen). In this design, the presence of AIEgen ensured that the nanoconjugate does not experience luminescence quenching due to aggregation in biological water environments. Instead, it can effectively convert the UCL of UCNPs into the fluorescence required to activate Ce6. Therefore, under 808-nm laser irradiation, UCNP-Ce6/AIEgen can activate Ce6 through two mechanisms. These two mechanisms work together to significantly increase the generation of ROS, thereby more effectively promoting tumor cell apoptosis. This study not only innovated the application methods of lanthanide—based inorganic materials and organic dyes but also provided new ideas for enhanced photodynamic therapy. The unique design of UCNP-Ce6/AIEgen enables it to maintain good biocompatibility while improving the therapeutic effect, bringing new possibilities to cancer treatment. This innovative nanoconjugate platform demonstrates great potential in the field of precision medicine, especially in optimizing photodynamic therapy.

To address the challenges of the high recurrence and metastasis of fatal tumors, Zhou et al. (2023) proposed an innovative combination therapy, aiming to overcome the limitations of single-modality treatment strategies such as surgery, PDT, and radiotherapy (RT). This approach fully exploited the complementary advantages of PDT and RT. They designed a near-infrared-induced PDT agent by combining lanthanide-doped UCNPs with red blood cell membrane vesicles embedded with Ce6 to achieve synchronous deep-depth PDT and low-dose RT while reducing radiation exposure (Fig. 2.19). In this nanoagent, gadolinium-doped UCNPs possess strong X-ray attenuation ability. They not only serve as highly efficient light transducers that can activate the loaded photosensitizer Ce6 under near-infrared light irradiation to implement PDT but also act as radiosensitizers to enhance the effect of RT. In this way, enhanced low-dose RT combined with PDT can not only produce ROS to destroy local tumor cells but also induce strong T-cell-dependent immunogenic cell death (ICD) to prevent systemic cancer metastasis, thus achieving synergistic inhibition of tumor growth. The advantage of this combination therapy lies in its ability to simultaneously utilize the targeting property of PDT and the deep-tissue penetration of RT, providing a more comprehensive treatment plan. Experimental results showed that this method can not only effectively kill local tumor cells but also stimulate the body's immune system to generate a lasting anti-tumor immune response, thereby inhibiting the development of distant metastases. Therefore, the combination of PDT and RT may become a highly attractive tumor-eradication strategy, providing new ideas and tools for cancer treatment.

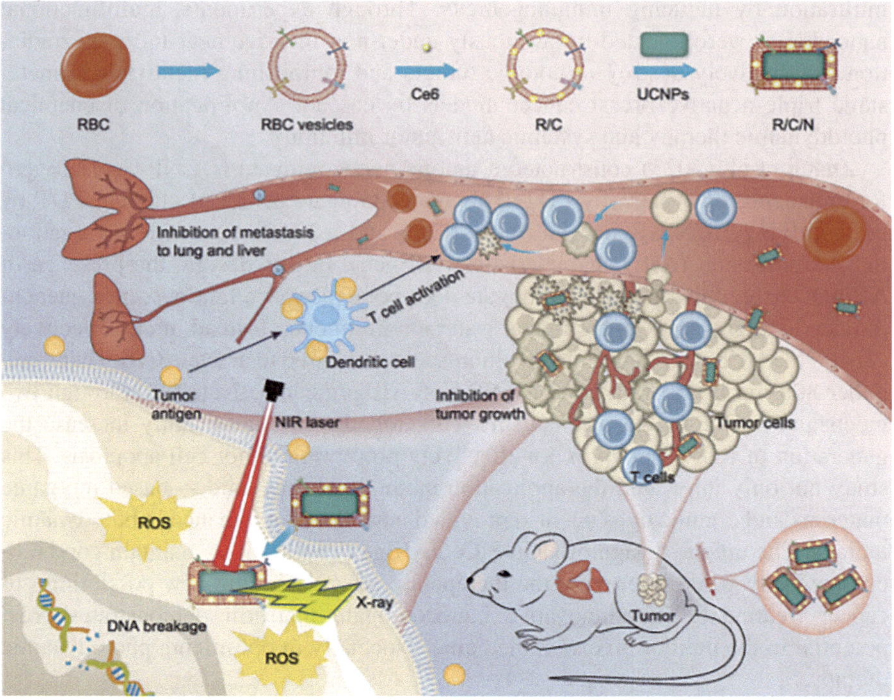

Fig. 2.19 Illustration of the synthesis of R/C/N and the synergetic anti-tumor effect of PDT combined with RT (Zhou et al. 2023)

Guo et al. (2024) combined two conjugated polymers with UCNPs through a coordination reaction to form UCNPs-CPs. The design of this composite structure fully utilizes the excellent overlap between the absorption spectrum of the conjugated polymers and the emission spectrum of UCNPs. When irradiated by NIR, energy transfer occurs between UCNPs and the polymers, generating a large amount of ROS. Additionally, sufficient heat energy is produced for PTT treatment. Therefore, UCNPs-CPs achieves the synergistic effect of PDT and PTT. In vitro cell experiments indicate that in a 4T1 breast cancer mouse model, the UCNPs-CPs nanomaterial shows a 100% anti-tumor growth rate, demonstrating excellent therapeutic effects. This study showcases how to optimize the synergy of multiple treatment mechanisms through the careful design of nanomaterials, providing new ideas for the development of more effective cancer treatment methods. The successful application of UCNPs-CPs highlights its significant value in the field of precision medicine, especially in terms of improving treatment efficiency and reducing side effects (Fig. 2.20).

The huge advantages and potential of collaborative therapy in improving anti-tumor efficacy have attracted researchers to quickly devote their attention to the design and construction of dual-mode collaborative therapy system, and certain

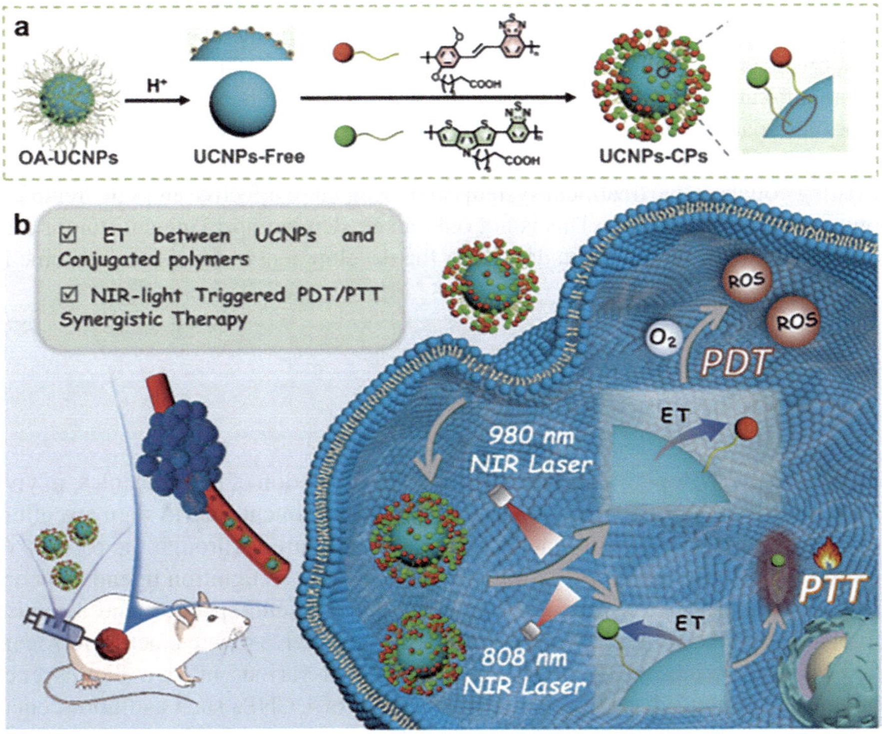

Fig. 2.20 Schematic illustration of (**a**) the preparation of UCNPs-CPs and (**b**) synergistic PDT & PTT anti-tumor mechanism (Guo et al. 2024)

research progress has been achieved. However, the synergistic therapy system based on simple nuclear or core–shell structure, loaded with anti-cancer drugs or photosensitizers through chemical bonding and physical adsorption, is easy to cause the leakage of photosensitizers and chemotherapy drugs, low load, and only can load photosensitizers or chemotherapy drugs with special structures, limiting the synergistic therapy effect. To overcome the above problems, researchers are exploring more complex and intelligent structural designs that aim to integrate multiple treatment modes into one, not only achieving a "1 + 1 > 2" synergistic effect, but also providing a universal drug delivery mode that can be applied to most anti-cancer drugs and photosensitizers to meet the treatment needs of different types of tumors. The ideal new nanoplatform should have the following characteristics: (1) *Efficient drug loading capability*: it can stably load multiple types of anti-cancer drugs and photosensitizers to ensure that sufficient drug concentrations reach the target. (2) *Intelligent response mechanism*: Responds to specific signals in the tumor microenvironment (such as pH, enzyme concentration, REDOX status, and so on) to precisely control drug release. (3) *Multifunctional integration*: Organic combination of multiple treatment modes (such as photodynamic therapy, photothermal therapy, chemotherapy, immunotherapy, and so on) to enhance the therapeutic effect through synergistic effect. (4) *Versatility and flexibility*: Provide a common loading mode for most anti-cancer drugs and photosensitizers to meet the needs of different tumor types and patient individualization. By developing more advanced and intelligent nanocarrier systems, researchers are expected to break through the limitations of existing collaborative treatment systems and bring more effective and safe treatment options to cancer patients. This is not only a key step in improving anti-tumor efficacy, but also lays a solid foundation for the development of precision medicine in the future.

2.3 Conclusions and Perspectives

At present, UCNPs play an important role in the detection of biomolecules, in vivo fluorescence imaging and tumor diagnosis and treatment due to their excellent optical properties and easily modified surface structures. Through the process of luminescence resonance energy transfer with specific recognition ligands, upconversion nanoparticles can be used to construct efficient, specific, and sensitive fluorescent probes of biological molecules such as active oxygen, active nitrogen, and active enzymes. Researchers have developed various in vivo fluorescence imaging techniques based on the characteristics of UCNPs such as fluorescence, magnetism, and easy modification. In the integration of tumor diagnosis and treatment, a drug delivery system is constructed by modifying silicon layers, polymers, and precious metals on the surface of upconversion nanomaterials, and various drug delivery systems based on tumor microenvironments and photocontrolled drug delivery systems are constructed to achieve controllable and accurate drug release. Upconverted nanomaterials as activators of photosensitizers can realize photodynamic and photothermal therapy of tumors. In addition, as gene and antigen transport carriers, upconversion nanomaterials can also realize gene therapy and immunotherapy. At present, the multi-mode collaborative anti-tumor system based on upconversion nanomaterials is still in its infancy, and there are still many problems to be solved in this field. If the treatment of tumor is combined with diagnosis, the disease type and lesion site can be accurately determined, and then targeted administration and the optimal treatment time are expected to reduce the number of doses and greatly improve the anti-tumor effect. Therefore, the design and construction of multi-diagnosis and treatment system with simple structure and multiple functions will have important clinical application prospects.

References

Atkins KM, Chaunzwa TL, Nayan L et al (2021) Association of left anterior descending coronary artery radiation dose with major adverse cardiac events and mortality in patients with non–small cell lung cancer. JAMA Oncol 7(2):206–219

Cen D, Qiang Z, Bingzhu Z et al (2023) A near-infrared light-responsive ROS cascade nanoplatform for synergistic therapy potentiating antitumor immune responses. Adv Funct Mater 33(9):2211402

Chan M-H, Yu-Ting P, Jung LI et al (2017) Minimizing the heat effect of photodynamic therapy based on inorganic nanocomposites mediated by 808 nm near-infrared light. Small 13(21):1700038

Chen G, Ohulchanskyy TY, Cheung LW et al (2011) Monodisperse NaYbF4: Tm3+/NaGdF4 core/shell nanocrystals with near-infrared to near-infrared upconversion photoluminescence and magnetic resonance properties. Nanoscale 3(5):2003–2008

Chen Q, Chao W, Liang C et al (2014) Protein modified upconversion nanoparticles for imaging-guided combined photothermal and photodynamic therapy. Biomaterials 35(9):2915–2923

Chen J, Dongya Z, Yan Z et al (2018) Developing a pH-sensitive Al(OH)3 layer-mediated UCNP@Al(OH)3/au nanohybrid for photothermal therapy and fluorescence imaging in vivo. J Mater Chem B 6(47):7862–7870

Chen Y, Huijing X, Xiaodan L et al (2024) Near-infrared laser-switching DNA phase separation nanoinducer for glioma therapy. ACS Nano 18(35):24426–24440

Cheng Z, Lin J (2015) Synthesis and application of nanohybrids based on upconverting nanoparticles and polymers. Macromol Rapid Commun 36(9):790–827

Cheng L, Kai Y, Yonggang L et al (2012) Multifunctional nanoparticles for upconversion luminescence/MR multimodal imaging and magnetically targeted photothermal therapy. Biomaterials 33(7):2215–2222

Cheng X, Jie Z, Jingyi Y et al (2022) Recent development in sensitizers for lanthanide-doped Upconversion luminescence. Chem Rev 122(21):15998–16050

Choi J, Kim SY (2023) Multi-shell structured upconversion nanocarriers that combine IDO inhibitor-induced immunotherapy with NIR-triggered photodynamic therapy for deep tumors. Biomater Sci 11(13):4684–4699

Cichos J, Karbowiak M (2014) A general and versatile procedure for coating of hydrophobic nanocrystals with a thin silica layer enabling facile biofunctionalization and dye incorporation. J Mater Chem B 2(5):556–568

Dai WB, Lei YF, Ye S et al (2016) Mesoporous nanoparticles Gd2O3@mSiO2/ZnGa2O4:Cr3+,Bi3+ as multifunctional probes for bioimaging. J Mater Chem B 4(10):1842–1852

Dong N-N, Marco P, Fabio P et al (2011) NIR-to-NIR two-photon excited CaF2:Tm3+,Yb3+ nanoparticles: multifunctional nanoprobes for highly penetrating fluorescence bio-imaging. ACS Nano 5(11):8665–8671

Dong H, Ling-Dong S, Chun-Hua Y (2020) Upconversion emission studies of single particles. Nano Today 35:100956

Du J, Tao J, Feng L et al (2024) MOF-coated upconversion nanoparticle agents enable synergistic photodynamic therapy and immunotherapy. Adv Funct Mater 34(36):2401272

Fan W, Bo S, Wenbo B et al (2013) Rattle-structured multifunctional Nanotheranostics for synergetic chemo-/radiotherapy and simultaneous magnetic/luminescent dual-mode imaging. J Am Chem Soc 135(17):6494–6503

Fang J, Chengyang Z, Yaru N et al (2023) Double local electromagnetic fields collaboratively enhanced triplet–triplet annihilation upconversion for efficient photocatalysis. Cat Sci Technol 13(7):2151–2159

Ge W, Zhang XR, Liu M et al (2013) Distance dependence of gold-enhanced upconversion luminescence in au/$SiO_2Y_2O_3$:Yb_{3+}, Er_{3+} nanoparticles. Theranostics 3(4):282–288

Guo X, Ling L, Wenhua J et al (2024) Composite nanomaterials of conjugated polymers and Upconversion nanoparticles for NIR-triggered photodynamic/photothermal synergistic cancer therapy. ACS Appl Mater Interfaces 16(16):19926–19936

Hu F, Bei L, Hongqian C et al (2019) Real-time monitoring of pH-responsive drug release using a metal-phenolic network-functionalized upconversion nanoconstruct. Nanoscale 11(18):9201–9206

Hu S, Lixian H, Liuyan Z et al (2023) Single-excitation triple-emission down-/up-conversion nanoassemblies for tumor microenvironment-enhanced ratiometric NIR-II fluorescence imaging and chemo-/photodynamic combination therapy. Anal Chem 95(7):3830–3839

Jiang C, Chen Y, Xiaolong L et al (2024) An intelligent NIR-IIb-responsive lanthanide@metal–organic framework core–shell nanocatalyst for combined deep-tumor therapy. J Mater Chem B 12(35):8626–8632

Jin F, Jing Q, Minxia Z et al (2020) NIR-triggered sequentially responsive nanocarriers amplified cascade synergistic effect of chemo-photodynamic therapy with inspired antitumor immunity. ACS Appl Mater Interfaces 12(29):32372–32387

Li L, Fangqiong T, Huiyu L et al (2010a) In vivo delivery of silica Nanorattle encapsulated docetaxel for liver cancer therapy with low toxicity and high efficacy. ACS Nano 4(11):6874–6882

Li Y, Wei L, Qian H et al (2010b) Copper sulfide nanoparticles for photothermal ablation of tumor cells. Nanomedicine 5(8):1161–1171

Li L-L, Ruobing Z, Leilei Y et al (2013a) Corrigendum: biomimetic surface engineering of lanthanide-doped upconversion nanoparticles as versatile bioprobes. Angew Chem Int Ed 52(32):8190–8190

Li C, Zhiyao H, Yunlu D et al (2013b) A facile fabrication of upconversion luminescent and mesoporous core–shell structured β-NaYF4:Yb3+, Er3+@mSiO2 nanocomposite spheres for anticancer drug delivery and cell imaging. Biomater Sci 1(2):213–223

Li D, Wen-Yong L, Qiyue S et al (2017) Multifunctional NaYF4:Yb3+,Er3+@SiO2@Au heterogeneous nanocomposites for upconversion luminescence, temperature sensing and photothermal conversion. RSC Adv 7(19):11491–11495

Li P, Yue Y, Binlong C et al (2018) Lanthanide-doped upconversion nanoparticles complexed with nano-oxide graphene used for upconversion fluorescence imaging and photothermal therapy. Biomater Sci 6(4):877–884

Li Y, Hongliang Q, Xin H et al (2022) Up-converted nano-gasholder with precise nitric oxide release remodels immunosuppressive microenvironment and potentiates tumor immunotherapy. Nano Today 42:101381

Li Y, Yingjie D, Yifan Z et al (2023) An engineered cascade-sensitized red-emitting upconversion nanoplatform with a tandem hydrophobic hydration-shell and metal-phenolic network decoration for single 808 nm triggered simultaneous tumor PDT and PTT enhanced CDT. Nanoscale 15(23):10067–10078

Liu J, Wenbo B, Shengjian Z et al (2012) Controlled synthesis of uniform and monodisperse upconversion core/mesoporous silica shell nanocomposites for bimodal imaging. Chemistry 18(8):2335–2341

Liu Y, Shengbin C, Shiping Z et al (2021a) Manipulating energy transfer in UCNPs@SiO2@ag nanoparticles for efficient infrared Photocatalysis. Inorg Chem 60(8):5704–5710

Liu B, Xinquan G, Qiannan S et al (2021b) Injectable in situ induced robust hydrogel for photothermal therapy and bone fracture repair. Adv Funct Mater 31(19):2010779

Liu B, Bing W, Ziyao W et al (2023a) Near-infrared light-controlled microRNA-21-loaded upconversion nanoparticles to promote bone formation in the midpalatal suture. ACS Appl Mater Interfaces 15(37):43503–43514

Liu Z, Zhiyuan F, Mohan C et al (2023b) An orthogonally activatable CRISPR-Cas13d nanoprodrug to reverse chemoresistance for enhanced chemo-photodynamic therapy. Chem Sci 14(15):4102–4113

Liu X-M, Zhen-Zhen Z, Xin-Rui H et al (2024) NIR light and GSH dual-responsive upconversion nanoparticles loaded with multifunctional platinum(IV) prodrug and RGD peptide for precise cancer therapy. ACS Appl Mater Interfaces 16(31):40753–40766

Molkenova A, Eun CH, Gibum L et al (2024) Cold-responsive hyaluronated upconversion nanoplatform for transdermal cryo-photodynamic cancer therapy. Adv Sci 11(19):2306684

Naccache R, Fiorenzo V, Venkataramanan M et al (2009) Controlled synthesis and water dispersibility of hexagonal phase NaGdF4:Ho3+/Yb3+ nanoparticles. Chem Mater 21(4):717–723

Nsubuga A, Korentin M, Nour F et al (2024) Sub 20 nm upconversion photosensitizers for near-infrared photodynamic theranostics. Adv Funct Mater 35:2410077

Park HS, Jongwoo K, Young CM et al (2020) Effectual labeling of natural killer cells with Upconverting nanoparticles by electroporation for in vivo tracking and biodistribution assessment. ACS Appl Mater Interfaces 12(44):49362–49370

Pegaz B, Elodie D, Jean-Pierre B et al (2006) Effect of nanoparticle size on the extravasation and the photothrombic activity of meso(p-tetracarboxyphenyl)porphyrin. J Photochem Photobiol B Biol 85(3):216–222

Qian Y, Weilin C, Man W et al (2024) Tumor microenvironment-specific driven nanoagents for synergistic mitochondria damage-related immunogenic cell death and alleviated immunosuppression. Small Methods 8(7):2301231

Ren W, Gan T, Shan J et al (2012) TWEEN coated NaYF4:Yb,Er/NaYF4 core/shell upconversion nanoparticles for bioimaging and drug delivery. RSC Adv 2(18):7037–7041

Schietinger S, Thomas A, Hai-Qiao W et al (2010) Plasmon-enhanced upconversion in single NaYF4:Yb3+/Er3+ codoped nanocrystals. Nano Lett 10(1):134–138

Shao Y, Bei L, Zhenghan D et al (2020) Engineering of upconverted metal–organic frameworks for near-infrared light-triggered combinational photodynamic/chemo–/immunotherapy against hypoxic tumors. J Am Chem Soc 142(8):3939–3946

Shi S, Cheng L, Yufu L et al (2024) Upconverting ultra-thin Bi2O2CO3 Nanosheets for synergistic photodynamic therapy and radiotherapy. Adv Healthc Mater 13(28):2401586

Song L, Nana Z, Fu-Jian X (2017) Hydroxyl-rich polycation brushed multifunctional rare-earth-gold core–shell nanorods for versatile therapy platforms. Adv Funct Mater 27(32):1701255

Song X, Huairong Z, Tao Y et al (2020) Controllable growth and assembling strategies based on nanomaterials for targeted and precise therapy of malignant cancers. Chin J Chem 38(12):1489–1496

Song X, Fengyan L, Feng T et al (2023) Upconversion nanoparticle-based optogenetic nanosystem for photodynamic therapy and cascade gene therapy. Acta Biomater 157:538–550

Song N, Xiaoting F, Xiaocui G et al (2024) A DNA/upconversion nanoparticle complex enables controlled co-delivery of CRISPR-Cas9 and photodynamic agents for synergistic cancer therapy. Adv Mater 36(15):2309534

Sun S, Qiang H, Siyuan T et al (2019) A magnetic, luminescent and mesoporous nanocomposite as protein drug carrier. Microporous Mesoporous Mater 277:261–266

Sun Y, Meihua C, Puxian X et al (2023) Integrated multi-mode glass ceramic fiber for high-resolution temperature sensing. Adv Powder Mater 2(4):100132

Tan C, Xingjun L, Zhuo L et al (2024) Near-infrared-responsive nanoplatforms integrating dye-sensitized upconversion and heavy-atom effect for enhanced photodynamic therapy efficacy. Nano Today 54:102089

Tian G, Wenyan Y, Junjiang J et al (2014) Engineered design of theranostic upconversion nanoparticles for tri-modal upconversion luminescence/magnetic resonance/X-ray computed tomography imaging and targeted delivery of combined anticancer drugs. J Mater Chem B 2(10):1379–1389

Wahab OJ, Daviddi E, Xin B et al (2023) Proton transport through nanoscale corrugations in two-dimensional crystals. Nature 620(7975):782–786

Wang C, Huiquan T, Liang C et al (2011a) Near-infrared light induced in vivo photodynamic therapy of cancer based on upconversion nanoparticles. Biomaterials 32(26):6145–6154

Wang C, Liang C, Zhuang L (2011b) Drug delivery with upconversion nanoparticles for multifunctional targeted cancer cell imaging and therapy. Biomaterials 32(4):1110–1120

Wang X, Kai L, Guangbao Y et al (2014a) Near-infrared light triggered photodynamic therapy in combination with gene therapy using upconversion nanoparticles for effective cancer cell killing. Nanoscale 6(15):9198–9205

Wang L, Xiaoying L, Jing W et al (2014b) Novel insights into combating cancer chemotherapy resistance using a plasmonic nanocarrier: enhancing drug sensitiveness and accumulation simultaneously with localized mild photothermal stimulus of femtosecond pulsed laser. Adv Funct Mater 24(27):4229–4239

Wang L, Chang G, Keyin L et al (2016) Cypate-conjugated porous upconversion nanocomposites for programmed delivery of heat shock protein 70 small interfering rna for gene silencing and photothermal ablation. Adv Funct Mater 26(20):3480–3489

Wang M, Jun S, Feifan Z et al (2019) NIR-triggered phototherapy and immunotherapy via an antigen-capturing nanoplatform for metastatic cancer treatment. Adv Sci 6(10):1802157

Wang R, Han Y, Rongxin F et al (2020) Biomimetic upconversion nanoparticles and gold nanoparticles for novel simultaneous dual-modal imaging-guided photothermal therapy of cancer. Cancer 12(11):3136

Wang M, Mengyu C, Chunxia L et al (2022) Tumor-microenvironment-activated reactive oxygen species amplifier for enzymatic Cascade cancer starvation/Chemodynamic/immunotherapy. Adv Mater 34(4):2106010

Wang J, Ping S, Ming L et al (2023) Dual-site Förster resonance energy transfer route of upconversion nanoparticles-based brain-targeted Nanotheranostic boosts the near-infrared phototherapy of glioma. ACS Nano 17(17):16840–16853

Wang J, Yi Z, Huadong C et al (2024) Motor-cargo structured Nanotractors for augmented NIR phototherapy via gas-boosted tumor penetration and respiration-impaired mitochondrial dysfunction. Adv Healthc Mater 13(30):2402063

Working Group on the Summit on Combination Therapy (2014) Combination pharmacotherapy to prevent cardiovascular disease: present status and challenges. Eur Heart J 35(6):353–364

Wu C, Meihui C, Li C et al (2022) NIR-responsive photodynamic nanosystem combined with antitumor immune optogenetics bacteria for precise synergetic therapy. ACS Appl Mater Interfaces 14(11):13094–13106

Xia L, Xianggui K, Xiaomin L et al (2014) An upconversion nanoparticle—zinc phthalocyanine based nanophotosensitizer for photodynamic therapy. Biomaterials 35(13):4146–4156

Xiang G, Menglin Y, Yongxi D et al (2023) Ratiometric nanothermometer $CaSc_2O_4$: Nd^{3+} operating in biological window for deep-tissue photothermal therapy. Ceram Int 49(9, Part A):13816–13822

Xiao Q, Xiangpeng Z, Wenbo B et al (2013) A Core/satellite multifunctional Nanotheranostic for in vivo imaging and tumor eradication by radiation/Photothermal synergistic therapy. J Am Chem Soc 135(35):13041–13048

Xu Z, Chunxia L, Ping'an M et al (2011) Facile synthesis of an up-conversion luminescent and mesoporous Gd_2O_3: Er^{3+}@$nSiO_2$@$mSiO_2$ nanocomposite as a drug carrier. Nanoscale 3(2):661–667

Xu W, Chen X, Hongwei S (2017) Upconversion manipulation by local electromagnetic field. Nano Today 17:54–78

Xu M, Bin X, Yue W et al (2021) Temperature-feedback nanoplatform for NIR-II Penta-modal imaging-guided synergistic photothermal therapy and CAR-NK immunotherapy of lung cancer. Small 17(43):2101397

Yang L, Ruipeng S, Ruoxi Z et al (2022a) Near-infrared upconversion mesoporous tin dioxide theranostic nanocapsules for synergetic cancer chemophototherapy. ACS Appl Mater Interfaces 14(2):2650–2662

Yang L, Sha Y, Yongcun Y et al (2022b) Upconversion nanoparticle@Au core–satellite assemblies for in situ amplified imaging of MicroRNA in living cells and combined cancer phototherapy. Anal Chem 94(19):7075–7083

Yang Z, Xichao W, Guohai L et al (2022c) Photocontrolled chondrogenic differentiation and long-term tracking of mesenchymal stem cells in vivo by upconversion nanoparticles. J Mater Chem B 10(4):518–536
Zhan Q, Jun Q, Huijuan L et al (2011) Using 915 nm laser excited Tm3+/Er3+/Ho3+-doped NaYbF4 upconversion nanoparticles for in vitro and deeper in vivo bioimaging without over-heating irradiation. ACS Nano 5(5):3744–3757
Zhang M-K, Xiao-Gang W, Jing-Yi Z et al (2018) Double-targeting explosible nanofirework for tumor ignition to guide tumor-depth Photothermal therapy. Small 14(20):1800292
Zhang X-C, Jun W, Guo-Guang S et al (2019a) Comprehensive genomic and immunological characterization of Chinese non-small cell lung cancer patients. Nat Commun 10(1):1772
Zhang Y, Yue Z, Guobin S et al (2019b) A DNA–azobenzene nanopump fueled by upconversion luminescence for controllable intracellular drug release. Angew Chem Int Ed 58(50):18207–18211
Zhang C, Wen-Jun Q, Xue-Feng B et al (2020) Nanomaterials to relieve tumor hypoxia for enhanced photodynamic therapy. Nano Today 35:100960
Zhang Y, Huaguang C, Ruiqi Z et al (2021a) Nanoparticulation of prodrug into medicines for cancer therapy. Adv Sci 8(18):2101454
Zhang Y, Weiwei C, Yanyun F et al (2021b) Activating a DNA nanomachine via computation across cancer cell membranes for precise therapy of solid tumors. J Am Chem Soc 143(37):15233–15242
Zhang L, Jun C, Huihui Y et al (2022) A chromogenic reaction-free distance-based paper device for facile detection of microRNA via viscosity amplification and surface hydrophobicity modulation. Sensors Actuators B Chem 359:131570
Zhang R, Lu Y, Yifei Z et al (2023) Utilizing dual-pathway energy transfer in upconversion nanoconjugates for reinforced photodynamic therapy. Nano Res 17(4):2941–2948
Zhang H, Yue C, Wanying L et al (2024) Engineering "Trojan horse" disguised by cancer cell membranes for mitochondrial targeting and amplified immunotherapy. Chem Eng J 480:147961
Zhao S, Rongrong T, Baiqi S et al (2019) Designing of UCNPs@bi@SiO2 hybrid theranostic nanoplatforms for simultaneous multimodal imaging and photothermal therapy. ACS Appl Mater Interfaces 11(1):394–402
Zhao W, Xinyi Z, Ruiting T et al (2023) The sensor platform combined with dual signal amplification and based on UCNPs and CRISPR/Cas12a for MiRNA-21 detection. Sensors Actuators B Chem 393:134238
Zhong H, Zhihua F, Taylor JM et al (2017) Covalent organic gels: inorganic acid-impregnated covalent organic gels as high-performance proton-conductive materials at subzero temperatures. Adv Funct Mater 27(32):1701465
Zhou J, Menghuan L, Yanhua H et al (2018) Engineering of a nanosized biocatalyst for combined tumor starvation and low-temperature photothermal therapy. ACS Nano 12(3):2858–2872
Zhou D, Yun G, Zhe Y et al (2023) Biomimetic Upconversion Nanoplatform synergizes photodynamic therapy and enhanced radiotherapy against tumor metastasis. ACS Appl Mater Interfaces 15(22):26431–26441
Zhu X, Sihan C, Xiuwen H et al (2023) Near-infrared nano-optogenetic activation of cancer immunotherapy via engineered bacteria. Adv Mater 35(8):2207198

Chapter 3
Application of Liquid Crystal-Based Material in Tumor-Related Sensing, Treatment, and Responsive Devices

Lishang Liu and Jinrong Chen

Abstract This article reviews and discusses the applications of liquid crystal-based materials in tumor-related sensing, treatment, and response devices. By taking advantage of the unique optical properties of liquid crystals, rapid and sensitive detection of tumor markers can be achieved. In addition, the role of liquid crystals in targeted drug delivery systems provides precision for cancer therapy. At the same time, combined with the liquid crystal response device, the intelligent wearable device is designed and developed to monitor and react to changes in human body signs and provide real-time feedback for personalized monitoring and treatment. These studies highlight the promise of liquid crystal materials in advancing the diagnosis and treatment of diseases such as cancer.

Keywords Liquid crystal · Sensing · Monitoring · Treatment · Cancer

3.1 Introduction of Liquid Crystal

Liquid crystals, as a state of matter that lies between isotropic liquids and anisotropic crystalline solids, have been widely applied across various fields their unique physical and chemical properties. In a crystal composed of anisotropic molecules, long-range order refers to both positional order (molecules occupying specific positions in the lattice) and directional order (molecular axes pointing in specific directions) (Woltman et al. 2007; Zhang et al. 2023). When such a crystal is heated from the solid phase, both types of order (positional and directional) may disappear simultaneously at a certain temperature, resulting in a new phase that is an isotropic liquid. If the directional order is lost while the positional order remains, the corresponding phase is known as a plastic crystal. Conversely, if the positional order is

L. Liu · J. Chen (✉)
Shandong Provincial Key Laboratory of Tumor Imaging Equipment Development and Tumor Diagnosis and Treatment Integration Technology, School of Medicine, Linyi University, Linyi, Shandong, People's Republic of China

X. Song et al. (eds.), *Functional Inorganic Materials for Tumor Diagnostics and Therapy*, https://doi.org/10.1007/978-981-95-0528-9_3

completely or partially lost while some degree of directional order is maintained, the resulting phase is called a liquid crystal (Chavda et al. 2022).

In 1888, Austrian botanist Friedrich Reinitzer accidentally discovered that a certain organic substance would change from a solid to a turbid paste and then to a transparent liquid as its temperature increased while studying the melting points of organic substances. Subsequently, German physicist Otto Lehmann embarked on the path of researching this substance, using a microscope with a heating function to observe the crystallization process as the substance cooled. With the research deepening, in 1889, Otto Lehmann named this substance with special properties "liquid crystal," and in 1922, Georges Friedel classified liquid crystals based on the partial order of elongated particles (Devadiga and Ahipa 2022). With the advancement of the semiconductor industry, the patterning of transparent electrodes and the micro-fabrication technology that integrates liquid crystals with semiconductor circuits gradually matured, leading to their widespread application in display devices (Zhang et al. 2023). As research progressed, liquid crystals, as dynamic functional materials, began to be applied in various fields such as optical instruments, lasers, diffraction gratings, spatial light modulators, dynamic lenses, and actuator devices, becoming an indispensable substance in scientific research and daily life. Liquid crystals possess many of the characteristics of liquids, such as high fluidity and the inability to support shear, while also exhibiting properties similar to crystals, such as birefringence and anisotropic magnetizability, and electric susceptibility (Lagerwall and Scalia 2012; Ula et al. 2018; Smalyukh 2018).

Liquid crystals can be categorized into two types based on their formation: thermotropic and lyotropic. Thermotropic liquid crystals undergo phase transitions influenced by temperature changes, and to date, most liquid crystal-based biosensing technologies have utilized thermotropic liquid crystals as the primary sensing medium, particularly rod-shaped 4-cyano-4′-pentylbiphenyl (5CB) (Lagerwall and Scalia 2012). The organic molecules that make up liquid crystal materials can vary, but to facilitate interactions that promote alignment, they must maintain an elongated shape. This requires that certain parts of the molecules are anisotropic or rigid to ensure the necessary directional order for the liquid crystal phase (Lyu et al. 2020). Lyotropic liquid crystals exist in solutions, where concentration is an important controllable parameter. Typically, lyotropic liquid crystal molecules have a hydrophobic group at one end and a hydrophilic group at the other. These amphiphilic molecules form ordered structures (micelles, vesicles, lamellar phases, or inverse phases) in both polar and nonpolar solvents, similar to soap and various phospholipids (Guardià et al. 2024).

Liquid crystals can exist in three main types based on the symmetry of their molecules: nematic, cholesteric, and smectic liquid crystals (Fig. 3.1). Nematic liquid crystals are the simplest and commonly used type. Molecules in the nematic phase exhibit long-range directional order, which can be altered by applying different surface anchoring conditions on the liquid crystal walls or by external electric or magnetic fields. The reorientation of nematic liquid crystal molecules under the influence of an external electric field is known as the Frederiks transition, which forms the theoretical basis for the application of liquid crystals in liquid crystal

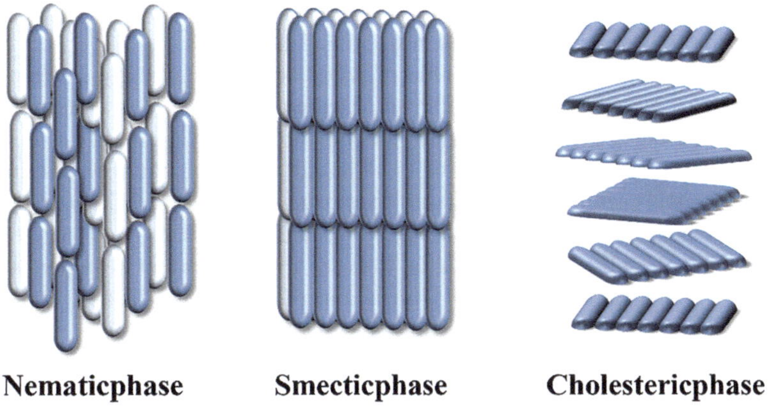

Fig. 3.1 Arrangement of liquid crystal molecules in the nematic, smectic, and cholesteric phases

displays. Smectic liquid crystals differ from other liquid crystal molecules in that they have a regular orientation but lack positional order, meaning that smectic liquid crystal molecules move within their layer, with layers sliding against each other. Because of this characteristic, smectic liquid crystals have a higher viscosity and are less sensitive to external electric and magnetic fields. As a result, they are widely used in the persistent image display of liquid crystal televisions, saving energy compared to traditional liquid crystal display methods. Cholesteric liquid crystals, like other liquid crystals, possess fluidity, deformability, visibility, and crystalline optical anisotropy, making them excellent nonlinear optical materials. Due to their unique helical twisted structure, they selectively reflect certain wavelengths of light, which is known as the selective reflection of cholesteric liquid crystals. This property has led to their widespread application in the development of optical devices, including circular polarizers and reflective displays (Zhang et al. 2023).

In the face of the increasingly severe threat of cancer, identifying effective methods for early detection, diagnosis, and precise treatment is the primary measure to extend the survival period of cancer patients and reduce mortality rates. In recent years, liquid crystal technology has not only made significant advances in display and optical equipment but has also shown tremendous potential in the field of biomedicine. Particularly in tumor diagnosis and treatment, liquid crystal technology has introduced new ideas and methods for accurate diagnosis and therapy. On the one hand, the adjustable optical properties of liquid crystal materials are utilized in tumor imaging. For instance, liquid crystal sensors can achieve label-free detection of tumor markers, thereby providing possibilities for early diagnosis. On the other hand, due to their special structure and properties, liquid crystal nanoparticles can serve as drug carriers, enhancing the targeting and therapeutic effects of medication. For example, the microenvironment formed by liquid crystals can alter the immunosuppressive state of tumors, thus improving the efficacy of antitumor immunity (Hu et al. 2020). Additionally, new materials like liquid crystal metals offer new options for photothermal therapy of tumors (Tucker et al. 2012; Turajlic and Swanton 2016;

Rastogi et al. 2019; Kim et al. 2023). These methods not only enable targeted treatment of tumors but also minimize damage to normal tissues.

However, despite the great potential of liquid crystal technology in tumor diagnosis and treatment, there are still many challenges and issues that need to be addressed. Ensuring the stability and biocompatibility of liquid crystal materials, enhancing the sensitivity and specificity of liquid crystal technology, and integrating liquid crystal technology with other therapeutic methods are all focal points of current research. This chapter reviews the research related to liquid crystals and tumor diagnosis and treatment, highlighting the application and advantages of liquid crystal technology in tumor diagnosis and therapy. We will also explore the challenges faced by liquid crystal technology and future research directions.

3.2 Liquid Crystal Based Material for Sensing

To date, the development of advanced detection and analysis technologies for tumor markers has become one of the important research topics internationally (Sorbye et al. 2023; Siegel et al. 2024). Tumor markers, as commonly used biological markers in clinical practice, are present in patients' tissues, body fluids, and excretions, and they hold significant value in cancer screening, diagnosis, and prognosis assessment (Wu and Qu 2015). A variety of detection techniques for these biomarkers have been documented by researchers, including radioimmunoassay (RIA), enzyme-linked immunosorbent assay (ELISA), chemiluminescence immunoassay (CLIA), and immunosensor technology. Despite their potential, these techniques often suffer from complexity in operation, high costs, and restricted use in clinical settings (Urruticoechea et al. 2010; Son et al. 2023; Kumar et al. 2023b).

The distinctive nature of liquid crystals, which can flow like liquids and exhibit anisotropic characteristics akin to crystals, makes them an excellent match for molecular recognition components. This compatibility enables the detection of molecular interactions on both solid and liquid surfaces. The advantages of speed, the absence of the need for labeling, and economic efficiency have highlighted the promising role of liquid crystal-based biosensing in the realms of diagnostic and clinical use (Tomilin et al. 2001; Woltman et al. 2007; Pani et al. 2023; Zhang et al. 2023). The Abbott group was the first to innovate by utilizing liquid crystal molecules for converting signals, with the integration of antibodies serving as the molecular detection agents, leading to the creation of a novel biosensing technology based on liquid crystals. This ground breaking study initiated a new era in the field of liquid crystal-based biosensing devices (Rouhbakhsh et al. 2022). Spanning over 20 years of advancement, liquid crystal-based sensing has shown notable success in chemical analyzes, encompassing the sensing of various substances like glucose, folic acid, and metal ions (Hussain et al. 2016; Luan et al. 2020; Devadiga and Ahipa 2022; Pani et al. 2023; Zhang et al. 2023). Meanwhile, the technology of liquid crystal biosensors has reaped substantial outcomes in bioanalytical applications, a multitude of investigators have harnessed the capabilities of liquid crystals for the identification and measurement of biomarkers associated with cancer, such as nucleic acid, antigens, proteins, enzymes, cancer cells, and cancer imaging (Hussain et al. 2016; Luan et al. 2020; Concellón 2023; Jiang et al. 2024).

3.2.1 Nucleic Acid Sensing

Nucleic acids, such as deoxyribonucleic acid (DNA) and ribonucleic acid (RNA), preserve genetic information for the next generation of organisms by storing it in nucleotides. Therefore, nucleic acid testing has extremely important application value in fields such as forensic science, medicine, and genetics. Recently, in DNA detection, liquid crystal droplet platforms have been extensively investigated by functionalized liquid crystal interfaces. Lai et al. reported a method for detecting DNA targets that hybridize to solid surfaces using liquid crystal (Fig. 3.2). The analytical principle of this method is based on the different interference colors displayed on the surface of liquid crystal droplets modified with single-stranded DNA (ssDNA) or double-stranded DNA (dsDNA). At the same time, streptavidin complex DNA was introduced into dsDNA to increase the surface mass density, which was used to distinguish ssDNA from dsDNA. By pre-mixing biotin-labeled DNA targets with streptavidin prior to DNA hybridization, branched-streptavidin complexes can be formed and clear liquid crystal signals observed. This liquid-crystal-based DNA detection principle is an important step toward developing a simple, instrumental and fluorescent carrier-free DNA detection method (Lai et al. 2011).

Verma et al. have also developed a DNA detection sensor based on liquid crystal droplets. They prepared positively charged polyl-lysine (PLL) coated droplets for significant absorption of negatively charged DNA, resulting in a change in the orientation of the droplet molecules (Verma et al. 2017). They effectively demonstrated that these DNA and PLL-decorated liquid crystal droplets, under physiological conditions, are gradually stimulated by DNA molecules to release propyl iodide (PI) dyes (Verma et al. 2017).

For continuous detection, Ma et al. revealed a unique, simple, and ultra-low sample consumption assay method using whispering gallery mode (WGM) together with liquid crystal droplet-based sensor. To create a fiber-optic probe, the sensor

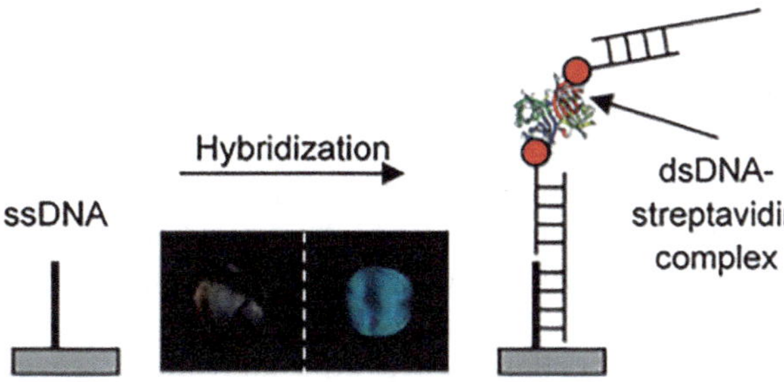

Fig. 3.2 Strategies for detecting DNA targets hybridized to solid surfaces using optical images of liquid crystals, reprinted from Lai et al. (2011), copyright 2011, with permission from American Chemical Society

combines liquid crystal droplets and a hollow capillary tube (HCT). The test solution maintained a stable suspension of the LC droplets. The DNA detection was then accomplished by measuring the liquid crystal droplet orientation change and WGM spectroscopy. When used to detect the target salmon sperm DNA, it produced a measurement range that could be adjusted from 3.75 to 11.25 g/mL with a sensitivity of 0.33 g/mL. The detection solution required only 3 nL of samples, the detection limit was 1.32 g/mL, and the effective DNA detection was only 3.96 pg (Ma et al. 2022).

3.2.2 Protein Sensing

The early identification, management, and prognostic assessment of cancer necessitate the creation of a highly specific and sensitive method for the detection of blood-borne tumor markers. Researchers Ren and Jang constructed an innovative sensor for the detection of tumor maker carcinoembryonic antigen (CEA), utilizing the sensitivity of liquid crystals. Aptamers, acting as molecular identifiers, were affixed to the surface of glass slides that had been treated with (3-aminopropyl)-triethoxysilane and dimethyltetradecyl-[3-(trimethoxysilyl)propyl]-chloride, using glutaraldehyde to secure them in place. These aptamers were designed to bind specifically to CEA. When CEA is present, the aptamers' looped structures, which initially covered the slides, unfolded, enabling them to engage with CEA and form a complex. This interaction resulted in a change in the liquid crystal molecules' orientation from an orderly to a disordered state. The resultant change in the optical appearance from dark to light can be simply observed with a polarized light microscope. The sensor offers the selectivity to differentiate CEA from other compounds with analogous protein structures (Ren and Jang 2019).

For the multiplexed detection multiple tumor markers, Qi and associates introduced, for the first time, an approach that leverages functional liquid crystal sensors and aptamer-based target-induced dissociation (TID) for the concurrent identification of several tumor markers in blood (Fig. 3.3). Magnetic beads (MBs) enveloped with specific aptamers (apt1) capture target proteins present in the bloodstream. Upon incubation with a double-stranded structure of another aptamer (apt2) and signal DNA, the protein-laden MBs form a sandwich complex of apt1/protein/apt2 due to apt2's specific targeting of the protein, prompting the release of signal DNA into the solution. This signal DNA is then specifically detected by an acutely sensitive liquid crystal sensor laden with DNA. Employing this methodology, a 3D-printed optical cell was utilized to achieve highly specific and sensitive concurrent detection of multiple tumor markers, including carcinoembryonic antigen (CEA), alpha-fetoprotein (AFP), and prostate-specific antigen (PSA). This high-efficiency, cost-effective multiplexing technique capitalizes on the straightforward isolability of MBs, the high specificity of aptamer identification, and the heightened sensitivity of the functional liquid crystal sensors. Additionally, its performance rivals that of commercial ELISA kits, and it remains unaffected by hemolysis,

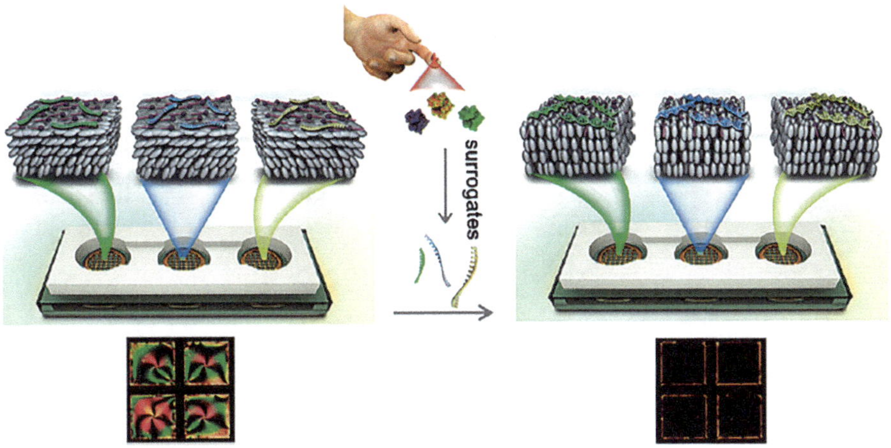

Fig. 3.3 Principle of simultaneous detection of multiple tumor markers in blood with functional LC sensors, reprinted from Qi et al. (2020), copyright 2020, with permission from American Chemical Society

indicating promising applicability in the clinical multiplexed detection of tumor markers (Qi et al. 2020).

In order to further improve the sensitivity of liquid crystal-based point-of-care testing (POCT) of protein markers Qi et al. developed an aptamer-CRISPR/Cas12a-regulated liquid crystal sensor (ALICS) that enables ultra-sensitive protein detection using a smartphone-coupled portable device. As shown in Fig. 3.4, the probe containing the protein target aptamer sequence and the Cas12a activation sequence is fixed on the substrate, and the Cas12a activation sequence is released in the presence of the target. Activate the CRISPR reaction and cut the basal link sequence, resulting in optical changes from light to dark in the DNA-functionalized liquid crystal interface. The result is an optical image captured by a smartphone that can be used to quantify the target concentration. The detection limits of ALICS for SARS-CoV-2 nucleocapsid protein (N protein) and carcinoembryonic antigen (CEA) model proteins were 0.4 and 20 pg/mL, respectively, which was higher than the typical sensitivity of SARS-CoV-2 and clinical CEA detection. At the same time, the detection effect of clinical samples is better than that of commercial ELISA and side-flow detection kits. Overall, ALICS is a highly sensitive and cost-effective POCT platform that shows great potential for pathogen detection and disease surveillance in resource-limited Settings (Qi et al. 2024).

Known as Mucin 16 (MUC16), CA125 stands among the most pioneering biomarkers in oncological studies. It functions as a glycoprotein on cellular surfaces and can also be detected in a soluble state within the blood plasma. CA125 is implicated in the immune response mediated by cells, with the hypothesis that tumor-origin CA125 may dampen the immune system's antitumor activity. An increase in CA125 levels serves as a marker for ovarian cancer at various stages and is also observed in malignancies like breast and stomach cancers (Charkhchi et al. 2020;

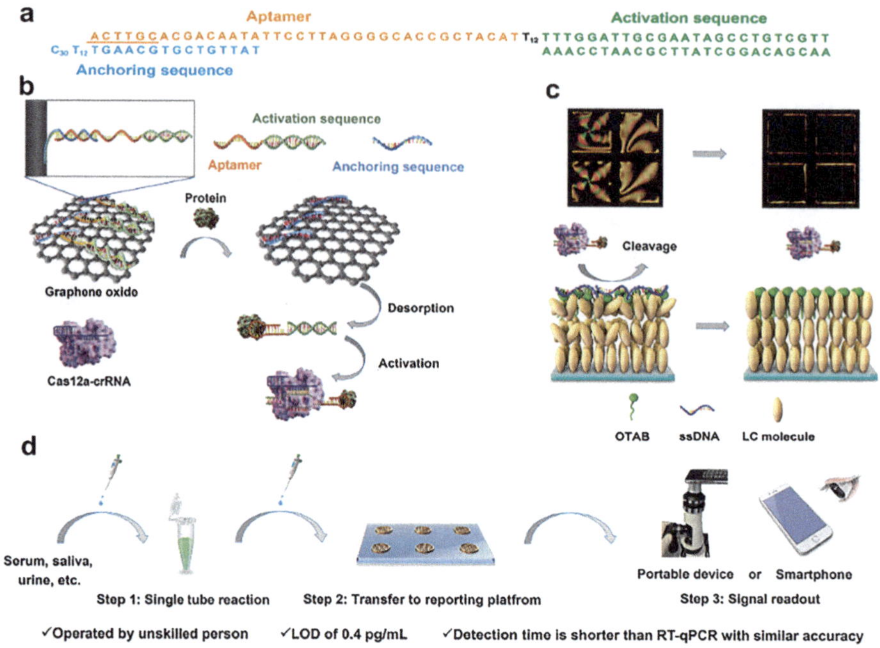

Fig. 3.4 Schematic illustration of aptamer-CRISPR/Cas12a-regulated liquid crystal sensor (ALICS), reprinted from Qi et al. (2024), copyright 2024, with permission from American Chemical Society

Zhang et al. 2021). Researchers, including Su, improved the sensitivity of immunoassay techniques that rely on liquid phase chromatography by performing surface modifications using a surfactant composed of dimethyltetradecyl-[3-(trimethoxysilyl) propyl]chloride (DMOAP). They affixed a range of antibodies specific to CA125 onto a monolayer of DMOAP that had been treated with ultraviolet light to detect the tumor biomarker CA125. The process of UV exposure was found to boost the binding affinity of the CA125 antibodies and the reliability of the immunoassay results, reaching a detection threshold for CA125 as low as 0.01 ng/ml. The optical patterns observed through a cross-polarization microscope correlated with the concentration of the target analyte. The findings suggest that the application of UV treatment to the calibration layer and the detection method using liquid crystals with significant birefringence can enhance the capabilities of label-free immunoassays based on liquid crystals. This method shows potential as a favorable substitute for conventional assays that rely on labeling techniques (Su et al. 2015).

Researchers Chiang and colleagues have crafted a method for quantifying proteins and performing label-free immunodetection using dye-doped liquid crystals (DDLC). The black dye added to the DDLC mixture as a dopant exhibits a pronounced dichroic ratio and an extensive absorption spectrum in the visible light range, facilitating the visual detection of perturbations in DDLC alignment due to biomolecules. A quantitative biosensing technique was developed based on DDLC, utilizing transmittance spectroscopy that capitalizes on the DDLC's absorption and anisotropy. It was found that as the concentration of bovine serum albumin (BSA), a standard protein, increases, the transmittance of DDLC diminishes, and employing linearly polarized light in spectroscopic assessments boosts the sensitivity of detection. Demonstrating clinical applicability, the immunoassay for the cancer biomarker CA125 was conducted by engaging CA125 with antibodies specific to CA125 that were immobilized on a surface. The transmittance of DDLC was observed to lower with rising CA125 concentrations, whereas the transmittance spectra remained consistent with the absence of CA125 when nonspecific proteins like BSA interacted with the anti-CA125 antibodies, indicating the specificity of the DDLC-based biosensing platform for immunodetection. The findings suggest that DDLC, due to its distinctive optical attributes and dichroism, holds promise as a biosensing material capable of qualitative detection for swift screenings and quantitative detection for cancer diagnostics (Chiang et al. 2018).

Sun have utilized a LC with high-birefringence properties to detect the cancer biomarker CA125, capitalizing on the large birefringence ($\Delta n = 0.33$ at 20 °C) to amplify the detection signal and markedly increase the sensitivity of the assay. To avoid signal loss typically associated with washing steps in immunoassays, the antigen and antibodies for CA125 are allowed to react on a substrate prior to its fixation, obviating the need for washing away unbound or unreacted biomolecules, thereby preventing signal loss. This method paves the way for a streamlined LC-based immunoassay that offers a heightened optical response when using a high-birefringence LC mixture over the traditional 5CB. Given that the variations in the optical textures of LCs only offer qualitative insights into the immunoreaction, this research also pioneers a quantitative approach to assessing the optical responses of LCs (Sun et al. 2015).

In contrast to thermotropic liquid crystals, the biocompatibility of lyotropic liquid crystals is enhanced by their affinity for water, but their biosensing capabilities have seldom been investigated. Researchers, including Shaban, have studied the nematic phase of disodium cromoglycate (DSCG), a phase shared by both thermotropic liquid crystals and lyotropic liquid crystals, employing it as an intermediate phase for sensing in lyotropic liquid crystals biosensor. The biosensing platform was designed such that the lyotropic liquid crystals was uniformly aligned by the planar anchoring strength of polyimide, yet this alignment was disturbed by the presence of bovine serum albumin (BSA) or the cancer biomarker CA125, which was captured by specific anti-CA125 antibodies. The extent of this disruption, and the consequent optical signal, was predominantly dictated by the quantity of the analyte. The optical responses, which are dependent on concentration and wavelength, were examined using transmittance spectroscopy within the visible light spectrum under conditions of parallel or cross-polarized light. The concentration of CA125 could be quantitatively determined through parameters derived from the spectrum on a linear calibration curve. The lyotropic liquid crystals-based biosensor demonstrated a detection limit for BSA and CA125 that was either superior or on par with that of biosensor technologies based on thermotropic liquid crystals. These findings offer the initial evidence that LCLC can be utilized for spectral quantitative biosensing applications (Shaban et al. 2021).

Alpha-synuclein (αS) has been proposed as a potential biomarker for the diagnosis of Parkinson's disease (PD). However, the detection of αS using a simple, rapid, and sensitive method remains challenging. Yang et al. combined the stimulus responsiveness of LC and the specific interaction of DNA aptamers with proteins to construct a novel biosensor for the detection of αS. In principle, the positively charged surfactant cetyltrimethylammonium bromide (CTAB) binds to negatively charged DNA aptamers by electrostatic interaction. In the presence of αS, the DNA aptamer binds specifically to αS and releases the amphipathic molecule CTAB, which subsequently assembled at the liquid crystal-aqueous interface, leading to the codirectional alignment of the liquid crystals under dark light signals. In the absence of αS, CTAB binding to DNA aptamers does not affect liquid crystal alignment, showing planar anchoring and a bright optical signal. The response time of liquid crystals to αS is fast and can be as short as a few minutes. The established liquid crystal biosensor showed good specificity for αS and recognized αS even in mixed proteins. The liquid crystal biosensor has a high sensitivity, and the detection limit for αS can be as low as 10 pM, which is comparable to ELISA. This study provides a simple, rapid, and sensitive new strategy for the detection of αS, which has good application prospects in early diagnosis and clinical application (Yang et al. 2021).

Researchers Zhou and colleagues presented a highly sensitive biosensor that leverages the targeted growth of silver nanoparticles on a liquid crystal platform for the multicolored visual detection of alkaline phosphatase (ALP) activity. The product of the enzyme's action gathers on the liquid crystal sensor film via a silver deposition method, leading to a marked amplification of the birefringence signal observable under polarized light microscopy. The incorporation of gold nanoparticles further amplifies the sensitivity by about four orders of magnitude. The intensity of the bright spots in the images captured by the polarizing optical microscope (POM) correlates with the level of ALP activity, with the signal strength quantified by calculating the product of the average brightness and pixel values using Photoshop. The assay's detection threshold for ALP is as low as 1.2 $nU{\cdot}mL^{-1}$, significantly more sensitive than other colorimetric or fluorescent detection methods by 5–7 orders of magnitude. This technique has been adapted for the highly sensitive detection of CEA isolated using immunomagnetic beads. This method bypasses the need for complex sample preparation procedures and is capable of analyzing complex samples with a detection limit down to 0.35 pg mL^{-1}. It represents an innovative approach to immunoassays using immunomagnetic bead separation (Zhou et al. 2018a).

Carboxyesterase (CES) is considered to be the most effective candidate biomarker of liver cancer cells and has achieved fruitful results in the early detection of liver cancer. Therefore, the establishment of a reliable and simple system for detecting CES is attractive for practical applications in medical diagnosis and treatment. Zhou et al. reported a novel CES monitoring strategy based on the water/liquid crystal interface. First, a crackable surfactant, N-octalkoxycarbonyl-methyl-N,N,N-trimethylammonium bromide (OTB), was synthesized to control the orientation of

4-cyano-4′-pentylbiphenyl (5CB) molecules at the water/lc interface. The generative process of an OTB self-assembled monolayer promotes the alignment of the liquid crystal, which is on average perpendicular to the surface (homotropy), while the corresponding black image is observed in polarized light microscopy (POM). Then the doping of CES induces the hydrolysis of OTB, leading to the destruction of the OTB monolayer and thus a distinct dark to light transition in the optical response. This optical transition provides a quick and easy way to detect CES. OTB-modified liquid crystal surfaces have been developed as sensors with a detection limit of 18 U/L for CES detection. The prepared liquid crystal-based sensor showed high sensitivity and specificity for the detection of CES and good long-term stability. Preliminary results show that the detection of CES in urine samples is successful, which also indicates that the liquid crystal sensor has good detection performance in practical situations. In generally, the prepared CES sensor has the advantages of low cost and simple fabrication, demonstrating its potential in medical adjuvant therapy of human samples (Zhou et al. 2019).

Breast malignancies rank among the most prevalent forms of cancer in women across the globe and account for the second highest cause of death from cancer. Researchers, Pouras and colleagues, have harnessed the biomarker human epidermal growth factor receptor 2 (HER-2) for the evaluation of breast cancer (Fig. 3.5). The sensor's operational mechanism relies on a surface treatment with dimethyltetradecyl[3-(trimethoxysilyl)propyl]chloride (DMOAP), which

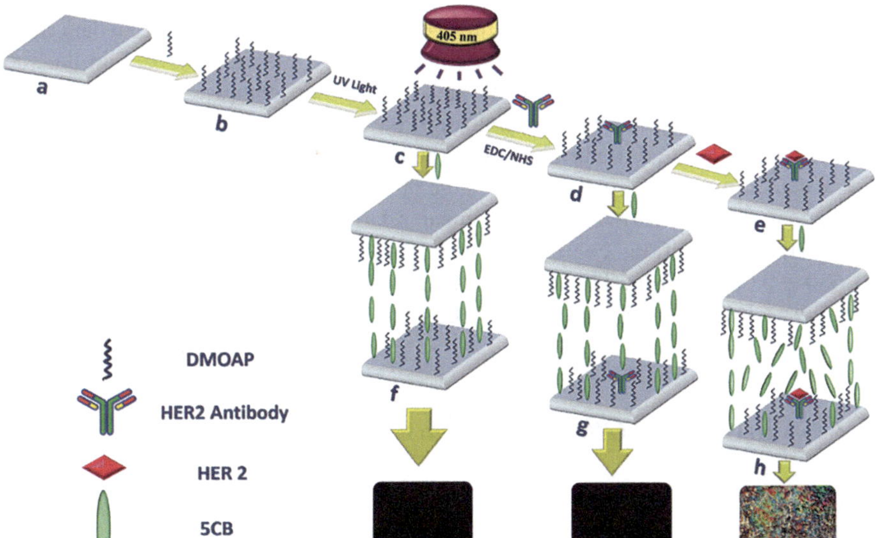

Fig. 3.5 Schematic preparation steps of liquid crystal-based biosensor for HER-2 detection, reprinted from Pourasl et al. (2023), copyright 2023, with permission from the authors and Springer Nature

encourages the formation of extended alkyl chains, thereby facilitating the uniform orientation of liquid crystal molecules at the interface. To further augment the integration of HER-2 antibodies on the liquid crystal aligning film, a straightforward ultraviolet-assisted technique was also utilized to escalate the functionality of the DMOAP-coated slides, enhancing the binding affinity and efficiency with HER-2 antibodies. The biosensor design leverages the specific interaction between the HER-2 protein and its corresponding antibodies, which perturbs the liquid crystal alignment. This perturbation leads to a shift in the optical appearance from non-birefringent to birefringent, allowing for the detection of HER-2. With a detection threshold reaching as low as 1 fg/mL, this innovative biosensor has been effectively applied in a proof-of-concept study to quantify HER-2 proteins in the context of breast cancer patients. Characterized by its high sensitivity, selectivity, and the absence of a need for labeling, this biosensor holds promise to broaden the scope of liquid crystal-based biosensing applications in the detection of a wide array of cancers (Pourasl et al. 2023).

Protein kinase C (PKC) is a protein that widely exists in all organisms and plays a key role in the pathogenesis of many diseases, including cancer. PKC has become a target for the diagnosis and treatment of many diseases. Liu et al. demonstrated a rapid PKC detection method based on liquid crystal biosensor and nanomodified signal amplification technology. PKC antibody/gold nanoparticles (anti-PKC/AuNPs) conjugates were immobilized on a self-assembled hybrid substrate of (3-aminopropyl)-triethoxysilane/N, N-dimethyl-N-octadecyl-(3-aminopropyl)-trimethoxysilyl chloride (APTES/DMOAP) using glutaraldehyde (GA). The above treatments keep liquid crystal molecules vertically aligned and produce a dark background under polarized light. When PKC is present, it is immobilized on the substrate by binding specifically to the anti-PKC of the complex. Specific binding events lead to changes in the morphology of the substrate, the arrangement of liquid crystal molecules, and the brightness of the liquid crystal membrane, enabling sensitive detection of PKC. In the linear range of 10–250 ng/mL, the average gray value of optical imaging showed a good linear relationship with the concentration of PKC, the correlation coefficient was 0.997, and the detection limit was (1.680 ± 0.005) ng/mL. The results of this study provide potential applications for highly sensitive and simple detection of protein kinases (Liu et al. 2021).

3.2.3 Toxins and Other Chemicals Sensing

Environmental pollutants such as endotoxins, aflatoxins, myricetin, pesticides, and antibiotic residues are key to a number of human diseases, including cancer. Therefore, sensitive detection of their content in food is very important.

MY detection has been suggested and demonstrated using a 5CB droplet-based sensing technique by Xiong et al. (2020). The 5CB droplets were generated via a syringe pump connected to a conical capillary microtubule and were functionalized in the aqueous phase with DTAB and DNA. Upon increasing the concentration of

MY, the structural configuration of the 5CB microdroplets underwent a transformation. Furthermore, the 5CB microdroplets were employed as optical microcavities to generate a typical whispering-gallery-mode (WGM) laser spectrum. As the MY concentration increased, a blue-shift phenomenon was observed in the WGM spectrum. Within the established detection range, the sensitivity of the system was determined to be 0.04 nm/μM.

The aminoglycoside antibiotic kanamycin is produced by Streptomyces kanamyceticus and is commonly administered to animals for the treatment of infections. However, excessive accumulation of kanamycin in the human body can lead to antibiotic resistance and adverse effects, including ototoxicity and nephrotoxicity, which may ultimately result in severe and potentially catastrophic harm (Jin et al. 2017). Therefore, the detection of kanamycin residues is of critical importance. There is an increasing demand for simple, practical, and cost-effective methods that enable rapid and label-free identification of kanamycin. Yin et al. (2021) proposed a novel approach for kanamycin detection based on surface-anchored liquid crystal droplets. In this method, the optical appearance of liquid crystal droplets transitions from a four-leaf clover shape to a uniformly dark cross shape with increasing surfactant concentration. The detection relies on the use of a kanamycin aptamer and the cationic surfactant cetyltrimethylammonium bromide (CTAB). Specifically, the addition of CTAB solution renders the liquid crystal droplets uniformly dark, while the CTAB/aptamer complex induces a four-leaf clover shape. The binding of kanamycin to its aptamer releases CTAB, resulting in a uniformly dark appearance of the droplets. A portable instrument was developed to measure the optical brightness of these droplets. This technique allows for the detection of kanamycin in complex matrices, such as milk and honey, with a detection limit as low as 0.1 ng/mL (or 0.17 nM). The development of this portable optical device represents a significant advancement in the creation of novel liquid crystal-based sensor platforms using surface-anchored liquid crystal droplets.

To elucidate the molecular interactions at the droplet interface, Pani et al. utilized aqueous liquid crystal dispersions to examine the interaction between mitochondrial cardiolipin (CL) and membrane-associated cytochrome c (Cyt c), the latter being a critical component of the apoptotic signaling cascade. In an aqueous environment, CL was found to encapsulate the surface of 5CB droplets and subsequently bind to Cyt c at the droplet interface (Fig. 3.6). For the first time, the lipid–protein interactions that induce reorientation of the liquid crystal droplet interface were demonstrated through an integrated approach combining atomic-level calculations, microscopic imaging, and spectroscopic analyzes (Pani et al. 2021).

Endotoxin, a component of the cell wall of Gram bacteria, consisted of two polysaccharide domains and a glycosphingolipid (lipid A) (Roberts 1966; Stephens and Mythen 2000). Jiang et al. (2021) demonstrated that liquid crystal droplet-based experimental measurements combined with machine learning techniques can effectively classify bacterial sources and quantify endotoxin concentrations derived from three distinct bacterial species in aqueous solutions. Their study highlighted the application of EndoNet, a machine learning model, to detect minute changes in the scattering field. This enabled the classification of bacterial sources and the

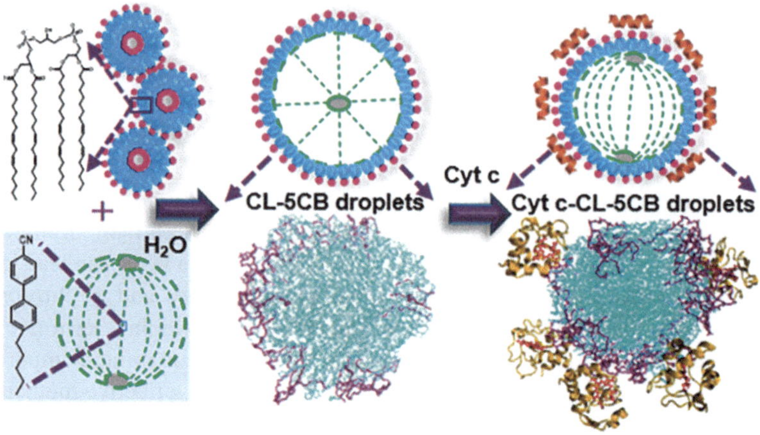

Fig. 3.6 Schematic illustration of lipid–protein interactions that control the reorientation at the liquid crystal droplet interface, reprinted from Pani et al. (2021), copyright 2021, with permission from American Chemical Society

measurement of endotoxin levels across an extensive range of eight orders of magnitude, from 0.01 $pg{\cdot}mL^{-1}$ to 1 $g{\cdot}mL^{-1}$.

Aflatoxin is recognized as one of the most toxic and carcinogenic mycotoxins. Even minimal consumption of food contaminated with Aflatoxin B1 (AFB1) can lead to cumulative effects, posing a significant risk to human health (Berthiller et al. 2013). Consequently, ensuring food safety against mycotoxins necessitates the development of highly sensitive and reliable detection methods to identify the causes of food poisoning (Liu et al. 2020; Dong et al. 2021). Recently, Cheng et al. demonstrated a novel approach using surface-anchored 5CB droplets (5CBSADrop) on a glass substrate coated with N,N-dimethyl-N-octadecyl(3-aminopropyl) trimethoxysilyl chloride (DMOAP) for the detection of AFB1 in food samples. The 5CBSADrop was formed by solvent evaporation. Under polarized optical microscopy, a black cross pattern corresponding to a radial orientation of the LC droplets was observed when CTAB was applied to the 5CBSADrop. In contrast, the 5CBSADrop exhibited a four-leaf clover shape when treated with a CTAB/AFB1 aptamer mixture, reflecting an escape-radial configuration of the LC droplets. Furthermore, a radial structure was observed when a mixture of CTAB, AFB1, and its aptamer was applied to the 5CBSADrop. These distinct patterns under POM enabled the identification of various AFB1 concentrations. The practical applicability of this method was demonstrated using peanut oil and rice samples. The 5CBSADrop technique is particularly promising for the food and agricultural industries due to its simplicity, minimal sample consumption (1 μL), high sensitivity, excellent stability, rapid on-field detection capability, and the elimination of the need for labeling and expensive equipment, all while maintaining low costs (Cheng et al. 2022).

The responsiveness of microdroplets floating in aqueous environments made the creation of new types of LC droplet environmental sensors. Based on enzymatic reactions, liquid crystal molecules can be employed to detect pesticides. Zhou et al. (2018b) developed a simple yet robust 5CB droplet sensor for accurate and practical detection of dichlorvos (DDVP) based on its hydrolysis by alkaline phosphatase (ALP). The sensor utilized sodium monododecyl phosphate (SMP), an ALP-cleavable surfactant, to modulate the orientation of 5CB liquid crystals (LCs) and thereby control their optical responses. The formation of an SMP monolayer at the water/liquid crystal droplet interface resulted in a dark cross optical image of 5CB. Upon addition of ALP and SMP, the optical appearance of 5CB transitioned to a bright fan-shaped pattern. Notably, when DDVP was preincubated with ALP, the liquid crystal maintained a dark optical image under polarized optical microscopy (POM). This sensor achieved a detection limit for DDVP of 0.1 ng/mL.

Volatile organic chemicals (VOCs) are often utilized in chemical synthesis as precursors. However, most VOCs are toxic and flammable, posing significant safety concerns for both human health and the environment (Kumar et al. 2015). Prolonged exposure to VOC vapors has been implicated in various diseases, including lung cancer and chronic respiratory inflammation (Phillips et al. 1999; Jareño et al. 2014; Zhu et al. 2020; Salimi and Milani Hosseini 2021), liquid crystal droplets could be a promising platform to detect various VOCs. For instance, An and Jang (2019) developed a sensing method based on liquid crystal droplet patterns to detect and monitor trace levels of organic aldehyde vapors. When liquid crystal droplets coated with a glycine solution were exposed to aldehyde vapors, the optical signal shifted from a bright sector to a dark cross pattern, as observed under polarized optical microscopy (POM). The results highlighted the excellent sensitivity and selectivity of the glycine/liquid crystal droplet pattern technique. Notably, the signal change was completed within a few minutes (2–7 min) when the sensor was exposed to aldehyde vapors in a real-world environment.

He et al. (2025) prepared surfactant stabilized W/LC/W emulsion with guest aqueous droplet loaded in the center of liquid crystal droplet (Fig. 3.7). Under the electrostatic interaction of surfactant with opposite charge, the encapsulated guest aqueous droplet was propelled to move from center of liquid crystal droplet toward the edge of liquid crystal droplet, finally disperse to surrounding continuous aqueous phase responsively. Through the characterization by BFM, CLSM, POM, and calculation of key interplay forces, the responsive release of guest aqueous droplet from liquid crystal droplet was depicted. With hCG encapsulated in the guest aqueous droplet located in the center of SDS stabilized double emulsion, amphiphile which carry positive charge triggered the release of hCG with positive concentration correlation. This response can be applied in the design of pregnancy test strip repurposing. Combined with the reduction effect of organophosphorus to enzyme, the hydrolysis of enzyme to myristoylcholine chloride and the trigger ability of myristoylcholine chloride, the responsive system was applied successfully in organophosphorus detection, as a proof-of concept to prove the application of responsive double emulsion in pregnancy test strip repurposing.

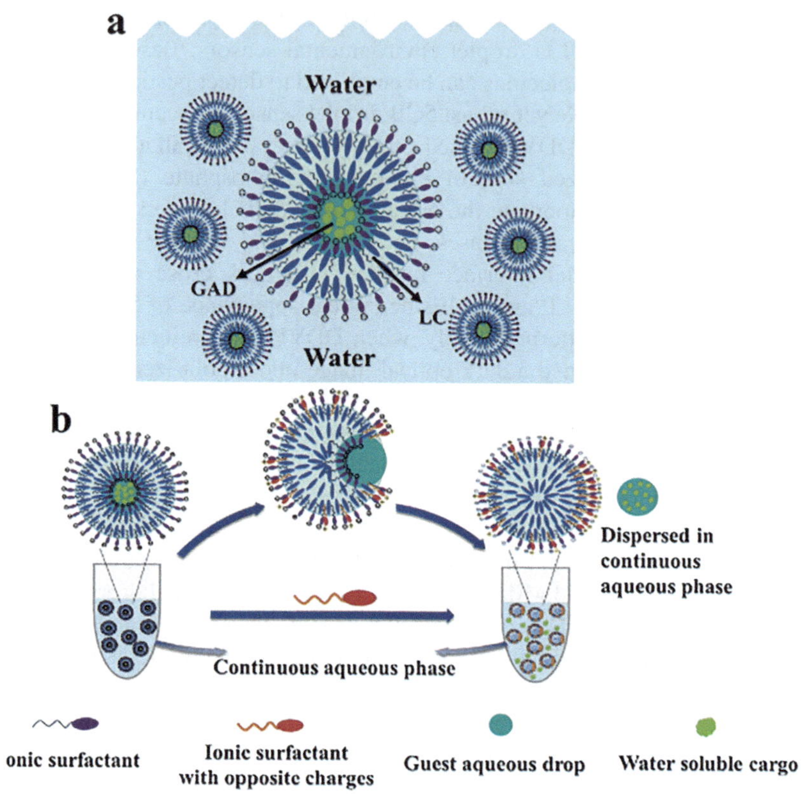

Fig. 3.7 Trapping and ejection of GAD in LC droplet, reprinted from He et al. (2025), copyright 2025, with permission from Elsevier

Deep convolutional neural networks (CNN) were investigated by Frazao et al. for using as pattern recognition systems to examine the dynamics of optical textures in liquid crystal droplets subjected to various VOCs (Frazão et al. 2021). Their study demonstrated that a single liquid crystal droplet could distinguish between 11 types of VOCs with subtle structural and functional differences. Through regression modeling, it was hypothesized that changes in the optical texture patterns of the droplets correlated with variations in VOC content. Consequently, the CNN-based approach provided a potential means for VOC detection based on the responses of individual liquid crystal droplets.

Bile acids (BAs) are crucial metabolites involved in the emulsification and digestion of dietary fats and lipophilic vitamins. However, excessive production of BAs is associated with disorders of the liver and intestines. Consequently, the diagnosis of hepatic and intestinal diseases has traditionally relied on the analysis of BAs in various bodily fluids, such as blood, liver tissue, gallbladder fluid, and urine. Cholic acid (CA) is a primary BA, accounting for approximately 31% of the total BAs produced in the liver. Secondary BAs include deoxycholic acid (DCA) and lithocholic acid (LCA). The correlation between CA and liver/intestinal diseases is stronger than that of other BAs. Host-guest chemistry has been employed to design liquid crystal droplet-based sensors. For instance, Deng et al. (2015) reported the selective detection of CA using a sensor composed of 5CB droplets coated with a β-cyclodextrin (β-CD)/tetradecyl trimethylammonium bromide (C_{14}TAB) complex. β-CD is a naturally occurring host molecule with a hydrophobic cavity that selectively recognizes CA and forms an inclusion complex with a high binding constant of 1:1. The interaction of CA with β-CD displaces C_{14}TAB from the β-CD cavity, inducing a conformational change in the liquid crystal droplet (Deng et al. 2015).

Detection by competitive adsorption was also a common method. To detect LCA, Fang's group changed the interface of 5CB droplets distributed in an aqueous solution by adsorbing surfactants on the 5CB/aqueous interface. The lower limit of detection of surfactant stabilized 5CB droplets may be changed in the range of 10–70 μM by altering the chain length of the surfactant (Bera and Fang 2013). Niu et al. developed a sensitive biosensor for cholic acid (CA) based on 5CB droplets dispersed in phosphate-buffered saline (PBS). The sensor relies on a radial-bipolar transition of the 5CB droplet, which occurs when CA competes with sodium dodecyl sulfate (SDS) for adsorption sites on the droplet surface. This liquid crystal droplet sensor is characterized by its rapid response, ease of use, and low cost. The detection limit of this method was determined to be 5 μM (Niu et al. 2019). Han et al. have reported a novel PDMS microchip based on the monodisperse droplet, wherein a single microfluidic device was used to create in situ monodisperse droplets and capture them. They could detect BAs quantitatively and quickly (<4 min) in real time with little sample consumption (~1.5 μL) by observing the form of the LC droplets under a POM. The detection limits for CA and DCA were 10 μM and 1 μM (Han et al. 2020). Gollapelli et al. recently introduced a novel and highly sensitive biosensor for detecting bile acids (CA and DCA) based on polyvinyl alcohol (PVA)-stabilized cholesteric liquid crystal (CLC) droplets using the surfactant SC12S. In this system, BAs can competitively displace other surface-active molecules from the LC/water interface through adsorption. The competitive adsorption of bile acids and SC12S on the CLC droplet surface induces a reconfiguration of the droplets from a homeotropic to a planar alignment. This innovative approach enables the detection of BAs using conventional optical microscopes, with a detection limit of 1 μM (Gollapelli et al. 2021).

3.2.4 Cell Sensing and Imaging

Circulating tumor cells (CTCs) serve as pivotal indicators for evaluating the efficacy of cancer treatments, given their propensity to detach from the primary lesion, circulate in the bloodstream, and initiate metastatic processes. Detecting CTCs is challenging due to their low prevalence in blood samples. Researchers, including Chang, have developed a label-free sensor that utilizes liquid crystals and aptamers targeting the epithelial cell adhesion molecule (EpCAM) (Fig. 3.8) to capture cancer cells that express this marker. The study delved into the optical and dielectric signals resulting from the interaction between the liquid crystals and captured breast cancer cells in varying quantities. By analyzing the dielectric constants—both real and imaginary—at frequencies of 1 kHz and 30 Hz, along with the relaxation frequency of the liquid crystals, the detection sensitivity was enhanced to the level of single CTCs. The liquid crystal-based sensor, specific for EpCAM, was effectively utilized to identify individual CTCs within human serum and blood samples. This approach underscores the capability of liquid crystal-based biosensing for detecting and quantifying at the cellular level, an essential advancement for the early identification of cancer metastasis and progression (Chang et al. 2022).

The conformational actions of liquid crystal microemulsions render them compelling for the detection of interfacial interactions involving chemicals or biological molecules within biological systems. Sidiq et al. proposed a simple method to generate biocompatible liquid crystal droplets by poly (L-lysine) (PLL)—liquid crystal interaction in situ for reporting the presence of cells and monitoring real-time cell-environment interactions mediated by topological defects in these droplets. In addition, responding PLL droplets were found to serve as a template for reporting annexin V-phosphatidylserine interactions, providing a simple measure of deleterious effects on cell health (Sidiq et al. 2017).

Researchers Yoon et al. have catalyzed a conformational shift in 5CB liquid crystal microemulsions by utilizing a medium composed of a folic acid-tethered block copolymer (PS-b-PAA-FA) and sodium dodecyl sulfate (SDS), to fabricate selective liquid crystal microemulsions for oropharyngeal epidermoid (KB) cancer cells (Fig. 3.9). These crafted microemulsions demonstrated a morphological change from a radial to a bipolar arrangement when in contact with KB cancer cells, whereas no such transition was detected during interactions with typical cell types such as fibroblasts and osteoblasts. The selectivity of these microemulsions for KB cancer cells is attributed to the presence of folic acid receptor-specific ligands (FA)

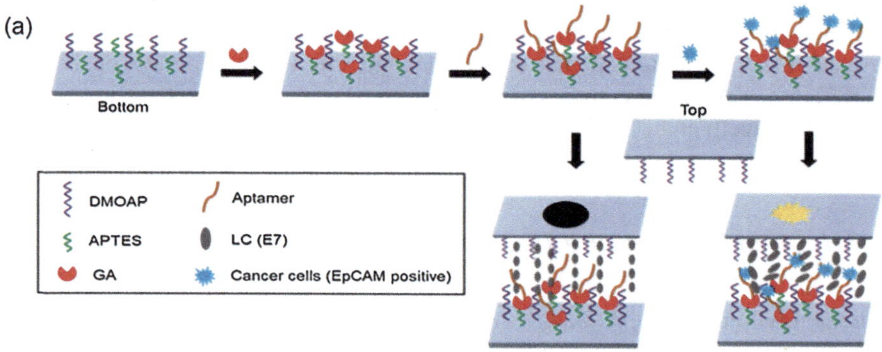

Fig. 3.8 Workflow for CTC detection by the label-free aptamer-based liquid crystal biosensor through optical measurements, reprinted from Chang et al. (2022), copyright 2022, with permission from Elsevier

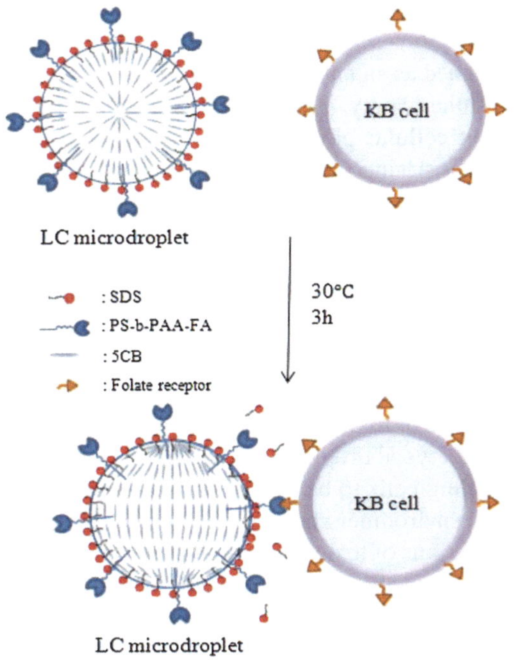

Fig. 3.9 KB cancer cells interactions with liquid crystal microdroplets containing folic acid-conjugated block copolymers, reprinted from Yoon et al. (2014), copyright 2014, with permission from American Chemical Society

on the microdroplet surface, facilitating a targeted interaction with KB cancer cells. It is the interaction between the ligand and receptor that is presumed to instigate the change from a radial to a bipolar configuration within the liquid crystal microemulsion. Consequently, the folic acid-ligand-tethered liquid crystal microemulsions exhibit promise for the ex vivo identification of KB cancer cells during the preliminary phases of tumorigenesis (Yoon et al. 2014).

Researchers, including Yuri-Choi, delved into the function of ligand–receptor interactions in detecting HepG2 cells through a biosensor based on liquid crystal microdroplets. They crafted microdroplets of liquid crystals that were modified with β-galactose and confirmed the significance of this β-galactose ligand by employing maltotriose as a mimic of the ligand. The findings from the experiments demonstrated that interactions between HepG2 cells and the β-galactose-modified liquid crystal microdroplets resulted in a reorientation of the liquid crystal molecules from a radial to a bipolar configuration, a change not observed with maltotriose-modified microdroplets. Subsequent studies revealed that the polystyrene component of the liquid crystal microdroplets plays an essential role in conveying the forces of ligand–receptor interactions to the microdroplet's core. The research has yielded a biosensor based on liquid crystal microdroplets that can detect HepG2 cells with high selectivity and sensitivity, offering a new approach for the detection of these cells in biological fluids outside of a living organism (Choi et al. 2015).

Gong et al. used a hybrid liquid crystal microlaser resonator to monitor the biochemical function of cells as well as the biological information of cells. This is a label-free laser emission imaging method for secreted molecules associated with various types of cell-environment interactions. Liquid crystal microdroplets are designed as signal amplifiers to report tiny molecular events sandwiched in an optical microcavity. Through integration with a raster scanner, dynamic information about cellular physiological processes is recorded through different laser wavelengths. Using a pancreatic cancer cell line, we demonstrated the ability to detect small molecules of REDOX oxygen, as well as larger molecules such as overexpressed proteins. It also illustrates the ability to monitor cellular responses to anticancer drugs. The proposed concept can be extended to multiple biological lasers for studying cell signaling, cell-to-cell interactions, and drug screening (Gong et al. 2022).

Manna et al. proposed that microscale thermotropic liquid crystal droplets can be localized inside living mammalian cells and used as chemical sensors to report the presence of toxins in the extracellular environment (Fig. 3.10). This method enables liquid crystal droplets to be enclosed in a semipermeable polymer capsule, thereby enabling cells to be internalized. Liquid crystal droplets are stable in the intracellular environment but undergo optical changes after cells are exposed to low concentrations of toxic amphiphilic substances. It is important to note that liquid crystal droplets in the intracellular environment respond to extracellular analytes that do not respond to liquid crystal in the absence of cellular internalization. It is also unresponsive to other chemical stimuli or processes related to cell growth or manipulation in culture. The results suggest that droplet activation involves transport and co-adsorption of amphiphilic toxins and other lipophilic cellular components to the surface of the internalized droplet. This work provides novel designs for bioabiotic

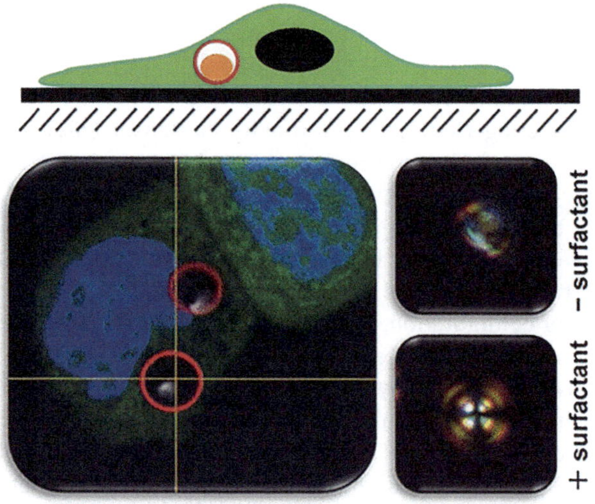

Fig. 3.10 Schematic of structured liquid droplets as chemical sensors that function inside living cells, reprinted from Manna et al. (2021), copyright 2021, with permission from American Chemical Society

systems that can sensitively and selectively report the presence of specific chemical factors outside the cell, and provides a basis for designing structured droplets that can sense and report on other biochemical or metabolic processes within the cell (Manna et al. 2021).

Harnessing the optical imaging attributes of liquid crystals, Aharon and other researchers devised an innovative differential spectroscopic assessment system grounded in their theoretical proposition. The system integrates liquid crystal devices, with one functioning as an adjustable filter and the other as a polarization modifier, both active across an expansive spectrum spanning visible light to the near-infrared region. Given that the alignment of collagen fiber textures is contingent upon their location within the skin and considering their lack of orderly structure, this system investigates the state of birefringence by progressively altering the orientation of the linearly polarized light projected onto the skin abnormality, and identifies disparities between a series of images while simultaneously gauging the aggregated output from a video camera. This method, which is non-intrusive, accentuates the regions of skin affected by tumors or lesions, effectively reducing the statistical fluctuations in polarization arising from light scattered by the lesion. It operates on the principle of non-invasive detection by evaluating regions of the skin that exhibit shifts in polarization statistics, induced by the scattering of light at the lesion site. When merged with fluorescent microscopy, this technology enables the in vivo visualization of skin tumors prior to surgical intervention. The system, when juxtaposed with outcomes from tissue histopathology, demonstrates that this differential spectroscopic polarization imaging for the skin presents a non-invasive technique, offering a new modality for tackling the challenges of precancerous diagnostics (Aharon et al. 2011).

Breast ultrasound and X-ray mammography serve as the benchmark for breast cancer diagnosis in conjunction with pathological diagnosis. Nonetheless, the limitations inherent in these techniques spur ongoing investigation and advancement in alternative diagnostic and screening methodologies for breast cancer. Contact thermal imaging technology, utilizing liquid crystal matrix technology that can be tailored to detect temperature ranges corresponding to the physiological temperature variances between malignant tumors and the adjacent healthy breast tissue, presents significant potential as a supplementary method to the existing clinical benchmarks (Biernat et al. 2016). Researchers, including Diana, have embarked on a prospective pilot study to evaluate the efficacy of liquid crystal contact thermal imaging in the early detection of breast cancer. Encompassing 274 females undergoing preventive breast assessments at the Polish Breast Center, the study involved thermal imaging examinations. Sensitivity and specificity were observed to be consistent across various age groups. The probability of breast cancer was notably higher in women with positive thermal imaging results, while a negative outcome diminished the likelihood of the disease. The findings indicate that liquid crystal contact thermal imaging is particularly effective for women below the age of 50, underscoring its potential as a supplementary diagnostic tool for the early identification of breast cancer (Hodorowicz-Zaniewska et al. 2020).

For individualized tumor therapy, the prospective design of nanocarrier should be a collection of targeting, imaging, and noninvasive therapeutic capabilities. Urandur et al. report the development of inverse hexagonal nanoliquid crystal (NLC) particles capable of simultaneously carrying formononetin (FMN) and tetrasylstyrene (TPE), a signature optical beacon with an aggregation induced emission (AIE) signature (Fig. 3.11). An ordered 3D mesoporous internal structure and a high liposome integral number of NLC nanoparticles (NLCNPs) frame an outer compartment for better payload deposition. Anisamide (AA), a novel sigma receptor targeting ligand, chemically modified these nanoparticles by carbodiimide conjugation, ensuring NLC as an excellent vehicle for potential application in monitoring tumor location as well as in delivering FMN by active AIE imaging. Confocal microscopy was used to demonstrate the ability of single live cell imaging conferred by NLC nanoparticles. Both in vitro and in vivo studies support that targeted nanoparticles (AA-NLC-TF) are more effective than non-targeted nanoparticles (NLC-TF) and free drugs. Clearly, such carefully designed multimodal NLCNPs

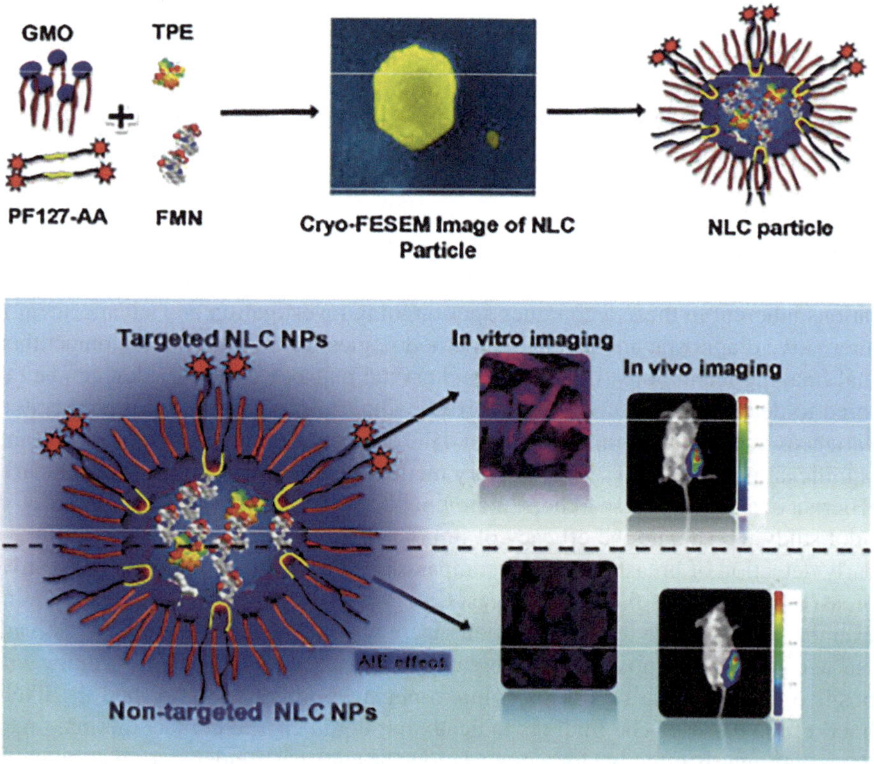

Fig. 3.11 Schematic illustration of synthetic steps for PF127-AA and schematic illustration of AIE based active bioimaging in vitro and in vivo, reprinted from Urandur et al. (2018), copyright 2018, with permission from American Chemical Society

may establish a promising platform for targeted and image-guided chemotherapy in breast cancer (Urandur et al. 2018).

Tumor cells exhibit significant metabolic alterations by adapting to the stress microenvironment. These cells have a high rate of glutamine breakdown and release the by-product ammonia (NH_3), which may act as a spreading signal between cancer cells and can reveal cellular heterogeneity (Fig. 3.12). Khan et al. selected a nematic liquid crystal (LC)E7 and doped with 4-pentyl-4′biphenyl carboxylic acid (PBA), encapsulated it in a polymer microcapsule (P-E7PBA) and then immobilized in a microfluidic channel on the cell. NH_3 release was studied in normal human umbilical vein endothelial cells (HUVECs) and myeloma, human primary glioblastoma (U87), human colon cancer (Caco-2), and human breast adenocarcinoma (MCF-7) cells. By polarized light microscopy, P-E7PBA was able to visualize the release of NH3 from the cells by changing the radial to bipolar (R-B) direction. The time required to respond to changes in R-B varied significantly among the various cell lines. The average reaction times of Caco-2, U87, and MCF-7 cells were 277, 155, and 121 s, respectively. NH_3 release from single cells captured in the microwell flow chip showed similar R-B changes. By using different probes to functionalize LC, the P-E7PBA droplet technique can be applied to multiple other targets (Khan et al. 2019).

To encapsulate, the profound studies conducted in recent years have propelled liquid crystals to emerge as innovative signal converters within the realm of bio and

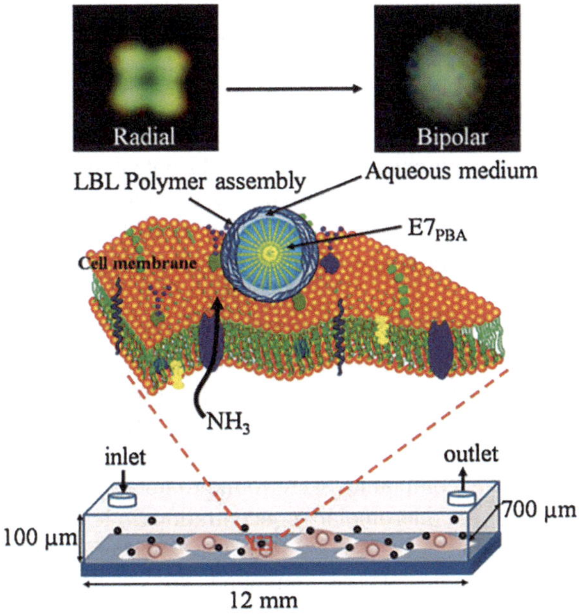

Fig. 3.12 Schematic of immobilized P-E7PBA droplets on cells cultured in a microfluidic channel, reprinted from Khan et al. (2019), copyright 2019, with permission from the Authors, published by WILEY-VCH Verlag GmbH & Co. KGaA, Weinheim

chemical molecule sensing. The novel sensing techniques predicated on liquid crystals have introduced a fresh visual detection approach for the preliminary identification of cancer, thereby broadening the horizons of liquid crystal applications. Nevertheless, the exploration of diverse pioneering biosensors founded on liquid crystals is far from exhausted, with the majority lingering in the phase of semi-quantitative assessments. In addition, research on the application of liquid crystal materials in medical imaging is also in its infancy. The combination of liquid crystal nanoparticles with drugs, fluorescence, and other substances for tumor imaging and treatment is also a hot topic in future research. Consequently, the development of mathematical models that link the color and luminosity of liquid crystal films to the concentrations of oncological markers, thereby facilitating precise analysis of these markers and the imaging of tumors, represents an avenue for future scholarly pursuits.

3.3 Liquid Crystal Materials for Tumor Therapy

3.3.1 Chemotherapy

Individuals battling cancer in its later stages often depend on chemotherapeutic agents. Yet, the majority of these drugs are characterized by their lack of precise targeting, severe side effects, and extended treatment durations, leading to a multitude of adverse effects on the patient's well-being. Consequently, it is imperative to innovate and bring to market antineoplastic medications that are both highly targeted and potent. Liquid crystal materials, with their unique properties, may offer new avenues in the development of such targeted therapies. Their anisotropic nature and responsive behavior to external stimuli could potentially be harnessed in the design of drug delivery systems that can more accurately target cancer cells while minimizing damage to healthy tissues. The exploration of liquid crystal materials in this context is a burgeoning field with much potential for future advancements in cancer treatment.

Researchers, including Fukushi, have put forth the concept that precursors to liquid crystals, known as liquid crystal-related compounds (LCRCS), can augment the process of hematopoietic differentiation even at low concentrations (Terasawa et al. 2009). Following this, an assessment of the biological activity of 16 distinct LCRCS on the K562 cell line, which is indicative of chronic myeloid leukemia, was carried out. Two specific compounds were found to notably suppress the proliferation of K562 cells at micromolar concentrations: 2-(4-butoxyphenyl)-5-(4-hydroxyphenyl)pyrimidine and 2-{4-(4-hexoxyphenyl) phenyl}-5-hydroxypyrimidine. These compounds are structurally akin, with a trio of aromatic rings at their core, comprising a pyrimidine ring with an alkyl chain and a hydroxyl group on both sides. The findings point toward the plausibility of LC RCS serving as novel antitumor pharmaceuticals (Fukushi et al. 2011). In congruence, Takahashi and other researchers explored the biological and pharmacological potency of amphiphilic liquid crystal compounds, focusing on phenyl-pyrimidine derivatives featuring D-glucosamine and cyanobiphenyl derivatives highlighted by terminal hydroxyl moieties, aiming to uncover new cancer-fighting capabilities of LCs. The cancer-inhibiting properties of LCs concerning cell proliferation, cell cycle distribution, and signaling pathways were scrutinized in A549 lung cancer cells and WI-38 normal fibroblasts, employing flow cytometry and Western blot techniques. The study's conclusions hint at selected benzimidazole and dibenzofuran derivatives that impede the growth of cancer cells while exhibiting no adverse effects on normal fibroblasts. Consequently, LCs could potentially be considered as emerging chemotherapy drugs (Takahashi et al. 2011).

The strategy of drug delivery via nanocarriers, which are taken up by cells, offers a distinct route for medicinal agents to penetrate into cancerous cells (Peer et al. 2020). These particulate carriers have the ability to accumulate in neoplastic tissues passively, which can boost the drugs' ability to infiltrate and remain within the tissue, thereby heightening the effectiveness of the treatment. Furthermore, the incorporation of specific ligands allows for the precise localization of these nanoparticles to the tumor site (Petrilli et al. 2013; Peer et al. 2020; Hani et al. 2023). Among the nanoparticles that have been commercialized for oncology treatments are liposomal formulations and conjugates of polymers with proteins. In addition, a variety of other nanoscale delivery systems hold promise for the advancement of cancer therapeutics, including dendrimers, nanoparticles made from silica, exosomes derived from cellular processes, plant-based delivery systems, and liquid crystals (Bonifácio et al. 2013; Wicki et al. 2015; Peer et al. 2020; Kemp and Kwon 2021; Hani et al. 2023).

Liquid crystals represent an intermediate state between solids and liquids, meaning they possess the crystalline order akin to solids and the fluidity similar to liquids (Mezzenga et al. 2019). The structural affinity of liquid crystals to biological membranes aids in the interaction between drugs and cell membranes, easing the penetration of drugs. In terms of structure, liquid crystals are advantageous for augmenting the charge of substances, which in turn boosts their thermodynamic stability and prolongs their service life (Bonifácio et al. 2013). Moreover, they possess the ability to release a spectrum of bioactive agents in a sustained manner (Guo et al. 2010; Bisoyi and Kumar 2010; Vamshi et al. 2020; Shan et al. 2022). As a result, the application of liquid crystals has broadened into the domains of biomedicine and pharmacy. The low toxicity and enhanced permeability of lyotropic liquid crystals position this technology as a viable alternative in cancer treatment. Characterized by amphiphilic properties, lyotropic liquid crystals consist of molecules with a hydrophilic head and a hydrophobic tail, such as those found in benzimidazole derivatives and cyanobiphenyl derivatives with terminal hydroxyl units, enabling the creation of compartments for the encapsulation of both lipid-soluble and water-soluble pharmaceuticals (Negrini et al. 2015; Vallamkondu et al. 2018; Wang et al. 2021). The crystalline forms of lyotropic liquid crystals include lamellar, cubic, and hexagonal phases, with the inverse cubic (Q2) and hexagonal (H2) phases being particularly noteworthy. These phases have been the focus of extensive research due to their capacity to facilitate the sustained release of a wide array of bioactive agents, from low-molecular weight compounds to proteins, peptides, and nucleic acids. Their nontoxicity, biodegradability, and bioadhesion also contribute to their utility in drug delivery (Shah et al. 2001). The inverse cubic and hexagonal phases, due to their capability for unlimited expansion, can be stably dispersed with an excess of water, resulting in colloidal dispersions that exhibit remarkable thermodynamic stability (Spicer 2005). Moreover, research by Negrini and colleagues has shown that the use of lyotropic liquid crystals to encapsulate antineoplastic drugs leads to a more precise release profile, with a release velocity a decade faster and a threefold increase in the efficacy of cancer cell destruction (Dang and Guan 2020). In particular, lyotropic liquid crystal nanoparticles (LLCNPs) have become a focal point as carriers for an array of hydrophobic and hydrophilic small molecule drugs, polypeptides, proteins, siRNA, DNA, and contrast agents (Zhai et al. 2019).

Researchers including Zhai have compiled an array of drug delivery systems founded on LCNPs, depicted in an idealized illustration (Fig. 3.13). The core framework is largely made up of monoolein (MO) and tricaprin (PT), encapsulating an assortment of supplementary agents and stabilizing compounds (Zhai et al. 2019). The routes of administration for LCNPs encompass a broad spectrum, from oral ingestion to intravenous injection, intraperitoneal delivery, ophthalmic applications, subcutaneous implantation, topical application, and targeted transdermal delivery to

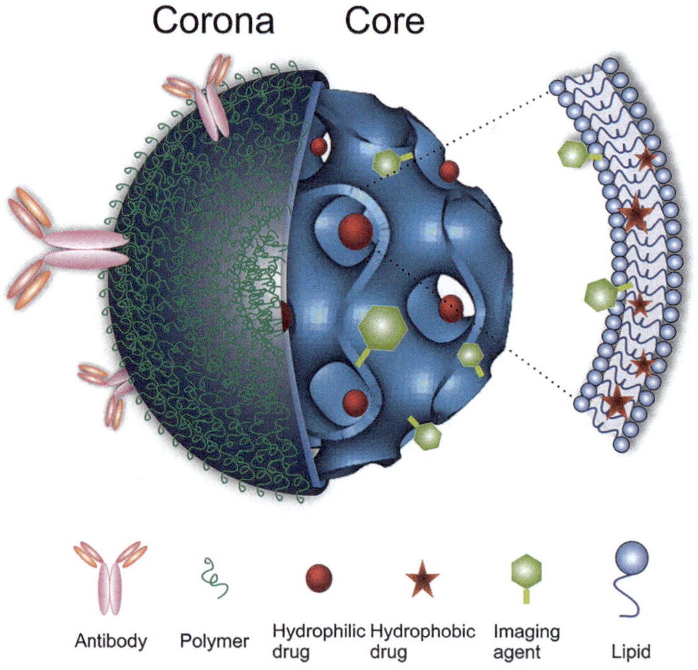

Fig. 3.13 "Idealized" LCNP drug delivery system, reprinted from Zhai et al. (2019), copyright 2019, with permission from American Chemical Society

specific regions such as internal organs, the brain, the ocular area, and the skin (Han et al. 2010; Li et al. 2013; Morsi et al. 2014; Thakkar et al. 2016; Liu et al. 2016). The demand for liquid crystal-based drug delivery in oncology treatment is increasingly garnering attention, highlighting the potential of these systems in the medical field.

Trevizan and colleagues have crafted a liquid crystals nano-dispersion (LCN) harboring docetaxel (DTX) and cetuximab, specifically designed for the treatment of prostate cancer. Given the overabundance of the epidermal growth factor receptor (EGFR) in prostate cancer, cetuximab (CTX), which is a monoclonal antibody against EGFR, is integrated into the formulation. The formulated system predominantly adopts a hexagonal phase, enabling nanodispersion with minimal polydispersity and a negatively charged zeta potential. The cytotoxic potency of the liquid crystal formulation, laden with DTX and CTX, has been shown to markedly diminish cellular viability in vitro at concentrations as low as 0.1 nM. Within the context of prostate cancer cells PC3, the EGFR-targeted LCN, functionalized with CTX, has illustrated enhanced cellular internalization relative to non-targeted DTX-laden LCN, culminating in heightened cytotoxic effects. Consequently, this research introduces a pioneering application of EGFR-directed liquid crystal nanodispersions for the selective conveyance of DTX to prostate cellular environments (Trevizan et al. 2021).

Researchers, Wang and colleagues, developed a cubic liquid crystalline formulation encapsulating curcumin, with the aim of discerning the specific phase and rheological traits inherent to the cubic liquid crystalline state (Wang et al. 2017). Subsequently, David and fellow researchers explored the efficacy of liquid crystalline systems in delivering curcumin. A range of surfactant concentrations was evaluated, including Tween 80, cremophor EL, and labrasol, to ensure the formulation would meet high safety standards and achieve a gel texture appropriate for vaginal applications. The utilization of these surfactants in the creation of liquid crystals marked a pioneering effort, and the resulting liquid crystalline systems were subjected to tests for their potential in treating cervical cancer. Outcomes demonstrated that the liquid crystalline systems loaded with curcumin were able to dispense the therapeutic dosage of the compound and exhibited efficacy against the cervical cancer cell lines under investigation (David et al. 2020).

Researchers Qin and colleagues investigated the potential of tricaprin-originating liquid crystals to serve as an innovative embolic agent for a controlled-release delivery system. They formulated a precursor solution of the tricaprin cubic phase laden with the chemotherapy agent hydroxycamptothecin (HCPT) and assessed its capabilities in terms of embolization, drug release outside of a biological context, and its cytotoxic effects. Within the dosage range examined, the presence of HCPT did not elicit any phase changes, and evaluations of the embolic impact in vitro confirmed that the isotropic solutions provided effective embolization. Findings from the in vitro release of the drug indicated a sustained release capability extending to 30 days, with the release kinetics being subject to the initial components and the drug payload. Assessments of cytotoxicity and the anticancer efficacy in vitro showed no marked lethality in normal cells after a 48-h period, whereas the formulation containing HCPT displayed pronounced anticancer effects on HepG2 and SMMC7721 cells, contingent upon the quantity of drug loaded and the duration of exposure. Collectively, these outcomes propose that a tricaprin-based cubic phase gel solution for embolization is a potential vector for the delivery of HCPT, enabling a prolonged release of the drug via vascular occlusion, and this methodology may hold promise for clinical utilization (Qin et al. 2016).

Monoolein and phytantriol, common lyotropic liquid crystal forming materials, may induce tissue cytotoxicity and adverse immunological responses, potentially limiting their broad application. To address these limitations, Shan et al. selected phosphatidylcholine and α-tocopherol as biocompatible carriers. By varying their ratios, the team investigated the resulting crystalline types, nanoscale structures, viscoelastic properties, release characteristics, and in vivo safety. Focusing on prostate cancer treatment, they explored the potential of an in situ lyotropic liquid crystal platform with injectable and sprayable properties for hormone-sensitive prostate cancer (HSPC) and castration-resistant prostate cancer (CRPC) (Fig. 3.14). For HSPC, spraying a lyotropic liquid crystal platform loaded with leuprolide and cabazitaxel onto the tumor bed post-resection significantly reduced tumor metastasis and extended survival time. For CRPC, the combination of leuprolide and cabazitaxel in the platform demonstrated superior tumor-inhibiting and anti-recurrence efficacy compared to cabazitaxel alone. This improvement was attributed to enhanced $CD4^+$ T cell infiltration and the presence of immune-potentiating cytokines in tumors.

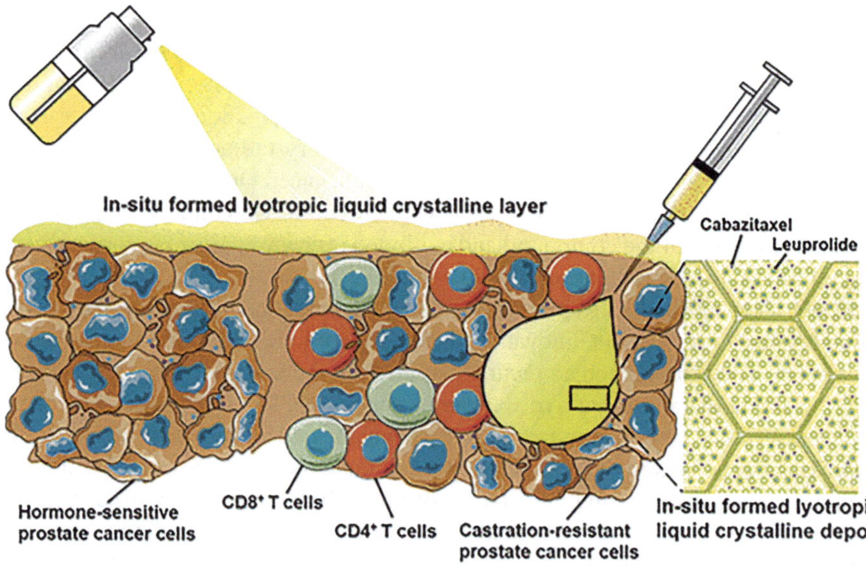

Fig. 3.14 Graphical illustration of injectable and sprayable in situ lyotropic liquid crystalline platform toward hormone-sensitive and castration-resistant prostate cancer, reprinted from Shan et al. (2023), copyright 2023, with permission from American Chemical Society

Overall, this dual-functional and clinically feasible strategy holds promise for treating both HSPC and CRPC (Shan et al. 2023).

Researchers, including Majumdar, have crafted a formulation of lyotropic liquid crystal nanoparticle encapsulating docetaxel, aiming to augment and precisely target the delivery of the medication to skin cancer cells. The emulsification-diffusion method was utilized to develop these lyotropic liquid crystal nanoparticles laden with docetaxel. The investigation encompassed studies on in vitro cellular responses and cellular uptake of the target. Findings demonstrated that the docetaxel-encapsulated lyotropic liquid crystal nanoparticle induced higher cytotoxic effects and an increased rate of cellular uptake compared to the standard drug formulations. By elevating the rate at which drugs are absorbed and sustaining therapeutic levels over a more extended timeframe, lyotropic liquid crystal nanoparticle emerge as a promising vehicular system for dermatological conditions, including cancer (Majumdar et al. 2019). Researchers, Tatiana and colleagues, have investigated the therapeutic efficacy of liquid crystal nanoparticles in conjunction with low-frequency ultrasound (LFU) for the treatment of skin cancer. They evaluated the effects of solid lipid nanoparticles (SLN) and liquid crystal nanogels as intermediaries for coupling with LFU to induce localized transport areas (LTR) in the skin and to facilitate the infiltration of doxorubicin (DOX) into the skin layers. The findings suggest that the refinement of nanoparticle formulations and the calibration of LFU parameters are crucial for the practical clinical utilization of LFU in oncology dermatology (Pereira 2023).

Researchers Li and colleagues have utilized the technique of small-angle X-ray scattering (SAXS) to meticulously orchestrate the in situ fabrication of a controlled-release vehicle for bufalin (BUF), an antitumor medication notorious for its poor solubility in the unencapsulated state and significant cardiac toxicity. A novel approach was taken by devising a pseudo-ternary plot for an oil (O)/surfactant (S)/water (W) system, integrating medium-chain caprylic/capric triglycerides (MCT), polyethylene glycol (15)-hydroxystearate, and sorbitan monooleate (Span 80). Through a process of gradual dilution, SAXS was instrumental in delineating the transitional phases from microemulsions (ME) to lamellar liquid crystalline structures and multilamellar vesicles. The medication bufalin was encapsulated in microemulsions of reduced viscosity, with the release of the drug occurring within the lamellar liquid crystal structures that formed in situ. A correlation was identified between the structural attributes of the LCME formulations, which emerged from the in situ phase transitions, and the variations in their viscosities and drug release patterns. Studies on cytotoxicity, cellular apoptosis, and pharmacokinetics have demonstrated enhanced bioavailability of bufalin subsequently to encapsulation. This research presents experimental substantiation for the clinical utilization of sustained-release formulations achieved through in situ phase transitions (Li et al. 2019).

Urandur et al. report the progress of a diagnostic lyotropic liquid crystal nanoparticle loaded with a unique MnO nanoparticle (MNP) for selective cancer imaging and therapy. Mnp acts as a fluorescent agent as well as a source of manganese (Mn^{2+}) and enables localized oxidative stress under cancer markers (acidosis, high

H_2O_2 levels). In pursuit of synergistic amplification of the antitumor activity of Mn^{2+}, betulinic acid (BA) was loaded in lyotropic liquid crystal nanoparticle. In this study, the nanostructure of the lyotropic liquid crystal nanoparticle phase interface was established by SAXS, Cryo-TEM and Cryo-FESEM. Interesting in vitro studies have shown that lyotropic liquid crystal nanoparticle triggers the production of hydroxyl radicals and exhibits greater selective cytotoxicity in cancer cells, ensuring the safety of normal cells. Compared with the control group, the tumor growth inhibition index of lyotropic liquid crystal nanoparticle in the experimental group reached 96.5%, achieving significant tumor ablation. Key insights into on-site drug release, local anticancer response, and tumor location were obtained by precise guidance of fluorescent MNPS. In addition, the safety, pharmacokinetics, and tumor distribution behavior of lyotropic liquid crystal nanoparticle were comprehensively evaluated in vitro and in vivo. This work highlights the promise of using intelligent endogenous stimulus-sensitive nanosystems to modulate the tumor microenvironment to enable advanced comprehensive cancer nanotherapeutics in the absence of any external stimuli (Urandur et al. 2020).

Researchers, including Saklani, have crafted an injectable system that forms a lipidic lyotropic liquid crystal (L3C) deposit in situ, designed for the precise and concurrent delivery of chemotherapy drugs directly into tumors. The L3C formulation consists of amphiphilic lipids and surfactants, such as monoolein, phosphatidylcholine, tocopherol acetate, and d-α-tocopherol polyethylene glycol-1000 succinate. The study showcased its transformative capability in vivo by devising a synergistic chemotherapy protocol ex vivo, with doxorubicin and paclitaxel serving as representative models for hydrophilic and hydrophobic medications, respectively. Post-injection into the tumor, it morphs into a hexagonal intermediate phase reservoir, which facilitates a sustained and precisely synchronized release of the encapsulated chemotherapy drugs. The results demonstrated that the L3C system, when utilized to deliver a meticulously timed combination of chemotherapeutic agents directly into the tumor, can markedly potentiate the antitumor efficacy and mitigate cardiotoxic side effects in comparison to desynchronized chemotherapy regimens or traditional intravenous administration methods. These insights underscore the L3C system's capacity to synchronize the co-delivery of chemotherapeutic agents within tumors, thereby enhancing the therapeutic impact of combined chemotherapy approaches (Saklani et al. 2024).

Researchers, Hu and colleagues, have created an array of amphiphilic poly amino acids featuring cholesterol side chains. The prototype pharmaceutical, DOX, was incorporated into the internal structure of a "core–shell" configuration, which manifests in spherical or elongated shapes, through physical entrapment and covalent bonding. These nanoparticles, laden with the drug, penetrated HeLa cell membranes through endocytic pathways, with cholesterol being instrumental in facilitating their translocation, as depicted in the figure. The system's capacity for drug delivery is characterized not only by the nanoparticles' cellular uptake but also by its pronounced responsiveness to variations in temperature and pH, predominantly owing to the incorporation of liquid crystal compounds. The controlled release of drugs is achieved by modulating the molecular arrangement of the macromolecules and altering the spatial conformation of the self-assembled structures. The synthesis of such liquid crystal materials offers innovative avenues for research in drug delivery systems (Hu et al. 2023).

Salinomycin (Sal) is a potent veterinary antibiotic known to be significantly toxic to a variety of tumor cells. Its therapeutic utility is limited due to its high lipophilicity (logP 7.5) and poor hydrophilicity. Liquid crystal nanoparticles provide a suitable delivery platform for these drugs. Nucleolin receptors are overexpressed on the cell surface and in the cytoplasm and can be used as targets for cancer therapy. LCNPs functionalized with F3 peptide (F3-Sal-nps) were designed by Kumar's team for the delivery of Sal to selectively target nucleolin receptors. The compound has excellent sustained release properties and is 4–5 times more cytotoxic than naive drugs. Meanwhile, F3-Sal-NPs reduced the gene expression and hemolytic toxicity in vitro ($6.5 \pm 1.5\%$), and improved the pharmacokinetics and bioavailability of Sal. F3-Sal-NPs significantly reduced tumor growth by 2.8-fold and induced tumor cell necrosis in vivo. The excellent performance of liquid crystal nanoparticles functionalized with F3 peptide in delivering Sal against breast cancer was clearly demonstrated (Kumar et al. 2023a).

Zahiri et al. prepared novel liquid crystal nanoparticles for coencapsulation of Bi_2O_3 and hydrophilic doxorubicin (DOX) and functionalization of folic acid (FA) for efficient targeting of melanoma for in vitro and in vivo CT scan imaging and chemotherapy. LCNPs Bi_2O_3 nanoparticles were prepared by glycerol mono-oleic acid-Pluronic F-127 (GMO/PF127/water). First, LC gels were prepared by homogenization of GMOs/water. Then, an aqueous stabilizer solution (PF127/Bi_2O_3/DOX) was added to the prepared LC gel and homogenized by homogenization and ultrasonic fragmentation. The prepared nanoparticles showed good stability and encapsulation efficiency. Higher cytotoxicity and cellular endocytosis of FA-Bi_2O_3-DOX-NPs were observed in folate receptor (FR)-overexpressing cells (B16F10) in vitro compared with Bi_2O_3-DOX-NPs and free DOX. In addition, FA-Bi_2O_3-DOX-NPs treated tumor-bearing mice exhibited ideal tumor suppression and improved survival compared with non-targeted drugs. On the other hand, the CT imaging ability of Bi_2O_3-DOX-NPs was examined in B16F10 tumor-bearing mice. The data obtained indicate that the developed FA-Bi_2O_3-DOXNPs have a high potential for the diagnosis and treatment of melanoma (Zahiri et al. 2024).

3.3.2 Photodynamic Therapy

To design multifunctional nanoparticles for cancer therapy, Jenni et al. coupled a diketone pyrrole-porphyrin-based photosensitizer to a block copolymer (Pluronic F108) and used it to stabilize water-lipid cubic liquid crystal nanoparticles (cubes) loaded with the antitumor drug docetaxel. Cell imaging experiments demonstrated that these nanoparticles localized lysosomes and mitochondria, and cytotoxicity tests demonstrated a slight but significant synergistic effect after irradiation that increased the toxicity induced by polyene paclitaxel, only allowing a further reduction in cell viability. The proposed fluorescent agent has some potential for combined chemical and photodynamic therapy of tumors (Jenni et al. 2020).

Researchers including Raquel have crafted nanoparticles of lyotropic liquid crystals encapsulating a novel photosensitizing agent, a chlorin derivative, specifically designed for photodynamic therapy (PDT) of skin cancer. Characterization of these systems was performed employing techniques such as fluorescence spectroscopy, dynamic light scattering, and small-angle X-ray diffraction (SAXRD) analysis. Studies of permeation in animal models were conducted both in vitro and in vivo, utilizing membranes from animal models. Moreover, findings from SAXRD suggested that the hexagonal liquid crystalline phase of the formulation was preserved following the encapsulation of the drug. The encapsulation rate exceeded 50% and assays for cell viability indicated that the nanodispersions posed no harm to L929 skin cells. The results verified that the nanodispersions with a hexagonal phase improved the rate of skin absorption of the drug when juxtaposed with the control group. Additionally, examinations utilizing fluorescence microscopy revealed that the distribution of the chlorin derivative within the skin layers of hairless mice was more pronounced compared to the control group. Collectively, these outcomes propose that the nanodispersions possess the capacity to convey photosensitizers to the skin, a critical prerequisite for the efficacious execution of topical PDT (Petrilli et al. 2013).

Oral malignancies rank among the most common head and neck cancers globally. PDT has shown significant promise in combating these cancers, leading to a decrease in chronic illness. Balian and colleagues explored the use of methylene blue (MB) within a mucoadhesive liquid crystal precursor system (LCPS) for the treatment of oral cancer. They focused on the photostability and the in vitro release, permeation, and retention characteristics of MB-loaded LCPS (MB-LCPS) as well as its in vitro PDT efficacy against both normal (HaCaT) and cancerous (HSC-3) cell lines. The LCPS was found to enhance the photostability of MB and provided a sustained release of MB. Moreover, LCPS significantly increased the retention of MB in porcine esophageal mucosa, which was approximately three times more than that of the MB solution alone. The retention of MB within LCPS was about twice its permeability, making it suitable for ensuring the therapy's effectiveness in the oral cavity. An in vitro cytotoxicity test revealed that MB-LCPS significantly boosted the antitumor efficacy of MB following a 20-min exposure to light at 660 nm with a fluence of 12.5 J/cm^2. The findings suggest that the developed formulation presents a promising approach for the potential use in PDT for oral cancer treatment (Balian et al. 2023).

The Nakayama laboratory has described the formation of colloidal liquid crystals inspired by the classic inorganic constituents found in biological minerals, such as calcium carbonate ($CaCO_3$) and hydroxyapatite (HAp), which assemble themselves alongside polyacrylic acid (PAA), as shown in Fig. 3.15, resulting in materials with unique optical and medical properties. The anisotropic stability of these colloids, which consist of inorganic crystalline cores with varying attributes and an enveloping organic polymeric shell, is achieved through the crystallization process of the polymer's amorphous precursors. Experiments conducted outside of a

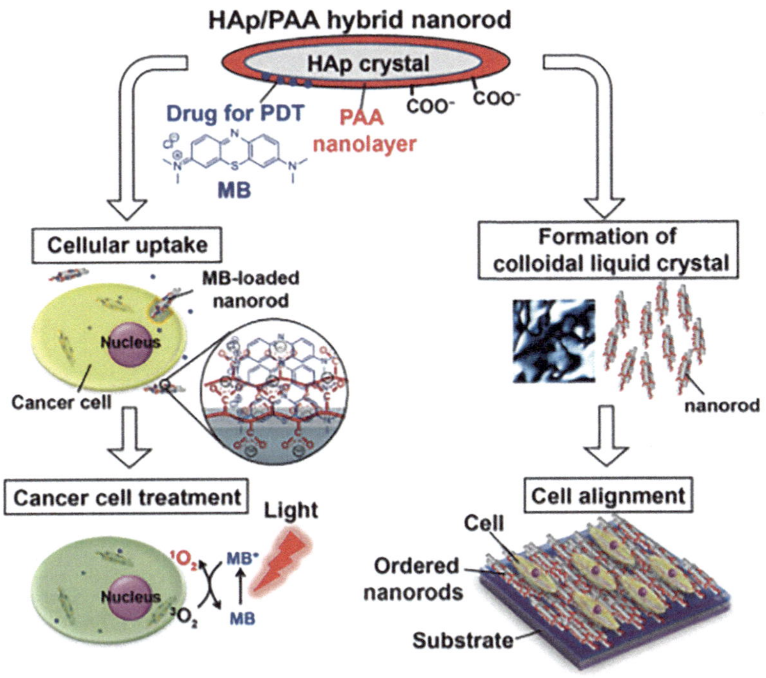

Fig. 3.15 Schematic illustration of HAp/PAA hybrid nanorods as a platform technology for the development of drug nanocarriers for PDT and bio-scaffolds for cell growth control, reprinted from Nakayama et al. (2019) copyright 2019, with permission from American Chemical Society

biological context have validated that these liquid crystalline colloids are characterized by elevated rates of cellular internalization, compatibility with biological systems, and efficacious destruction of cancer cells, thus highlighting their capacity to serve as nanocarriers for photodynamic therapy (PDT) applications. The synthesis of these biomineral-inspired colloidal liquid crystals, which occurs under gentle conditions, positions them as a novel category of sustainable nanomaterials endowed with self-assembling and responsive properties to environmental stimuli (Nakayama et al. 2015, 2018, 2019, 2020; Nakayama and Kato 2022).

3.3.3 Gene Therapy

Kim et al. developed a method based on organic nanorods (cellulose nanocrysts [CNC]) and inorganic nanodiscs (Laponite) [LAP] indocyanine green (ICG), glucose oxidase (GOx), and copper sulfate (Cu)-loaded hybrid gel for combination starvation therapy (ST), chemodynamic therapy (CDT), and photothermal therapy (PTT) for localized cancer treatment (Fig. 3.16). A hybrid CNC/LAP network with

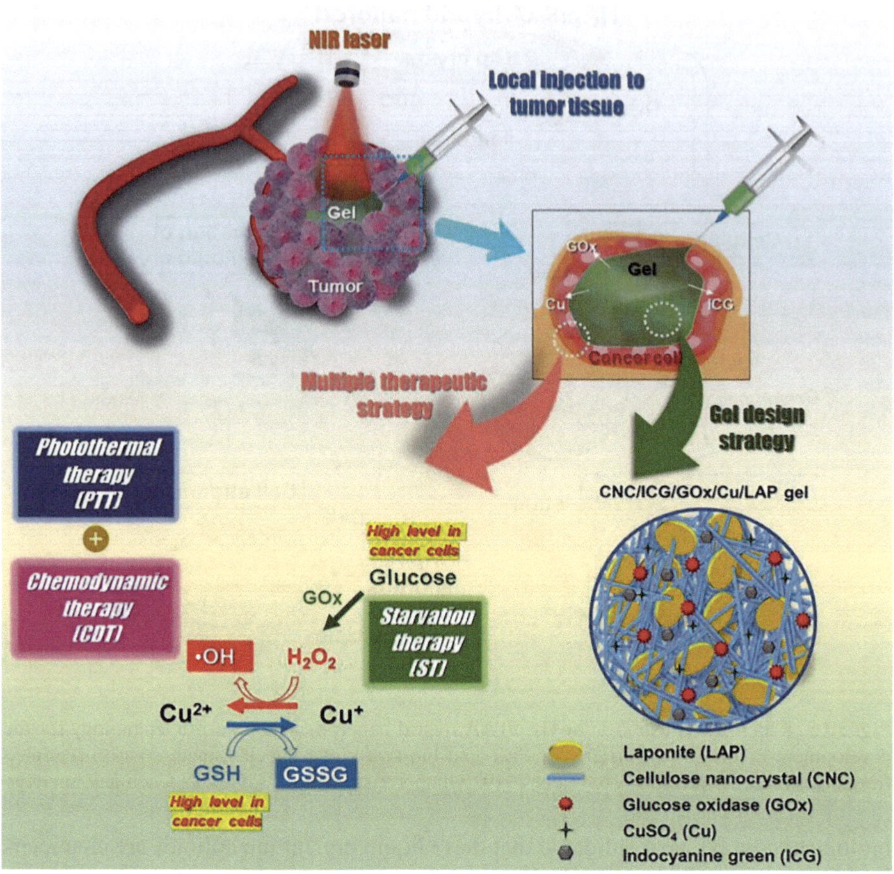

Fig. 3.16 Schematic of nano-aggregated gel network for multiple local cancer therapy, reprinted from Kim et al. (2023), copyright 2023, with permission from Bioengineering and Translational

nematic phase was designed to achieve rapid gelation, controlled viscoelasticity, syringe injectability, and longer in vivo retention time. In addition, ICG was introduced into the CNC/LAP gel system to induce hyperthermia in tumor tissues to amplify the CDT effect. GOx for glucose deprivation (associated with Warburg effect); Cu is introduced into cancer cells for the generation of hydroxyl radicals (based on Fenton-like chemistry) and the degradation of glutathione (GSH). The ICG/GOx/Cu-mounted CNC/LAP gel combined with near-infrared (NIR) laser achieved enhanced anti-proliferation, cellular reactive oxygen species (ROS) production, cellular glutathione degradation, and apoptosis induction in colorectal cancer (CT-26) cells. In addition, local injection of CNC/ICG/GOx/Cu/LAP gel into implanted CT-26 tumors combined with near-infrared laser irradiation showed strong tumor growth inhibition. In conclusion, the designed hybrid nanorod/

nanodisc gel network can be effectively applied to local PTT/ST/CDT of cancer cells (Kim et al. 2023).

Zhang et al. demonstrated that α-helical peptides are also nanofillers for the synthesis of anisotropic hydrogels. α-helical poly (γbenzyl-L-glutamic acid) (PBLG) chains are arranged into nematic liquid crystals by shear force. The ordered structure is then stabilized by embedding in a cross-linked polymer matrix. The liquid crystal PBLG composite gel was then transformed into a poly (Lglutamic acid) (PLGA) composite gel, which expanded anisotropically with pH. Thus, ph-triggered anisotropic expansion can be achieved even with non-responsive hydrogel matrix. More importantly, high expansion anisotropy was obtained. Because PLGA composite gel has large pH triggered anisotropic swelling, it can be used as an embolic agent for embolization therapy (Zhang et al. 2024b).

RNA interference (RNAi)-based therapeutics have emerged as a promising strategy to silence the expression of targeted pathological genes, primarily through the use of small interfering RNAs (siRNAs) and microRNA (miRNA) inhibitors. However, challenges such as short half-life, poor cellular uptake, and nonspecific distribution of small RNAs necessitate the development of novel delivery systems to enhance RNAi efficacy.

Gentile et al. developed a cationic liquid crystalline nanoparticle (CLCN) to efficiently deliver synthetic siRNAs and miRNAs (Fig. 3.17). CLCNs were fabricated via high-speed homogenization and assembled with siRNA or miRNA molecules in nuclease-free water to form CLCN/siRNA or miRNA complexes. These complexes exhibited homogeneous and stable characteristics, with a diameter of approximately 100 nm and a positively charged surface. CLCNs demonstrated nontoxicity and were efficiently internalized by human cells through endocytosis. Significant gene expression inhibition was observed in transiently transfected lung cancer H1299 cells treated with CLCN/anti-GFP complexes 24 h post-transfection. Biodistribution studies revealed that CLCNs and CLCN-RNAi complexes were successfully delivered to various organs and subcutaneous H1299 tumor xenografts in mice 24 h after systemic administration. These findings indicate that CLCNs represent an advanced

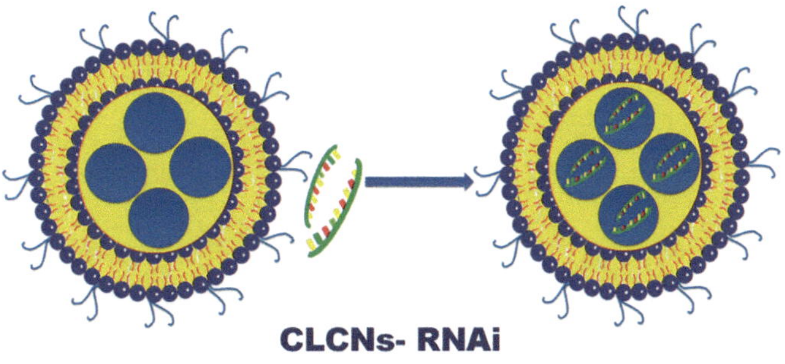

Fig. 3.17 Schematic representation of CLCNs-RNAi binding, reprinted from Gentile et al. (2017), copyright 2017, with permission from Gentile et al. and Oncotarget journal

delivery system capable of protecting RNAi molecules from degradation and facilitating efficient delivery both in vitro and in vivo (Gentile et al. 2017).

RNAi technology has shown great promise in the treatment of cancer and other genetic diseases. Despite efforts to increase the target tissue distribution of siRNAs, safe and effective delivery of siRNAs to diseased cells with adequate cytoplasmic transport is another key factor for successful clinical application of RNAi. He et al. constructed lipid-based liquid crystal nanoparticles, called nano-transformers, that can change structure in an acidic intracellular environment and enable efficient siRNA delivery for cancer therapy (Fig. 3.18). The developed nanoconverters showed good siRNA loading efficiency and low cytotoxicity. Unlike conventional cationic nanocarriers, the fusion of the inner membrane caused by the lipid conformational switch allows the siRNA to easily dissociate from the nanocarrier and be released directly into the cytoplasm. It was shown that transfection of cyclin-dependent kinase 1 (CDK1)-siRNA-loaded nano-transformers reduced the relevant mRNA by 95% in vitro and significantly inhibited tumor growth in vivo without causing any immunogenic response. This work highlights the potential of lipid-based nano-transformers as a next-generation siRNA delivery system with increased efficacy and improved safety (He et al. 2018).

Pereira et al. was conducted to explore the effectiveness of combining low-frequency ultrasound (LFU) with nanotechnology-enhanced formulations for the topical treatment of skin tumors. They used the effects of solid lipid nanoparticles (SLN) and hydrophilic nanogels as carriers on LFU-induced localized transport

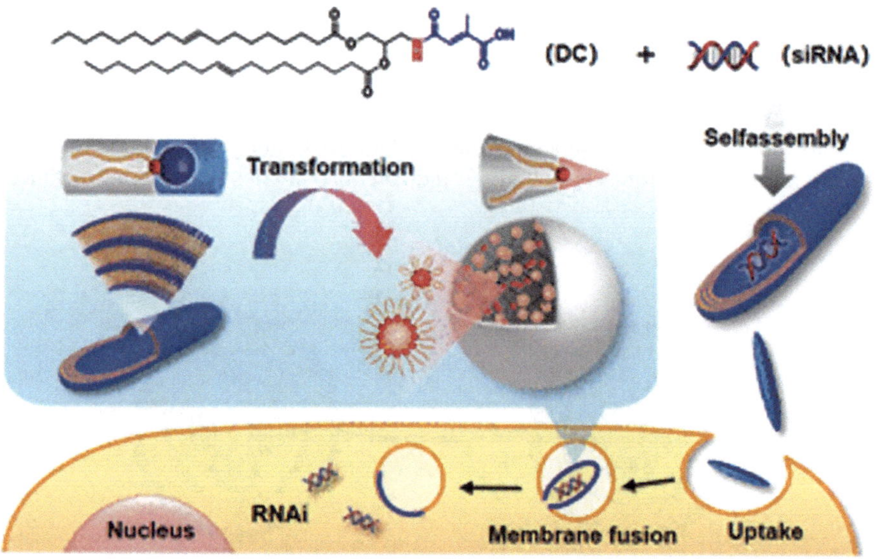

Fig. 3.18 Schematic illustration of the synthesis of DCLC nano-transformers and the delivery mechanism, reprinted from He et al. (2018), copyright 2018, with permission from American Chemical Society

regions (LTR) in the skin and the subsequent absorption of doxorubicin (DOX) in LFU-prepared skin. The SLN were created using the microemulsion method, while liquid crystalline nanogels were prepared with Poloxamer. In laboratory tests, the skin was subjected to LFU to achieve a resistivity of approximately 1 KΩ cm^2, after which the penetration of DOX from DOX-nanogel and SLN-DOX was measured in different skin layers. Mice with induced squamous cell carcinoma (SCC) were treated with LFU using nanogels, with the LFU applicator positioned either 5 mm or 10 mm from the tumor surface, followed by the application of DOX-nanogel. The results showed that LFU combined with nanogel as a coupling medium produced larger LTR areas compared to LFU with SLN. In skin prepared with LFU, DOX-nanogel notably enhanced drug penetration into the viable epidermis, whereas SLN-DOX impeded drug movement through LTR. In live animal testing, LFU with nanogel and a 10 mm applicator distance led to significant tumor suppression and a reduction in tumor cell count and necrosis. These results highlight the significance of fine-tuning nanoparticle formulations and LFU settings for the effective clinical use of LFU in treating skin tumors (Pereira et al. 2023).

3.3.4 Immunotherapy and Synergy Therapy

The risk of postoperative recurrence and metastasis in cancer patients is still the main cause of death (Turajlic and Swanton 2016). Tumor immunotherapy has been shown to be a promising treatment to achieve a durable antitumor response by enhancing the host immune response, selectively against residual tumor and circulating tumor cells (Tucker et al. 2012). Investigators Hu et al. reported a local and continuous drug delivery system, liquid crystal forming system (LCFS) loaded with doxorubicin (DOX) and remquimod (R848) (D/R@LCFS), to provide effective chemoimmunotherapy with reduced systemic toxicity. After local administration, D/R@LCFS converted the tumor into an in situ vaccine via DOX-triggered immunogenic cell death effect, combined with the immune stimulatory effect of R848. At the same time, the combination therapy of D/R@LCFS can promote the recruitment of effector $CD8^+$ T cells, as well as the polarization of myeloid-derived suppressor cells and immunosuppressive type 2 polarized macrophages to killer antigen-presenting cells, which is conducive to antigen-specific T cell immune response and induce more immunogenic phenotypes of tumors. The in situ vaccine prepared by D/R@LCFS and the remodeled TME combined with immune checkpoint blockade can induce systemic immune responses and long-term immune memory effects, and significantly prevent postoperative recurrence and metastasis of B16F10 or 4T1 tumors. Thus, this combined strategy of spatiotemporal TME modulation is expected to provide a clinically available option for effective chemoimmunotherapy after surgery (Hu et al. 2020).

3.4 Smart Liquid Crystal Materials

3.4.1 Liquid Crystal Based Materials for Responsive Release

Liquid crystals (LCs) are anisotropic fluids that combine the long-range order of crystals with the mobility of liquids 1, 2. This combination of properties has been widely used to create reconfigurable materials that optically report information about their environment, such as changes in electric fields (smartphone displays), temperature (thermometers) or mechanical shear, and the arrival of chemical and biological stimuli (sensors). Responsive liquid crystal-based materials not only report their environment but also transform loadings through self-regulated interactions.

3.4.1.1 Thermal Responsive Release

Targeted, on-demand delivery has been of interest using materials responsive to environmental stimuli. Beyazkilic et al. introduced a delivery technique that enables targeted, on-demand release of aqueous microdroplets from a liquid crystal medium through contactless stimulation. This method employs a nematic liquid crystal doped with a photothermal dye, which generates heat upon exposure to near-infrared (NIR) light. The heat overcomes the elastic constraints within the liquid crystal phase, facilitating the release of entrapped water droplets into the surrounding aqueous solution. By designing the geometry of liquid crystal-based emulsions and adjusting light intensity and position, the release can be controlled in two distinct modes: pulsatile and continuous. In the pulsatile mode, water droplets are transiently released from a water-in-liquid crystal emulsion layer as the isotropic-nematic phase boundary, controlled by light, sweeps across the emulsion. In this mode, water droplets are ejected steadily from a droplet-shaped water-in-liquid crystal emulsion due to heating-induced internal flow regulated by light. This contactless stimulation enables on-demand dosing of chemicals such as dopamine and its oxidizing reagent for in situ hydrogen peroxide assays. The dual-mode release system was further demonstrated for the controlled delivery of bioactive small molecules, including tetracycline and dexamethasone, which exhibit antibacterial and immunomodulatory activities. These features make the system promising for applications in chronic wound healing and biomedical implant coatings. The use of photothermal liquid crystals in this dual-mode release platform holds potential for drug delivery, controlled mixing, and photothermal therapy (Beyazkilic et al. 2022).

The controlled release of drugs and chemicals is crucial for theranostics and material synthesis. Borbora et al. highlighted that both pulsatile and continuous release modes are essential for various applications, including programmed chemical reactions, mechanical actuation, and the treatment of diverse diseases. Continuous release is vital for managing conditions such as glaucoma, cancer, asthma, and cardiovascular disease, where sustained drug efficacy over an extended period is required without frequent high-dose administration. In contrast, pulsatile release is critical for treating diseases like inflammatory conditions, acute gastritis, diabetes, and rheumatoid arthritis, where specific drug doses are needed at defined intervals to ensure timely therapeutic intervention (Borbora et al. 2023).

3.4.1.2 Field Indeced Release

LCNs possess unusual anisotropic optical and dielectric properties and have been extensively investigated and promoted in the field of soft robotics and actuators (Gelebart et al. 2018; Zhan et al. 2020a). Liu's group proposed an approach to manipulating liquid at self-contained porous coating surfaces. Dynamic surfaces composed of liquid crystal polymer networks (LCNs) have been developed for on-demand liquid secretion and uptake in coatings (Fig. 3.19a, b). The release of liquid can be controlled globally or locally by carefully designing the molecular architecture of the LCNs. An alternative approach involves using patterned interdigitated electrodes (IDEs) to apply an electric field, enabling liquid transport both through the thickness and in three dimensions (Fig. 3.19c, d).

These LCNs are nanoporous and filled with a liquid crystal fluid that acts as a porogen during polymerization. The networks are copolymerized with a light-responsive azobenzene derivative. The coating is applied to a glass substrate patterned with IDEs (Fig. 3.19e). Liquid secretion and subsequent reallocation are achieved by sequentially applying UV light and an alternating current (AC) field. UV light induces uniform liquid secretion at the coating surface, which is then laterally transported and collected at predefined locations by the AC field. These dual triggers can be used sequentially or simultaneously to provide a gating effect, enabling precise control over the timing, location, and remote steering of liquid secretion (Zhan et al. 2022).

3.4.2 Liquid Crystal Based Materials for Interskin Communication

A significant portion of recent research in "smart materials" focuses on integrating sensing and actuation capabilities that respond to environmental changes. However, to truly advance these materials to a "smart" level, we face the challenge of enabling them to exchange chemical information with humans, devices, or among materials themselves. Additionally, developing mechanisms to extend the lifetime of materials and devices, such as by optimizing their tribological properties, is crucial. These advancements can also serve as communication and feedback mechanisms in applications like robotics, virtual reality, and adaptive drug delivery (Zhan et al. 2020b; Li et al. 2022). For example, taking mammal skin as inspiration, coatings that chemically alter their surface properties by secreting a chemical substance interacts with

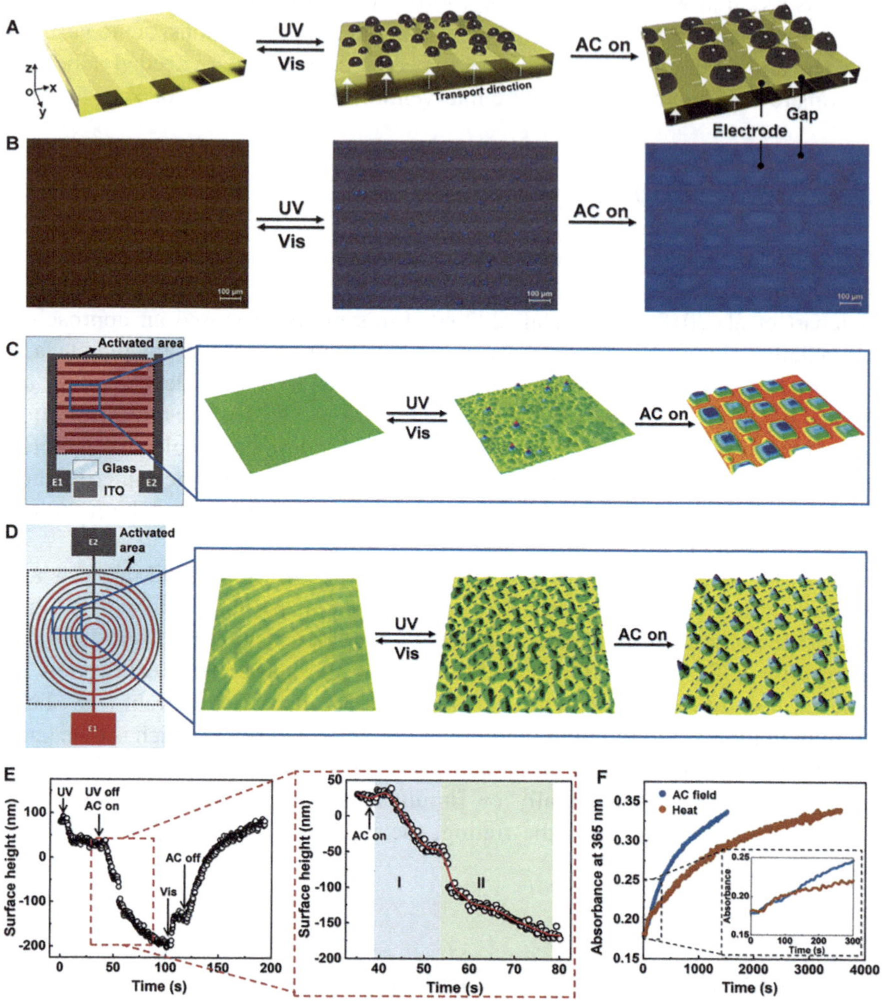

Fig. 3.19 3D-liquid manipulation. (**a**) Liquid secretion under UV illumination (365 nm, 11 mW cm^{-2}) and the subsequent reallocation process driven by the AC field (5.0 V μm^{-1}, 200 kHz) at the LCN coating surface, as observed by (**b**) optical microscope and (**c**), (**d**) digital holographic microscope. IDEs with (**c**) square pattern and (**d**) circular pattern are used. (**e**) Surface height change of the LCN coating subjected to sequential light and AC field stimulation. Surface height change occurs between 35 s and 80 s is zoomed in, which was indicated by stage I and II. Solid red line is eye guide. (**f**) Thermal relaxation (at 42 °C) of cisisomer and the accelerated process under the influence of the AC field. Single point kinetics analysis is performed at the absorbance peak at 365 nm. Insert is the zoom-in of the first 5 min, reprinted from Zhan et al. (2022), copyright 2022, with permission from the Authors. Angewandte Chemie International Edition published by Wiley-VCH GmbH

their environment by affecting adhesion, wettability, temperature, or touch experience. Liu et al. designed an artificial skin capable of triggered fluid release (Fig. 3.20). This system utilizes a radio frequency (RF)-actuated liquid crystal polymer network that expels initially trapped liquid through its nano- and micro-sized pores. The study demonstrated that this mechanism can also release chemical reagents or medicinal compounds upon fluid secretion. The polymer skin can be

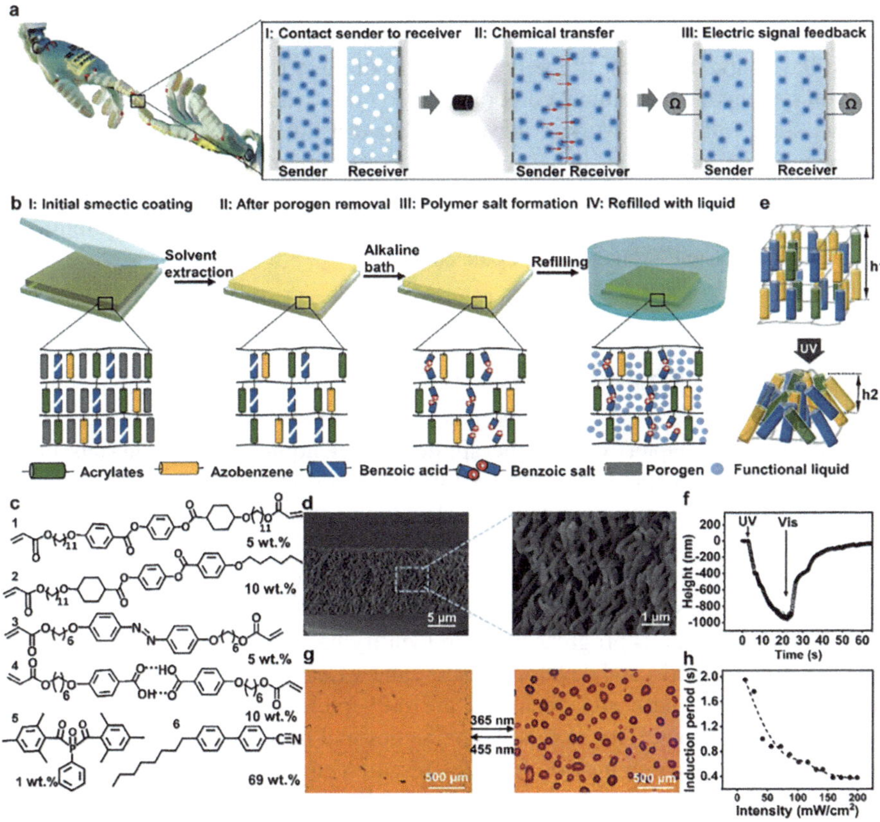

Fig. 3.20 Design concept of LCN polymer coatings communicating for information transfer. (**a**) Schematic illustration of the process of chemical communication, including information transfer, reception, and sensing. (**b**) Fabrication procedure of the LCN coating on a glass substrate. (**c**) Material composition of the communicating coating. (1, 2 LC monomers, 3 photoresponsive constituent, 4 benzoic acid dimer, 5 photoinitiator, 6 porogen). (**d**) Cross-sectional scanning electron microscope (SEM) image of the porous LCN coating and its zoomed-in view of the SEM image in highlighting the porous structure. (**e**) Scheme illustrating the smectic LC network and its contraction due to order decrease (h2 < h1). (**f**) Coating surface height changes over time upon UV and visible light irradiation. (**g**) Optical microscope images showing the secretion and reabsorption of PEG solution under light illumination at different wavelengths. (**h**) Induction period of secretion upon UV illumination as a function of illuminating intensity, reprinted from Zhang et al. (2024a), copyright 2024, with permission from the Authors. Angewandte Chemie International Edition published by Wiley-VCH GmbH

refilled through controlled absorption. The integration of this trigger-induced liquid secretion into soft robotics and microfluidics is anticipated to enable applications such as controlling chemical reactions, facilitating drug delivery, and modulating surface wettability and adhesion (Zhang et al. 2024a).

Chemical communication among individuals is a complex and dynamic process involving the production, release, and recognition of chemical signals (Llopis-Lorente et al. 2017). In many cases, chemical signals are used by organisms to communicate information about their identity, location, reproductive status, or physiological states. Chemical communication is a ubiquitous process in nature, and it has sparked interest in the development of electric-sense-based robotic perception systems with chemical components. To mimic the sweating functionality of human skin, Liu and coworkers have developed a liquid crystal (LC) polymeric network based artificial skin with programmable liquid secretion and excretion driven by radio frequency (RF) electrical fields. This approach allows for the transfer of cargo between two polymer coatings, and the transfer can be monitored through an electrical signal (Zhan et al. 2020b).

3.5 Conclusions and Perspectives

To sum up, this chapter reviews the application of liquid crystal materials in the diagnosis and treatment of cancer in recent years, focusing on the use of liquid crystals and their related compounds in the development of new cancer detection and therapeutic drugs. Although a variety of technologies have been developed into viable cancer diagnosis and treatment plans, the exploration of new technologies based on liquid crystals for advanced cancer diagnosis has always attracted interest. Despite the tremendous progress made in identifying cancer biomarkers, establishing detection methods, and developing new cancer treatment drugs over the past few decades, most of these advancements are only preliminary work demonstrating potential applications. There are still challenges and obstacles to overcome before these technologies can transition from the research environment to clinical applications. For cancer diagnosis, biosensors must have the sensitivity to transduce recognition events into readable signals, which is essential. Therefore, ongoing efforts are dedicated to enhancing existing strategies or discovering selective methods to quantify cancer biomarkers.

So far, research on liquid crystal-based immunoassay methods has yielded promising results for the development of cancer screening technology and has shown potential to replace traditional immunoassay methods. Leveraging the unique optical properties of liquid crystals, as well as their temperature sensitivity, many label-free, visual tumor marker detection methods have been developed. Undoubtedly, in the near-future, liquid crystal-based immunoassays will undergo further development and improvement to meet the requirements of clinical diagnostics.

Liquid crystals also have great potential in cancer treatment; some types of liquid crystals exhibit the ability to inhibit cell growth. However, the mechanism by which they inhibit cell growth is not yet clear. Therefore, detailed research on the molecular mechanisms and structure-activity responses of liquid crystals is needed to promote tumor-specific and highly potent liquid crystal-based chemotherapy drugs. Moreover, the cubic phase liquid crystal morphology is conducive to greater encapsulation of drugs and simultaneous transport of different substances, providing controlled and sustained release of large molecular weights, and, compared to other intermediate phases, it also has greater dilution stability. Preparing nanocarriers encapsulated with liquid crystal-drug can increase the bioavailability and targeting of the drug, reduce side effects, and strive for higher therapeutic efficiency. However, the cost of preparing and characterizing this nanotechnology is high, and it involves high energy requirements, which may be a barrier to future development. In view of this, future research should focus on new formulations aimed at reducing preparation costs and conducting more in vivo and toxicity evaluations so that such formulations can move toward multicenter clinical screening to assess their benefits in different populations, bringing such technology closer to the market.

With the significant demand for the development of clinical diagnostics, modern medicine, and the detection and treatment of cancer biomarkers, liquid crystal-based cancer diagnosis and treatment will certainly receive increasing attention. Future research will focus on developing high-throughput detection methods with higher sensitivity and selectivity, miniaturization, and multifunctionality as well as more economical and efficient cancer treatment drugs. Through interdisciplinary collaboration between chemists, biologists, physicists, clinicians, material scientists, and engineers, further development and application of liquid crystals in cancer biology and modern medicine may be realized.

References

Aharon O, Abdulhalim I, Arnon O et al (2011) Differential optical spectropolarimetric imaging system assisted by liquid crystal devices for skin imaging. J Biomed Opt 16:086008

An Z, Jang C (2019) Fabrication of liquid crystal droplet patterns for monitoring aldehyde vapors. ChemPlusChem 84:1554–1559

Balian GMFC, Luiz MT, Filippo LDD, Chorilli M (2023) Mucoadhesive liquid crystal precursor system for photodynamic therapy of oral cancer mediated by methylene blue. Photodiagn Photodyn Ther 44:103739

Bera T, Fang J (2013) Optical detection of lithocholic acid with liquid crystal emulsions. Langmuir 29:387–392

Berthiller F, Crews C, Dall'Asta C et al (2013) Masked mycotoxins: a review. Mol Nutr Food Res 57:165–186

Beyazkilic P, Akcimen S, Elbuken C et al (2022) Contactless pulsed and continuous microdroplet release using photothermal liquid crystals. Adv Funct Mater 32:2205385

Biernat M, Trzyna M, Byszek A, Jaremek H (2016) Liquid crystal foil for the detection of breast cancer. In: Photonics applications in astronomy, communications, industry, and high-energy physics experiments. SPIE, Bellingham, pp 873–882

Bisoyi HK, Kumar S (2010) Liquid-crystal nanoscience: an emerging avenue of soft self-assembly. Chem Soc Rev 40:306–319

Bonifácio BV, Silva PB, Ramos MA et al (2013) Nanotechnology-based drug delivery systems and herbal medicines: a review. IJN 9:1–15
Borbora A, Xu Y, Dey S et al (2023) Lubricated interfaces enabling simultaneous pulsatile and continuous chemical release modes. Adv Mater 35:2302264
Chang T-K, Tung P-C, Lee M-J, Lee W (2022) A liquid-crystal aptasensing platform for label-free detection of a single circulating tumor cell. Biosens Bioelectron 216:114607
Charkhchi P, Cybulski C, Gronwald J et al (2020) CA125 and ovarian cancer: a comprehensive review. Cancer 12:3730
Chavda VP, Dawre S, Pandya A et al (2022) Lyotropic liquid crystals for parenteral drug delivery. J Control Release 349:533–549
Cheng S, Khan M, Yin F et al (2022) Surface-anchored liquid crystal droplets for the semi-quantitative detection of aflatoxin B1 in food samples. Food Chem 390:133202
Chiang Y-L, Lee M-J, Lee W (2018) Enhancing detection sensitivity in quantitative protein detection based on dye-doped liquid crystals. Dyes Pigments 157:117–122
Choi Y, Lee K, Gupta KC et al (2015) The role of ligand–receptor interactions in visual detection of HepG2 cells using a liquid crystal microdroplet-based biosensor. J Mater Chem B 3:8659–8669
Concellón A (2023) Liquid crystal emulsions: a versatile platform for photonics, sensing, and active matter. Angew Chem Int Ed 62:e202308857
Dang Y, Guan J (2020) Nanoparticle-based drug delivery systems for cancer therapy. Smart Mater Med 1:10–19
David R, Akmar Binti Anwar N, Yian KR et al (2020) Development and evaluation of curcumin liquid crystal systems for cervical cancer. Sci Pharm 88:15
Deng J, Lu X, Constant C et al (2015) Design of β-CD–surfactant complex-coated liquid crystal droplets for the detection of cholic acid via competitive host–guest recognition. Chem Commun 51:8912–8915
Devadiga D, Ahipa TN (2022) A review on the emerging applications of 4-cyano-4′-alkylbiphenyl (nCB) liquid crystals beyond display. Mater Sci Eng 275:115522
Dong Y, Wen C, She Y et al (2021) Magnetic relaxation switching immunoassay based on hydrogen peroxide-mediated assembly of ag@au–Fe3O4 nanoprobe for detection of aflatoxin B1. Small 17:2104596
Frazão J, Palma SICJ, Costa HMA et al (2021) Optical gas sensing with liquid crystal droplets and convolutional neural networks. Sensors 21:2854
Fukushi Y, Hazawa M, Takahashi K et al (2011) Liquid crystal-related compound-induced cell growth suppression and apoptosis in the chronic myelogenous leukemia K562 cell line. Investig New Drugs 29:827–832
Gelebart AH, Liu D, Mulder DJ et al (2018) Photoresponsive sponge-like coating for on-demand liquid release. Adv Funct Mater 28:1705942
Gentile E, Oba T, Lin J et al (2017) Cationic liquid crystalline nanoparticles for the delivery of synthetic RNAi-based therapeutics. Oncotarget 8:48222–48239
Gollapelli B, Tatipamula AK, Dewanjee S et al (2021) Detection of bile acids using optical biosensors based on cholesteric liquid crystal droplets. J Mater Chem C 9:13991–14002
Gong C, Sun F, Yang G et al (2022) Multifunctional laser imaging of cancer cell secretion with hybrid liquid crystal resonators. Laser Photon Rev 16:2100734
Guardià J, Reina JA, Giamberini M, Montané X (2024) An up-to-date overview of liquid crystals and liquid crystal polymers for different applications: a review. Polymers 16:2293
Guo C, Wang J, Cao F et al (2010) Lyotropic liquid crystal systems in drug delivery. Drug Discov Today 15:1032–1040
Han S, Shen J, Gan Y et al (2010) Novel vehicle based on cubosomes for ophthalmic delivery of flurbiprofen with low irritancy and high bioavailability. Acta Pharmacol Sin 31:990–998
Han X, Han D, Zeng J et al (2020) Fabrication and performance of monodisperse liquid crystal droplet-based microchips for the on-chip detection of bile acids. Microchem J 157:105057

Hani U, Gowda BHJ, Haider N et al (2023) Nanoparticle-based approaches for treatment of hematological malignancies: a comprehensive review. AAPS PharmSciTech 24:233
He S, Fan W, Wu N et al (2018) Lipid-based liquid crystalline nanoparticles facilitate cytosolic delivery of siRNA via structural transformation. Nano Lett 18:2411–2419
He Q, Su H, Jiang H et al (2025) Responsive release of guest aqueous droplet in liquid crystal droplet assists the detection of organophosphorus using commercial pregnancy test strip. Sensors Actuators B Chem 423:136777
Hodorowicz-Zaniewska D, Zurrida S, Kotlarz A et al (2020) A prospective pilot study on use of liquid crystal thermography to detect early breast cancer. Integr Cancer Ther 19:153473542091577
Hu M, Zhang J, Yu Y et al (2020) Injectable liquid crystal formation system for reshaping tumor immunosuppressive microenvironment to boost antitumor immunity: postoperative chemoimmunotherapy. Small 16:2004905
Hu Z, Wang G, Zhao S et al (2023) Sustained release properties of liquid crystal functionalized poly (amino acid)s nanoparticles. J Mol Liq 369:120891
Hussain Z, Qazi F, Ahmed MI et al (2016) Liquid crystals based sensing platform-technological aspects. Biosens Bioelectron 85:110–127
Jareño J, Munoz MA, Wagner C et al (2014) Volatile organic compounds (VOC) in exhaled breath in patients with lung cancer. Chest 145:334A
Jenni S, Picci G, Fornasier M et al (2020) Multifunctional cubic liquid crystalline nanoparticles for chemo- and photodynamic synergistic cancer therapy. Photochem Photobiol Sci 19:674–680
Jiang S, Noh J, Park C et al (2021) Using machine learning and liquid crystal droplets to identify and quantify endotoxins from different bacterial species. Analyst 146:1224–1233
Jiang L, Shui Q, Han B et al (2024) Recent advances in liquid crystal droplet-based biosensors. TrAC Trends Anal Chem 180:117920
Jin N, Paraskevaidi M, Semple KT et al (2017) Infrared spectroscopy coupled with a dispersion model for quantifying the real-time dynamics of kanamycin resistance in artificial microbiota. Anal Chem 89:9814–9821
Kemp JA, Kwon YJ (2021) Cancer nanotechnology: current status and perspectives. Nano Convergence 8:34
Khan M, Li W, Mao S et al (2019) Real-time imaging of ammonia release from single live cells via liquid crystal droplets immobilized on the cell membrane. Adv Sci 6:1900778
Kim S, Hwang C, Jeong DI et al (2023) Nanorod/nanodisk-integrated liquid crystalline systems for starvation, chemodynamic, and photothermal therapy of cancer. Bioeng Transl Med 8:e10470
Kumar P, Deep A, Kim K-H, Brown RJC (2015) Coordination polymers: opportunities and challenges for monitoring volatile organic compounds. Prog Polym Sci 45:102–118
Kumar H, Gupta NV, Jain R et al (2023a) F3 peptide functionalized liquid crystalline nanoparticles for delivering Salinomycin against breast cancer. Int J Pharm 643:123226
Kumar RR, Kumar A, Chuang C-H, Shaikh MO (2023b) Recent advances and emerging trends in cancer biomarker detection technologies. Ind Eng Chem Res 62:5691–5713
Lagerwall JPF, Scalia G (2012) A new era for liquid crystal research: applications of liquid crystals in soft matter nano-, bio- and microtechnology. Curr Appl Phys 12:1387–1412
Lai SL, Tan WL, Yang K-L (2011) Detection of DNA targets hybridized to solid surfaces using optical images of liquid crystals. ACS Appl Mater Interfaces 3:3389–3395
Li J, Wu L, Wu W et al (2013) A potential carrier based on liquid crystal nanoparticles for ophthalmic delivery of pilocarpine nitrate. Int J Pharm 455:75–84
Li Y, Angelova A, Liu J et al (2019) *In situ* phase transition of microemulsions for parenteral injection yielding lyotropic liquid crystalline carriers of the antitumor drug bufalin. Colloids Surf B Biointerfaces 173:217–225
Li J, Jamieson WD, Dimitriou P et al (2022) Building programmable multicompartment artificial cells incorporating remotely activated protein channels using microfluidics and acoustic levitation. Nat Commun 13:4125

Liu R, Wang S, Fang S et al (2016) Liquid crystalline nanoparticles as an ophthalmic delivery system for tetrandrine: development, characterization, and in vitro and in vivo evaluation. Nanoscale Res Lett 11:254
Liu D, Li W, Zhu C et al (2020) Recent progress on electrochemical biosensing of aflatoxins: a review. TrAC Trends Anal Chem 133:115966
Liu H, Su X, Zhang J et al (2021) Highly sensitive and rapid detection of protein kinase C based on liquid crystal biosensor. Colloids Surf A Physicochem Eng Asp 628:127346
Llopis-Lorente A, Díez P, Sánchez A et al (2017) Interactive models of communication at the nanoscale using nanoparticles that talk to one another. Nat Commun 8:15511
Luan C, Luan H, Luo D (2020) Application and technique of liquid crystal-based biosensors. Micromachines 11:176
Lyu X, Xiao A, Shi D et al (2020) Liquid crystalline polymers: discovery, development, and the future. Polymer 202:122740
Ma Z, Xu M, Zhou S et al (2022) Ultra-low sample consumption consecutive-detection method for biochemical molecules based on a whispering gallery mode with a liquid crystal microdroplet. Opt Lett 47:381–384
Majumdar A, Dubey N, Dubey N (2019) Dermal delivery of docetaxel loaded nano liquid crystals for the treatment of skin cancer. Int J App Pharm 11:188–193
Manna U, Zavala YM, Abbott NL, Lynn DM (2021) Structured liquid droplets as chemical sensors that function inside living cells. ACS Appl Mater Interfaces 13:42502–42512
Mezzenga R, Seddon JM, Drummond CJ et al (2019) Nature-inspired design and application of lipidic lyotropic liquid crystals. Adv Mater 31:1900818
Morsi NM, Abdelbary GA, Ahmed MA (2014) Silver sulfadiazine based cubosome hydrogels for topical treatment of burns: development and in vitro/in vivo characterization. Eur J Pharm Biopharm 86:178–189
Nakayama M, Kato T (2022) Biomineral-inspired colloidal liquid crystals: from assembly of hybrids comprising inorganic nanocrystals and organic polymer components to their functionalization. Acc Chem Res 55:1796–1808
Nakayama M, Kajiyama S, Nishimura T, Kato T (2015) Liquid-crystalline calcium carbonate: biomimetic synthesis and alignment of nanorod calcite. Chem Sci 6:6230–6234
Nakayama M, Kajiyama S, Kumamoto A et al (2018) Stimuli-responsive hydroxyapatite liquid crystal with macroscopically controllable ordering and magneto-optical functions. Nat Commun 9:568
Nakayama M, Lim WQ, Kajiyama S et al (2019) Liquid-crystalline hydroxyapatite/polymer nanorod hybrids: potential bioplatform for photodynamic therapy and cellular scaffolds. ACS Appl Mater Interfaces 11(19):17759–17765
Nakayama M, Kajiyama S, Kumamoto A et al (2020) Bioinspired selective synthesis of liquid-crystalline nanocomposites: formation of calcium carbonate-based composite nanodisks and nanorods. Nanoscale Adv 2:2326–2332
Negrini R, Fong W-K, Boyd BJ, Mezzenga R (2015) pH-responsive lyotropic liquid crystals and their potential therapeutic role in cancer treatment. Chem Commun 51:6671–6674
Niu X, Ye K, Wang L et al (2019) A review on emerging principles and strategies for colorimetric and fluorescent detection of alkaline phosphatase activity. Anal Chim Acta 1086:29–45
Pani I, Sharma M, Pal SK (2021) Probing nanoscale lipid–protein interactions at the interface of liquid crystal droplets. Nano Lett 21:4546–4553
Pani I, Sil S, Pal SK (2023) Liquid crystal biosensors: a new therapeutic window to point-of-care diagnostics. Langmuir 39:909–917
Peer D, Karp JM, Hong S et al (2020) Nanocarriers as an emerging platform for cancer therapy. In: Nano-enabled medical applications. Jenny Stanford Publishing, New Delhi
Pereira TA (2023) Liquid crystalline nanogel targets skin cancer via low-frequency ultrasound treatment. Int J Pharm 646:123431
Pereira TA, Ramos DN, Sobral LM et al (2023) Liquid crystalline nanogel targets skin cancer via low-frequency ultrasound treatment. Int J Pharm 646:123431

Petrilli R, Praca F, Carollo A et al (2013) Nanoparticles of lyotropic liquid crystals: a novel strategy for the topical delivery of a chlorin derivative for photodynamic therapy of skin cancer. Curr Nanosci 9:434–441
Phillips M, Gleeson K, Hughes JMB et al (1999) Volatile organic compounds in breath as markers of lung cancer: a cross-sectional study. Lancet 353:1930–1933
Pourasl MH, Vahedi A, Tajalli H et al (2023) Liquid crystal-assisted optical biosensor for early-stage diagnosis of mammary glands using HER-2. Sci Rep 13:6847
Qi L, Liu S, Jiang Y et al (2020) Simultaneous detection of multiple tumor markers in blood by functional liquid crystal sensors assisted with target-induced dissociation of aptamer. Anal Chem 92:3867–3873
Qi L, Liu J, Liu S et al (2024) Ultrasensitive point-of-care detection of protein markers using an aptamer-CRISPR/Cas12a-regulated liquid crystal sensor (ALICS). Anal Chem 96:866–875
Qin L, Mei L, Shan Z et al (2016) Phytantriol based liquid crystal provide sustained release of anticancer drug as a novel embolic agent. Drug Dev Ind Pharm 42:307–316
Rastogi P, Njuguna J, Kandasubramanian B (2019) Exploration of elastomeric and polymeric liquid crystals with photothermal actuation: a review. Eur Polym J 121:109287
Ren H, Jang C-H (2019) A simple liquid crystal-based aptasensor using a hairpin-shaped aptamer for the bare-eye detection of carcinoembryonic antigen. Biochip J 13:352–361
Roberts RS (1966) Preparation of endotoxin. Nature 209:80
Rouhbakhsh Z, Huang J-W, Ho TY, Chen C-H (2022) Liquid crystal-based chemical sensors and biosensors: from sensing mechanisms to the variety of analytical targets. TrAC Trends Anal Chem 157:116820
Saklani R, Yadav PK, Tiwari AK et al (2024) Synchronized codelivery of combination chemotherapies intratumorally using a lipidic lyotropic liquid crystal system. ACS Appl Mater Interfaces 16:29098–29111
Salimi M, Milani Hosseini SMR (2021) Smartphone-based detection of lung cancer-related volatile organic compounds (VOCs) using rapid synthesized ZnO nanosheet. Sensors Actuators B Chem 344:130127
Shaban H, Lee M-J, Lee W (2021) Label-free detection and spectrometrically quantitative analysis of the cancer biomarker CA125 based on lyotropic chromonic liquid crystal. Biosensors 11:271
Shah JC, Sadhale Y, Chilukuri DM (2001) Cubic phase gels as drug delivery systems. Adv Drug Deliv Rev 47:229–250
Shan X, Luo L, Yu Z, You J (2022) Recent advances in versatile inverse lyotropic liquid crystals. J Control Release 348:1–21
Shan X, Li X, Luo Z et al (2023) A clinically-achievable injectable and sprayable *in situ* lyotropic liquid crystalline platform in treating hormone-sensitive and castration-resistant prostate cancer. ACS Nano 17:6045–6061
Sidiq S, Prasad GVRK, Mukhopadhaya A, Pal SK (2017) Poly (L-lysine)-coated liquid crystal droplets for cell-based sensing applications. J Phys Chem B 121:4247–4256
Siegel RL, Giaquinto AN, Jemal A (2024) Cancer statistics, 2024. CA Cancer J Clin 74:12–49
Smalyukh II (2018) Liquid crystal colloids. Ann Rev Condensed Matter Phys 9:207–226
Son MH, Park SW, Sagong HY, Jung YK (2023) Recent advances in electrochemical and optical biosensors for cancer biomarker detection. Biochip J 17:44–67
Sorbye H, Grande E, Pavel M et al (2023) European Neuroendocrine Tumor Society (ENETS) 2023 guidance paper for digestive neuroendocrine carcinoma. J Neuroendocrinol 35:e13249
Spicer PT (2005) Progress in liquid crystalline dispersions: cubosomes. Curr Opin Colloid Interface Sci 10:274–279
Stephens R, Mythen M (2000) Endotoxin immunization. Intensive Care Med 26:S129–S136
Su H-W, Lee M-J, Lee W (2015) Surface modification of alignment layer by ultraviolet irradiation to dramatically improve the detection limit of liquid-crystal-based immunoassay for the cancer biomarker CA125. JBO 20:057004
Sun S-H, Lee M-J, Lee Y-H et al (2015) Immunoassays for the cancer biomarker CA125 based on a large-birefringence nematic liquid-crystal mixture. Biomed Opt Express 6:245–256

Chapter 4
Nanozymes and Their Applications in Tumor Imaging and Treatment

Dan-Dan Wang

Abstract Artificial nanoenzymes with enzyme-like catalytic activity have gradually become substitutes for natural enzymes due to their low production cost, high stability, and good tolerance. In recent years, with the rapid development of nanotechnology, various enzyme mimics have emerged. With unique properties such as high specific surface area, high porosity, tunable morphology, and excellent biocatalytic properties, nanoenzymes have broad application prospects in the fields of biosensing, biocatalysis, biomedical imaging and therapy. This chapter systematically summarizes the latest research progress of nanoenzymes, including the catalytic principle, classification, and related applications of nanoenzymes, which are classified into two categories: tumor imaging and therapy. Finally, the current challenges and future perspectives of nanoenzymes are also discussed. It is hoped that this review of this important field will provide meaningful guidance for tumor therapy, biosensing, and so on.

Keywords Nanozymes · Catalytic principle · Classification · Tumor imaging · Tumor treatment

4.1 Introduction

Natural enzymes, which are mainly composed of proteins, are promising, widely used and efficient biocatalysts, which are widely used in industrial, pharmaceutical, food, and bioengineering fields due to their excellent catalytic activity and substrate specificity (Breaker 1997; Wolfenden and Snider 2001; Bornscheuer et al. 2012) The main difference compared to inorganic catalysts, which increase the rate of reaction, is that inorganic catalysts are involved in chemical reactions that can

D.-D. Wang (✉)
Shandong Provincial Key Laboratory of Tumor Imaging Equipment Development and Tumor Diagnosis and Treatment Integration Technology, School of Medicine, Linyi University, Linyi, Shandong, People's Republic of China

X. Song et al. (eds.), *Functional Inorganic Materials for Tumor Diagnostics and Therapy*, https://doi.org/10.1007/978-981-95-0528-9_4

usually be carried out at high temperatures or pressures, whereas enzymes mostly catalyze biological reactions under mild conditions and therefore under higher reaction conditions such as temperature and pH (Lin et al. 2014; Liang and Yan 2019). To address the inherent limitations of natural enzymes, such as high production costs, low stability, and difficulties in storage and recycling, a series of enzyme mimics have been inspired, developed and explored.

In recent years, nanomaterials such as magnetic nanoparticles, gold nanoparticles, and rare-earth nanoparticles have been found to have excellent enzyme-like activities and have been named as nanoenzymes. Nanoenzymes are a class of artificial enzymes synthesized on the basis of nanomaterials, which have both the unique properties of nanomaterials and high catalytic efficiency (Wei and Wang 2013; Lin et al. 2014; Liang and Yan 2019; Wu et al. 2019). Based on the rapid development of nanotechnology and the deep understanding of nanoscience, a series of studies on nanoenzymes have emerged. By mimicking the catalytic sites of traditional enzymes or through multivalent reactions, nanoenzymes have been greatly developed to replace natural enzymes. Synthetic nanoenzymes have the advantages of high catalytic efficiency, economy, stability and easy synthesis, and have been widely used in the fields of medicine, bioengineering and chemistry (Wang et al. 2018; Liu et al. 2021a). The catalytic properties of nanoenzymes are mainly due to the similarity of their active center and electron transfer structure to that of natural enzymes, where the active center confers the enzyme or catalytic activity.

Artificial nanoenzymes with enzyme-like catalytic activity are gradually becoming substitutes for natural enzymes due to their low production cost, high stability, and good tolerance. This chapter systematically summarizes the recent advances on nanoenzymes, including synthesis, design strategies, and related applications. This chapter is divided into three major categories, namely, classification of nanozymes and their catalytic principles, preparation of nanozymes, and the applications of various classifications of nanozymes for diagnosis and therapy of tumors are introduced. Finally, the current challenges and future prospects of nanoenzymes are also discussed. It is hoped that this review of this important field will provide meaningful guidance for tumor therapy, biosensing, and many more.

4.2 Catalytic Principle and Classification of Nanozymes

4.2.1 General

Nanoenzymes are a class of artificial enzymes synthesized on the basis of nanomaterials, which have both the unique properties of nanomaterials and high catalytic efficiency. Compared with natural enzymes, nanoenzymes have the advantages of high catalytic activity, mild reaction conditions, good stability, low cost, and easy mass production. Since then, nanoenzymes have entered the vision of researchers. In order to meet the basic research, application research, and industrial promotion needs of nanozymes, scientists have established a standard for quantitatively comparing the catalytic activity and kinetics of various nanozymes, and defined the nanozyme unit (U) of nanozyme catalytic activity, which means that under the optimal reaction conditions, the amount of nanozyme required to catalyze the conversion of 1 μ mol substrate to product per minute is 1 U. The specific activity (U/mg) of nanozymes is defined as the catalytic activity per unit mass of nanozymes, and further specifies the detection conditions and calculation formulas for nanozyme specific activity. The standard further facilitates the development of nanoenzymatic-based detection and diagnostic technologies (Gao et al. 2007). Nanozymes follow enzyme kinetics (Mie's equation) and catalytic mechanisms (ordered cis, stochastic cis, or ping-pong reactions). In this case, the Mie equation represents the starting rate of an enzymatic reaction as a function of substrate concentration, the equation can be expressed as $V_0 = V_{max}[S]/(K_m + [S])$, where V_{max} is the reaction rate when the enzyme is saturated with the substrate, K_m value is the Mie's constant (K_m refers to the substrate concentration in mol/L when the reaction rate V reaches 1/2 V_{max} (K_m value can be used to identify enzymes of different nature).

Natural enzymes in living organisms are proteins consisting of hundreds of amino acid molecules with an internal activation center and a binding group. Typically, the binding group recognizes and binds a substrate molecule, which is induced by the binding group to gradually approach the activation center and catalyze a change in the substrate molecule, which is then detached from the enzyme. Their catalysis is related to their unique structure. Nanoenzymes are composed of variable-valent metals, for example, iron tetraoxide nanoenzymes have both divalent and trivalent iron on their surface, and these variable-valent metal ions can transfer electrons through valence conversion, thus catalyzing chemical reactions efficiently.

To date, a large number of nanomaterials, such as metal oxide nanoparticles, carbon-based nanomaterials, noble metal nanoparticles, and MOFs, have been shown to act as catalysts in the same way as natural enzymes (Sun et al. 2018; Wang et al. 2018; Lou-Franco et al. 2020; Ma et al. 2020; Niu et al. 2020; He et al. 2021a; Liu et al. 2021b; Cai et al. 2021). Based on these new characteristics, nanozymes have been widely used in biosensing, environmental remediation, disease diagnosis and treatment, antibacterial, and intracellular biomolecule cell protection, etc. (Shang et al. 2020; Fan et al. 2021; Lopez-Cantu et al. 2022). Usually, enzymes can be classified into multiple types based on their different mechanisms. Most nanozymes can simulate the activity of peroxidase (POD), peroxidation, catalase (CAT), superoxide dismutase (SOD) dismutase (Fig. 4.1).

4.2.2 *Oxidase Activity*

Oxidases are enzymes with important physiological functions, which reduce peroxides such as hydrogen peroxide and lipid peroxides, and use substrates as electron donors for oxidation. OXD can catalyze the oxidation of substrates (electron donors)

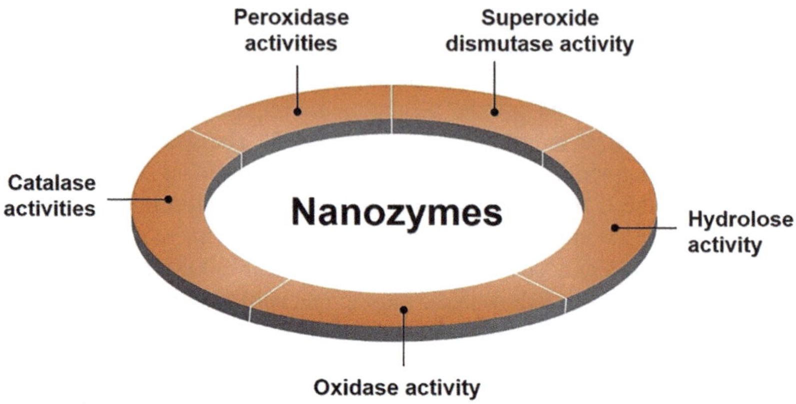

Fig. 4.1 The main enzyme-like catalytic activities of nanozymes

to corresponding oxidation products in the presence of O_2 (electron acceptor), while producing H_2O or H_2O_2. As is well-known, the majority of O_2 consumption in the biosphere is catalyzed by the heme copper oxidase superfamily, such as cytochrome c oxidase. Recently, it has been found that many inorganic nanomaterials can catalyze the oxidation of one or more substrates with the help of O_2, exhibiting OXD like activity, such as cerium-based oxides, precious metals (Au, Pt, Pd, and many more), manganese-based oxides, MOFs, and many more (Liu et al. 2012; Xiong et al. 2015; Cai et al. 2016; Cheng et al. 2016).

4.2.3 Peroxidase Activity

Peroxidase (POD) is a class of enzymes with important physiological functions, which reduce of oxides and oxidation using substrates as electron donors (Attar et al. 2019; Liu et al. 2019). Most natural PODs are heme proteins that can activate H_2O_2 and produce intermediates with high valence states and can extract electrons from different substrates. In recent years, discovery of a large number of iron-based nanomaterials with POD-like catalytic activity. For example, Gao et al. first reported the intrinsic catalytic performance of Fe_3O_4 nanoparticles on typical POD substrates in 2007, laying the foundation for the development of nanomaterials with POD like activity. Nowadays, nanomaterials, such as magnetic nanoparticles (Fe_3O_4, $CoFe_2O_4$, $ZnFe_2O_4$, etc.), carbon nanomaterials (graphene oxide, fullerene-C_{60}, carbon nanodots, nitrogen-doped carbon dots, etc.), metal oxide nanomaterials (Co_3O_4, CeO_2, MnO_2, etc.), metal sulfide nanomaterials (ZnS, MoS_2, etc.), metal nanomaterials (Au, Ag, Pt, etc.), and nanocomposites (GO-Fe_3O_4, Cu Ag/Rgo, MoS_2/GO, etc.), have been found to have inherent POD like properties.

4.2.4 Catalase Activity

Catalase (CAT) is one of the most important oxidoreductases in the human body. CAT can catalyze the decomposition of H_2O_2 into H_2O molecules and O_2, which plays a crucial role in catalyzing the elimination of ROS, preventing cell membrane damage, and inhibiting the growth of tumor cells (Glorieux and Calderon 2017). In addition, CAT is the hallmark enzyme of the peroxisome, accounting for about 40% of the total peroxisomal enzymes and is widely present in the peroxisome of cells. However, in practice, it also has the disadvantages inherent in natural enzymes, such as poor stability and difficulty in preservation. Nanozymes, which emerged as a result of the times, are also designed to mimic the activity of CAT. So far, metals, metal oxides, MOFs, carbon-based nanomaterials, metal sulfides, and Prussian blue have been shown to have CAT-like enzymatic activity and have been used in a wide range of applications such as tumor therapy and immunoassays (Huang et al. 2019; Xu et al. 2022).

4.2.5 Hydrolase Activity

It is a general term for enzymes that catalyze hydrolysis reactions and can also be said to be a special type of transferase that uses water as the receptor for the transferred group. Hydrolytic enzymes can catalyze the hydrolysis of a range of unique chemical bonds. For example, nucleases can cleave the phosphodiester bonds of nucleotide chains, while esterases can hydrolyze esters into acids and alcohols. Hydrolytic nanozymes are nanomaterials with hydrolytic enzyme like activity, which are classified into nucleases, esterases, phosphatases, proteases, and so on according to the different hydrolytic substrates (Gabrielli et al. 2020; Hou et al. 2022). In recent years, a large number of hydrolytic nanozymes have been reported, such as micelles, vesicles, and nanoparticles (Pasquato et al. 2000; Fernández et al. 2015; Ding et al. 2019). The state of micelles and vesicles depends mainly on the nature of the solvent, the ionic strength of the solution and the presence of additives, with limited control over the topology and catalytic sites. In contrast, for nanoparticles, the topology and catalytic sites can be well controlled.

4.2.6 Superoxide Dismutase Activity

Superoxide dismutase (SOD) is an antioxidant metalloenzyme present in living organisms. It catalyzes the decomposition of superoxide anion radicals, generating oxygen and hydrogen peroxide, and plays an important role in the balance between oxidation and antioxidation in the body. It is closely related to the occurrence and development of many diseases SOD is a group of metalloenzymes that catalyze the dismutation of O_2^- into H_2O_2 and O_2 (Fridovich 1978). Natural SOD is found in most prokaryotic cells and some eukaryotic cells, and is widespread in a variety of organelles. Depending on the cofactor, natural SOD enzymes can be classified into four types: manganese SOD (MnSOD), iron SOD (FeSOD), copper zinc SOD (CuZnSOD), and nickel SOD (NiSOD) (Zhao et al. 2021). As the first line of defense against ROS-induced oxidative damage, SOD is expected to play an important role in the treatment of oxidative stress-related diseases. However, due to its unstable structure and high cost, the clinical application of SOD has been severely limited. In 1985, Krato et al. reported that C_{60} exhibited O_2^- scavenging activity, which was early evidence that nanomaterials could simulate SOD like activity. However, the concept of nanoenzymes had not been proposed at that time. Since then, various nanomaterials mimicking SOD have been gradually developed. So far, about 100 nanozymes have been discovered to exhibit SOD like activity, among which fullerene and CeO_2 are the most well-known (Kroto et al. 1985; Ali et al. 2004; Korsvik et al. 2007).

4.3 Application of Nanozymes in Tumor Diagnosis and Treatment

4.3.1 General

Cancer is formed due to the unlimited proliferation of cells after losing control. It has a high incidence rate and mortality, which seriously threatens human health. Early diagnosis and treatment of cancer have always been crucial for improving disease prognosis. Scientists have been exploring effective methods for early diagnosis and treatment of tumors. The traditional treatment methods for cancer include radiotherapy, chemotherapy, and surgery. Although these methods play a positive role in cancer treatment, they have many drawbacks such as high toxicity and poor treatment effectiveness. Therefore, nanoenzymes with enzyme like activity provide new opportunities for the treatment of tumors. On the one hand, nanoezymes have unique enzyme like properties that can directly promote tumor cell apoptosis or inhibit tumor growth. On the one hand, nanoenzymes have unique enzyme-like properties that can directly promote apoptosis or inhibit tumor growth. On the other hand, in addition to catalytic therapy using individual nanoenzymes, cascade catalytic nanoenzymes systems can be constructed for cancer therapy. In addition, nanoenzymes are widely used in various tumor therapies, such as adjuvant radiotherapy, chemotherapy, photothermal therapy, chemodynamic therapy, immunotherapy, and many more, to improve the therapeutic efficacy through the combined strategy.

As a new type of nanomaterial with high catalytic efficiency and good stability, nanoenzymes have received widespread attention since their birth and have been applied in many fields, such as environmental protection, food testing, and chemical production (Fig. 4.2). In particular, anti-cancer nanomaterials and diagnostics have been extensively researched in the past decade.

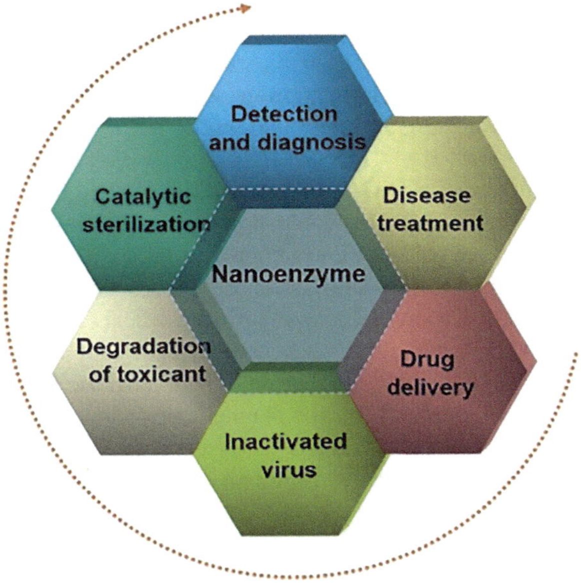

Fig. 4.2 Biomedical applications of nanozymes

4.3.2 Naoenzyme in Diagnosis of Tumors

Early diagnosis of cancer can significantly improve the survival rate of cancer patients (Linehan et al. 2010). Therefore, developing highly sensitive and selective early diagnostic methods is crucial. High precision and high selectivity tumor imaging is crucial for different cancer treatments such as surgery, RT, phototherapy, and many more. With the advantages of high catalytic activity, low cost, high stability, and multifunctionality, nanoenzymes have been successfully applied to the detection of cancer-related genes, molecules, and cells. Biomarkers can be tumor-specific DNA, RNA, or proteins. Nanozymes have excellent enzyme mimicking ability and have been used to detect different biomarkers (Wang et al. 2014). Nowadays, more and more nanozymes are developed and widely used for tumor imaging through various imaging techniques, such as fluorescence, US, magnetic resonance (MR), photoacoustic (PA), photothermal, and computed tomography (CT) imaging (Yang et al. 2012; Zhong et al. 2019; Feng et al. 2020; Wang et al. 2020a, d; Cao et al. 2021; Ma et al. 2021).

NIR-II photons have been shown to significantly improve the resolution of in vivo imaging in heavily scattering tissue environments, such as NIR-II photoluminescence imaging of intact cerebral vessels in mice. Chemiluminescence (CL) imaging, on the other hand, is thought to provide a better signal-to-noise ratio because it does not involve the background noise associated with the excitation beam. Chang et al. developed A BSAZn-based nanoenzyme, which was demonstrated for glucose sensing, tumor therapy, and in vivo H_2O_2 imaging (Fig. 4.3) (Chang et al. 2023). With the BSAZn nanoenzyme, the Gox-Glucose could introduce significant damage on the tumor tissue. A burn scar could be observed on the first day after administration, and the tumor almost shrunk away within 8 days (Fig. 4.3b). Furthermore, in vivo NIR-II CL was performed via subcutaneous injection of 1 M H_2O_2 (Fig. 4.3c). This efficient NIR-II CL nano-system was demonstrated for in vivo H_2O_2 imaging and the sensing of glucose. The proposed electron transfer mechanism would provide a reasonable guideline for the design of novel single-dye NIR-II CL systems.

The use of nanomedicine and nanomaterials to diagnose and treat diseases earlier and more accurately is a key element of current medical science and technology. Many new nanomaterials can fulfill these needs, but at the cost of serious side effects. Diagnostic imaging techniques currently in clinical use include photoacoustic (PA), computed tomography (CT), and magnetic resonance imaging (MRI), with MRI being particularly specialized for its high anatomical and spatial resolution. Translated with www.DeepL.com/Translator (free version) He et al. constructed a tumor/NIR light-responsive nanoplatform PDA/ICG/NONOate-MnO_2 (PINM), in which the biocompatible and pyrolytic polydopamine (PDA) was prepared to load photosensitizer ICG, NONOate, and MnO_2 NPs as nanoenzyme to fabricate PDA/

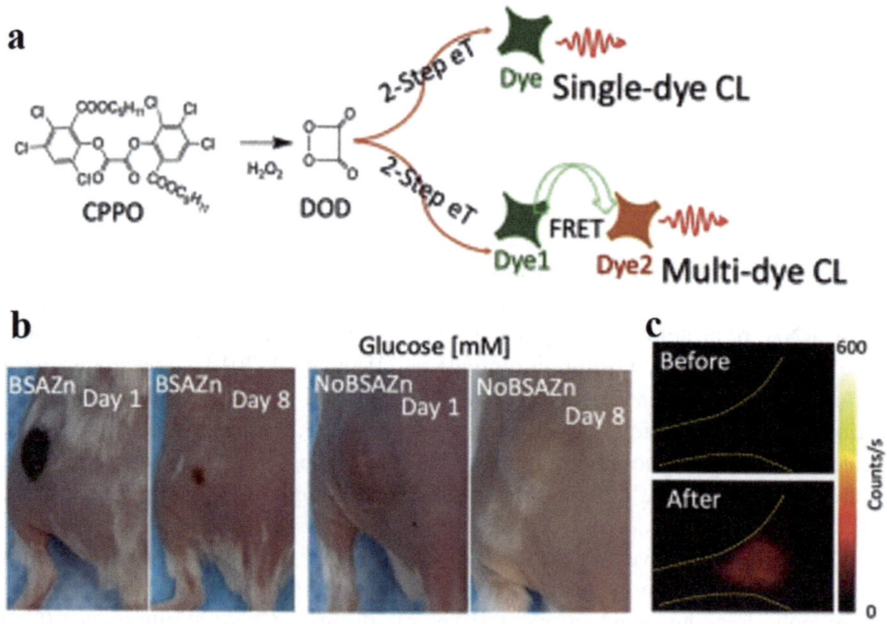

Fig. 4.3 (**a**) The CL mechanism of single-dye CL and multi-dye CL. (**b**) The image of the Gox-Glucose therapyeffect for treating with and without BSAZn nanoenzyme. (**c**) The in vivo CL image of mouse left hind limb before and after injection of 10 mL 1 M H_2O_2, with exposure time 1 s (reproduced from Chang et al. 2023)

ICG/NONOate-MnO_2 (PINM) (Fig. 4.4) (He et al. 2021b). MnO_2 NPs are a common agent for MRI. Therefore, in this study, the tumor was observed by MRI after accumulating PINMH at the tumor site through the HAmediated target. As shown in (Fig. 4.4b), T2-weighted MRI signals in tumors gradually enhanced with time, and PINMH showed time-dependent tumor accumulation. PINMH The tumor/near-infrared (NIR) light-responsive nanoplatform not only can generate O_2 and NO for

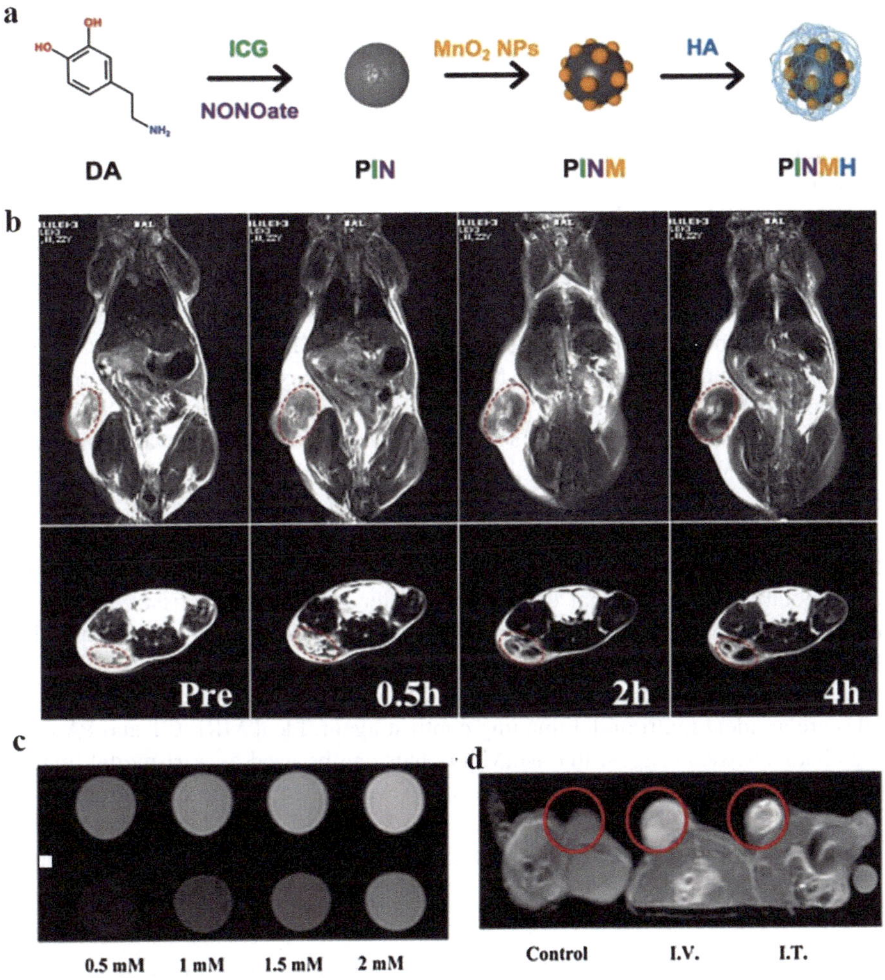

Fig. 4.4 (**a**) The synthetic route of PINMH. (**b**) MR images of PINMH for 0 h, 0.5 h, 2 h, and 4 h. In vitro T1-weighted MRI images of PPy@MnO_2-BSA(Ce6) with different Mn concentrations in citrate buffer solution and GSH solution; (**c**) the T1-MRI signal intensities of PPy@MnO_2-BSA(Ce6) aqueous solutions in citrate buffer solution and GSH solution; (**d**) in vivo T1-weighted MRI images of HT29 tumor-bearing mice before injection, and 6 h after I.V. injection and I.T. injection with PPy@MnO_2-BSA(Ce6). Red circles highlight the tumor site (reproduced from He et al. 2021a; Guo et al. 2020)

remodeling the GME of tumors and phototherapy but also exhibited good biosafety for clinical transformation. Guo et al. also developed a novel nanoenzyme theranostic nanoplatform, named PPy@MnO_2-BSA (bovine serum albumin (BSA)-modified manganese dioxide (MnO_2)/polypyrrole (PPy)) was invented for T1-MRI-guided combined photothermal therapy (PTT) and photodynamic therapy (PDT) of tumor (Fig. 4.4c, d) (Guo et al. 2020). Also, MnO_2, which could reduce the relaxation time of nearby protons, served to enhance the tumor T1-MRI ability. Furthermore, owing to the unique tumor microenvironment-responsive MRI enhancing ability of the MnO_2, the as-synhesized PPy@MnO_2-BSA could serve as an intelligent molecular probe to monitor and guide the precise tumor treatment. The in vitro and in vivo experiments both affirmed the great potential of PPy@MnO_2-BSA(Ce6) nanoparticle in MRI-guided combined PDT and PTT for tumor. This study lays a solid foundation for the theoretical design and clinical application of future therapeutic nanoprobes.

Zhang et al. prepared an "all-in-one" nanoenzyme by thermal injection of Mn ion into the MoOx matrix. MnMoOx showed oxidase (OXD)-like activity and catalase (CAT)-like activity in the tumor microenvironment (Fig. 4.5) (Zhang et al. 2022). Mn^{2+} can be used as a safer contrast agent due to its paramagnetism. The results of the in vitro experiment suggested that the MnMoOx NUs could provide clear contrast under T1-weighted MRI imaging (Fig. 4.5b). Besides, Mo can be used as a CT imaging contrast reagent since it is a heavy element. Their results showed that after injection of the MnMoOx NUs, the CT number was significantly higher than that before injection, and the signal did not completely disappear even after 6 h, proving the long holdup time in the tumor site (Fig. 4.5c). In fact, photoacoustic (PA) imaging has the advantages of deep tissue penetration and high sensitivity. A strong PA signal was detected in the tumor after injection of the MnMoOx Nus. Notably, the PA signal disappeared faster in the muscle tissue than in the tumor, and the NUs were almost completely metabolized after 6 h. The results show that the MnMoOx NUs have an appropriate residence time at the tumor site to promote tumor imaging and treatment (Fig. 4.5d). These data suggest that MnMoOx NUs are an ideal multimodal imaging contrast agent. Ideal MRI, CT and PA imaging characteristics suggest that MnMoOx Nus can be used as a tri-modal imaging contrast agent. This study comprehensively investigated the material structure and integrated therapeutic effect of nanoenzymes designed by introducing point defects, providing a theoretical basis for constructing the next generation of nanoenzymes.

4.3.3 *Enzyme Activity-Based Nanozyme Therapy System*

In living organisms, numerous biochemical processes occur concurrently within cells via a series of enzymatic reactions. These processes encompass various life functions. Enzyme cascade reactions in biological systems enhance the catalytic efficacy of enzyme reactions due to diminished diffusion barriers, elevated local concentrations of intermediates, and superior mass transfer efficiency. Nonetheless,

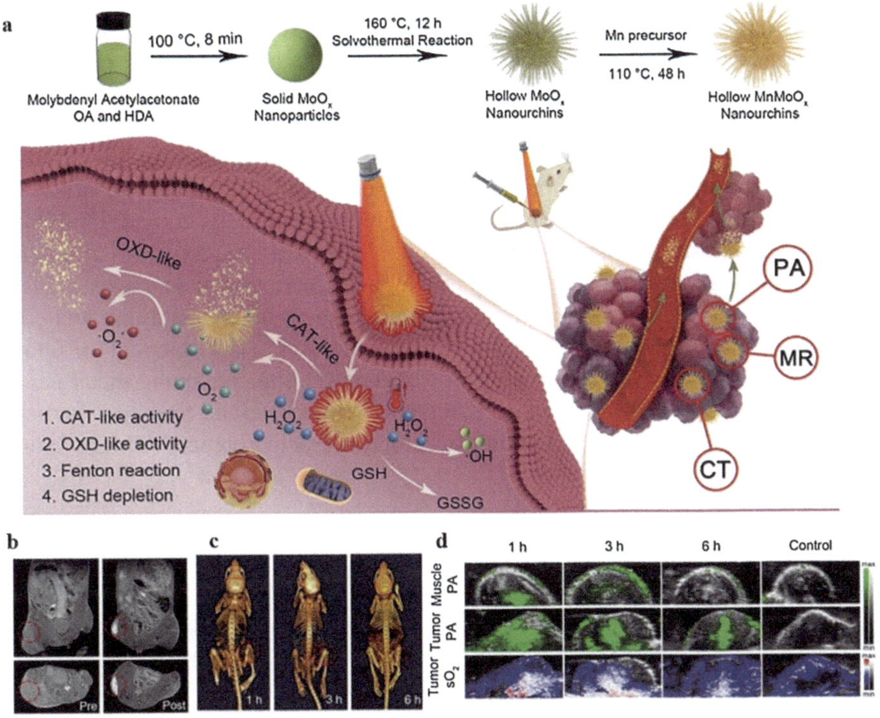

Fig. 4.5 (**a**) Illustration of the preparation procedure and working mechanisms of the MnMoOx NUs. (**b**) T1-weighted MRI beforeand after intratumor injection. (**c**) CT images of MnMoOx NUs solutions with different concentrations in PBS. (**d**) PA imaging of tumor oxygenation change and PA images of the muscle of tumor-bearing mice after being in situ injected with MnMoOx NUs at different time points (reproduced from Zhang et al. 2022)

natural enzymes' inherent limitations, such as stringent reaction conditions and poor stability, significantly impede their broader application in the biomedical field. As an alternative to natural enzymes, nanozymes exhibit diverse enzymatic activities. By incorporating multiple nanozymes into a platform through various methodologies, a cascade catalytic system can be established to effectively boost catalytic efficiency. Additionally, certain nanozymes can emulate various enzyme activities, offering new prospects for developing biomimetic self-cascade reaction systems. Cascade pathways generally follow a sequential pattern where the reaction product of one enzyme serves as a substrate for another enzyme. For instance, GOx or GOx-like nanozymes can initiate a reaction by catalyzing the production of substantial H_2O_2 molecules from glucose within tumors. Subsequently, the nanozyme participates in the catalytic reaction using H_2O_2 molecules as substrates to generate OH through POD-like activity. The strong oxidative OH can be utilized for antitumor therapy or biosensing. Building upon this foundation, an increasing number of cascaded nanodiagnostic and therapeutic systems have been developed in recent years to amplify tumor effects.

Iron apoptosis is an iron-dependent non-apoptotic cell death. Combining both apoptotic and non-apoptotic modes of cell death kills sufficient numbers of tumor cells to prevent them from developing intrinsic resistance. Wang et al. prepared a FeCo bimetallic MOF-derived Co_8FeS_8@Co1-xS nanocatalyst. Nanocatalysts exhibit peroxidase (POD)-like activity and oxidative activity in reducing glutathione (Fig. 4.6a) (Wang et al. 2024). The dual active sites of Co_8FeS_8@Co1-xS serve as a self-propagating platform, promoting the production of significant quantities of •OH while simultaneously depleting GSH, thus exerting a synergistic inhibitory effect on apoptosis and ferrotoxicity in tumor cells. Co_8FeS_8@Co1-xS exhibited remarkable synergistic antitumor effects through the induction of apoptosis and ferroptosis in vivo. This compound demonstrates excellent biocompatibility and efficacy as a standalone therapeutic agent, indicating its significant potential for future clinical applications in cancer treatment. Metal-active site-containing metal nanocatalysts have shown excellent performance in various catalytic reactions, particularly their specific response to the unique tumor microenvironment (TME). Zhong

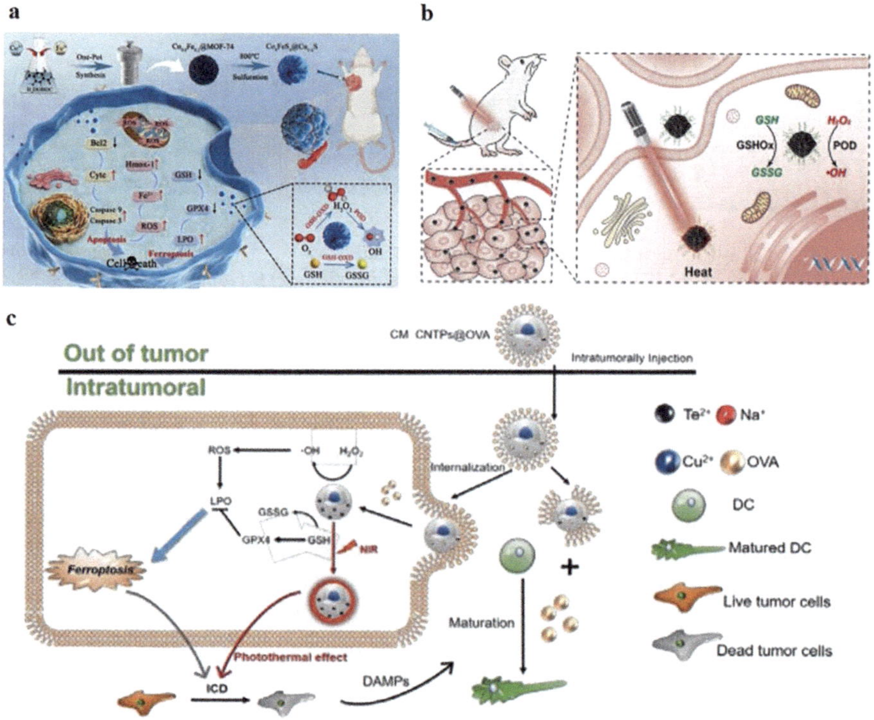

Fig. 4.6 (**a**) Schematic illustration of the successive synthetic procedures of high peroxidase-like and glutathione oxidase activities of Co8FeS8@Co1-xS nanozyme for apoptosis—ferroptosis synergistic tumor therapy; (**b**) Schematic illustration of Bi-NC@HA for cancer therapy through synergetic catalytic and photothermal therapy; (**c**) Scheme of CM CTNPs@OVA behaviors in vivo for catalytic immunotherapy combined with PTT (reproduced from Wang et al. 2024; Zhong et al. 2023; Fang et al. 2023)

et al. fabricated bismuth nanoclusters loaded onto nitrogen-doped porous carbon (Bi-NC) as a cancer therapy nanocatalyst (Fig. 4.6b) (Zhong et al. 2023). The Bi-NC@HA nanocatalyst demonstrated outstanding dual peroxidase (POD) and glutathione oxidase (GSHOx)-like enzymatic activity. Cellular studies revealed that Bi-NC@HA led to increased intracellular reactive oxygen species (ROS) levels and mitochondrial damage, thereby enhancing cancer cell apoptosis. Furthermore, in vivo studies showed that intravenous administration of Bi-NC@HA in mice resulted in significant tumor suppression under near-infrared light irradiation. This research offers a promising approach for developing high-performance metal nanocatalysts for cancer catalytic therapy.

Tumor cell vaccines represent a crucial and widely employed category of cancer immunotherapy, utilizing tumor cells as the basis for vaccination and offering a promising strategy for cancer treatment. By simulating peroxidase and glutathione peroxidase with tellurium copper nanoparticles, artificial nanocatalase was created to trigger ferroptosis. Based on this, a vaccine (called CM CTNPs@OVA) was designed and established to catalyze immunotherapy (Fig. 4.6c) (Fang et al. 2023). The cascade reaction initiated by the release of copper ions from cuprous telluride nanocatalytic enzymes promotes ferroptosis through the generation of reactive oxygen species (ROS) and depletion of glutathione (GSH), leading to immunogenic cell death (ICD). This process results in the release of damage-associated molecular patterns (DAMPs) and enhances dendritic cell (DC) maturation. The fluorescence intensity of ROS and lipid peroxidation (LPO) in all formulation groups was significantly higher compared to the control group, with the sequence of signal intensity being CM@CTNPs@OVA > CTNPs@OVA > CTNPs > Control. This demonstrates that CTNPs-based nanoparticles can effectively induce ROS and LPO production in B16-OVA cells. Additionally, while the stimulation of tumor antigens released by ICD due to ferroptosis alone was inadequate, the combination with exogenous antigens significantly enhanced the antigen-presenting capacity of mature DC cells, thereby eliciting a robust and prolonged anti-cancer immune response and facilitating systemic antitumor immunity. Moreover, photothermal therapy (PTT)-mediated maturation and anti-cancer immune activation of ICD provided substantial support. The PTT-integrated strategy utilizing ferroptosis-activated vaccines has successfully mitigated the immunosuppressive tumor microenvironment, presenting a promising and attractive approach for cancer treatment.

Triple-negative breast cancer exhibits resistance to conventional radiation and chemotherapy regimens, and currently, there is a paucity of molecularly targeted therapies for this aggressive malignancy. Wang et al. developed a novel amphiphilic pH-sensitive polyethylenimine (PEI) modified with stearic acid and hexahydrophthalic anhydride to encapsulate hydrophobic superparamagnetic iron oxide (SPIO) and IR780, thereby creating an innovative therapeutic diagnostic nanostructure (Wang et al. 2020b). The study revealed that the SPA-SPIO-IR780 group exhibited significant phototherapeutic efficacy in vivo. By improving delivery efficiency and generating hydroxyl radicals, reactive oxygen species (ROS), and temperature elevation through SPIO, it achieved notable therapeutic results. Additionally, localized hyperthermia at the tumor site enhances oxygenation-a crucial factor in amplifying photodynamic therapy effects. This research highlights the considerable potential of this therapeutic diagnostic nanomaterial system in cancer diagnosis and treatment.

4.3.4 The Combination of Nanozyme and Photodynamic Therapy

Photodynamic therapy (PDT) involves the energy transition of a photosensitizer upon absorbing light at specific wavelengths, leading to electron generation. When these electrons interact with surrounding oxygen, reactive oxygen species (ROS) are produced, which subsequently oxidize cellular biomolecules like lipids, proteins, and DNA, ultimately causing tumor cell apoptosis. Recognized for its precision, control, non-invasiveness, and efficiency, PDT has emerged as a preferred therapeutic approach in medical research. To enhance PDT efficacy, particularly by mitigating hypoxia, numerous nanomaterials have been employed. For instance, perfluorocarbons, known for their high oxygen-carrying capacity, facilitate oxygen delivery to tumor sites. Additionally, chemically active materials such as MnO_2, $MnFe_2O_4$, Pd@Pt, Prussian blue nanoparticles, and metal-organic frameworks (MOFs) have been utilized to decompose intratumoral hydrogen peroxide into oxygen.

Li et al. synthesized an organic semiconductor nanocatalyst (OSPE) that can be remotely activated by near-infrared light (Li et al. 2019). Near-infrared light activation not only generates $^1O_2$ for photodynamic therapy, but also specifically initiates the cascade activation of the enzyme progenitor. The OSPE-mediated photolysis of RNA leads to downregulation of transfer-related protein expression and ultimately helps suppress post-treatment metastasis (Fig. 4.7a). By further shifting the light response to the second near-infrared window or integrating it with endoscopic light transmission systems, it is highly probable that deep-seated tumors can be effectively treated using this pre-nanocatalyst. The development of therapeutic agents that regulate hypoxia to enhance treatment efficacy is considered an attractive strategy for photodynamic therapy (PDT) of cancer. Zhang et al. synthesized a multifunctional nanoplatform comprising Pt and BODIPY-modified COF, wherein the in situ loaded Pt imparts peroxidase-like properties to the porous nanoscale COF, facilitating the continuous decomposition of H_2O_2 into O_2 (Fig. 4.7b) (Zhang et al. 2024). The mouse treatment group (IV) showed the best therapeutic effect because the Pt@COF-BDP nanocatalyst enhanced the PDT effect by catalyzing the production of oxygen through the oxidation of H_2O_2, effectively addressing the problem of hypoxia-induced weakened PDT effect in solid tumors. The nanomedicine showed a marked photodynamic effect under laser irradiation, making it an efficient antitumor nanoplatform. Li et al. constructed a biomimetic metal-organic framework (FA-EM@MnO_2/ZIF-8/ICG) drug delivery system that can control the release of the tumor microenvironment (Fig. 4.7c) (Li et al. 2022). The MnO_2 coating on the ZIF-8/ICG nanoplatform surface enables self-sustained oxygen generation within

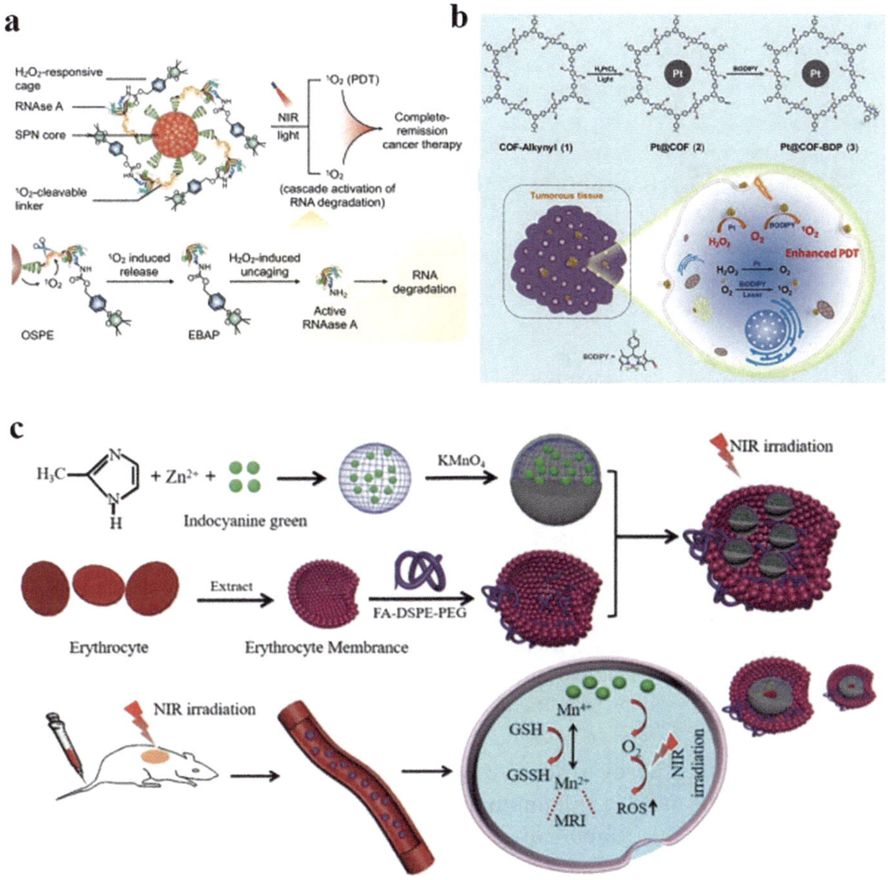

Fig. 4.7 (**a**) Illustration of the proposed mechanism for the photoactivated synergistic therapeutic action of OSPE including PDT and intracellular RNA degradation. (**b**) Schematic illustration of the preparation and application of 3 to enhance PDT. Illustration for the preparation of a FA-EM@ MnO2/ZIF-8/ICG nanoplatform for MRI imaging-guided cancer therapy. (**c**) The construction of FA-EM@MnO_2/ZIF-8/ICG nanoplatform through drug loading, MnO_2 decoration and erythrocyte membrane coating and proposed action mechanism of FA-EM@MnO2/ZIF-8/ICG nanoplatform in a mouse tumor model (reproduced from Li et al. 2019, 2022; Zhang et al. 2024)

the tumor microenvironment (TME) under hydrogen peroxide conditions, thereby augmenting phototherapy efficacy. Both in vitro and in vivo cellular experiments demonstrated that the FA-EM@MnO_2/ZIF-8/ICG + H_2O_2 + NIR group exhibited the highest levels of reactive oxygen species (ROS). This system presents promising potential for photodynamic tumor therapy and holds significant clinical application prospects.

Photodynamic therapy (PDT) for cancer is constrained by tumor hypoxia. Liu et al. devised a liposome system named nano-Pt/VP@MLipo and introduced a novel nanoparticle delivery approach that leverages CAT-like activity and

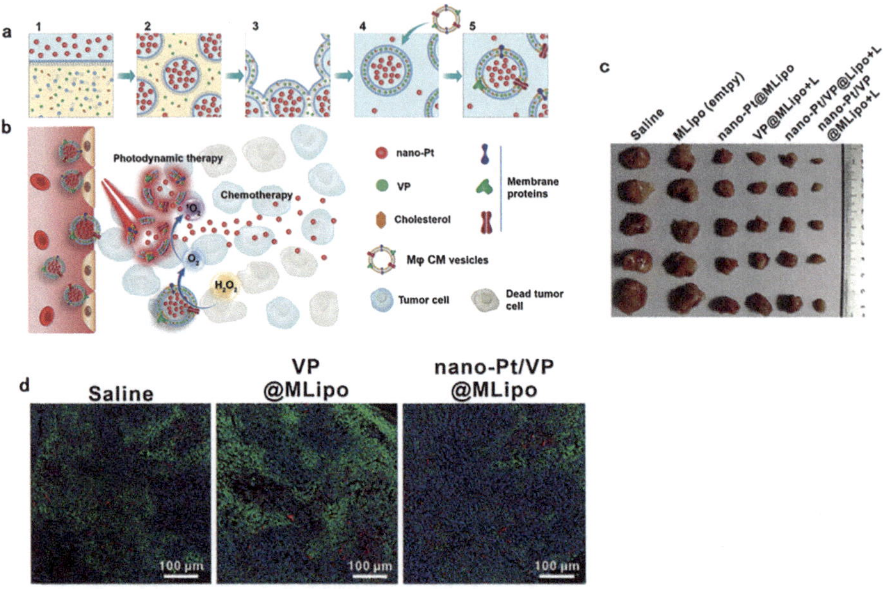

Fig. 4.8 Schematic illustration of the fabrication (**a**) of nano-Pt/VP@MLipo and (**b**) chemophototherapy performance in tumors. Antitumor therapy in vivo. (**c**) Photographs of the tumors harvested on day 20. (**d**) Pimonidazole positive hypoxic areas in 4 T1 tumors (reproduced from Liu et al. 2021b)

chemotherapeutic efficacy to achieve synergistic chemophototherapy (Fig. 4.8) (Liu et al. 2021a). The nano-Pt (platinum nanoparticles) function as a nanocatalyst akin to hydrogen peroxide, augmenting photodynamic therapy (PDT) by facilitating oxygen supply. The oxygen produced through nano-Pt catalysis enhances the photodynamic activity mediated by VP. This process renders the liposome membrane more permeable, facilitating the release of nano-Pt. These ultra-small particles, measuring 3–5 nm, can penetrate deeper into tumor tissues, and the generated oxygen further promotes penetration, thereby improving chemotherapeutic efficacy.

Develop a multifunctional effective carrier with enhanced oxidative stress and GSH depletion capacity to break the redox equilibrium in the tumor microenvironment. Feng et al. prepared polyethylene glycol (PEG)-modified mesoporous carbon (MC-PEG) as an ICG carrier and compared it with PEG-modified graphene oxide (GO-PEG) (Fig. 4.9) (Feng et al. 2023). The study revealed that MC-PEG demonstrated threefold greater oxidative capacity via a POD-like mechanism compared to GO-PEG, along with superior and prolonged GSH depletion capabilities, thereby amplifying oxidative stress. Under near-infrared light irradiation, MC-PEG, which exhibits higher photothermal conversion efficiency and augmented enzyme-like activity, significantly enhances the synergistic effect of PDT-PTT in tumors. MC holds significant promise for PTT-PDT tumor therapy, offering valuable insights for the advancement of PDT-based cancer treatments.

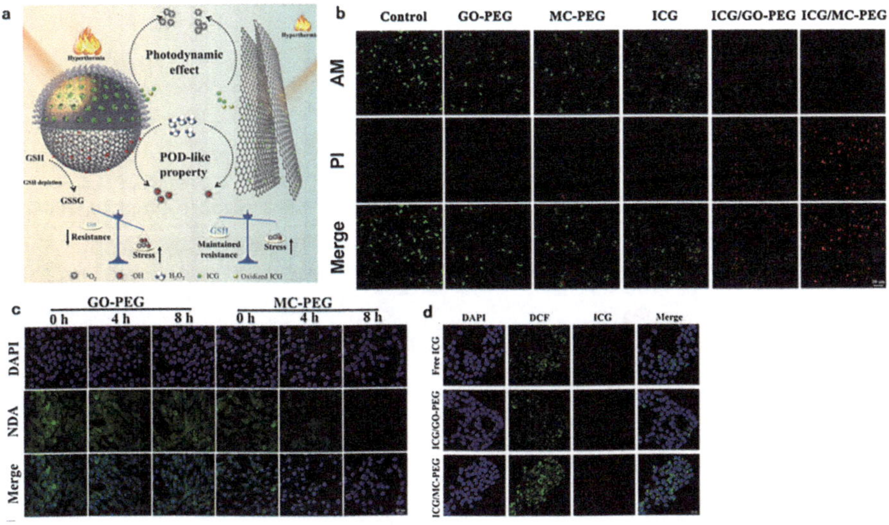

Fig. 4.9 (**a**) Schematic illustration of photothermal-photodynamic synergistic effect of ICG/MC-PEG and ICG/GO-PEG. (**b**) Fluorescence images of 4 T1 cells incubated with GO-PEG, MC-PEG, ICG, ICG/GO-PEG, and ICG/MC-PEG under NIR irradiation at 0.75 W/cm^2 and co-stained with Calcein AM/PI (scar bar: 20 μm). (**c**) Cellular GSH depletion of 4T1 cells incubated with GO-PEG and MC-PEG. (**d**) Fluorescence density of free ICG, ICG/GO-PEG, and ICG/MC-PEG incubated with DAPI and DCF (reproduced from Feng et al. 2023)

Hypoxia in malignant tumors not only induces tumor angiogenesis, invasion, and irreversible metastasis, but also reduces the efficacy of many treatment methods, including photodynamic therapy (PDT). They developed a phototherapy system based on tumor homing penetration and autogenous oxygenated graphene (GMtC) for highly efficient and tumor microenvironment-sensitive hypoxic solid tumors (Fig. 4.10) (Wu et al. 2022a, b). The 4T1 cancer cells were subjected to incubation with GMtC and their respective controls under hypoxic conditions. The oxygen levels within the cells were monitored utilizing an indicator. A comparative analysis of the red fluorescence intensity revealed that the GMPC and GMtC groups exhibited significantly lower red fluorescence intensity compared to the GOtC group. This observation indicates that the MnO_2 nanocatalyst can effectively generate oxygen within the cellular environment. The results of the MTS assay demonstrated the superiority of GMtC as a highly penetrating and effective nanophotosensitizer for selective PDT of deep-seated tumor cells. The hypoxic distribution in the GMPC-treated group was also significantly reduced compared to the GMtC-treated group. The addition of manganese dioxide significantly increased the production of oxygen, thereby enhancing the generation of reactive oxygen species (ROS) in photodynamic therapy (PDT). In conclusion, this study underscores the substantial potential of the two-dimensional treatment and diagnostic system for addressing deep-seated hypoxic tumors.

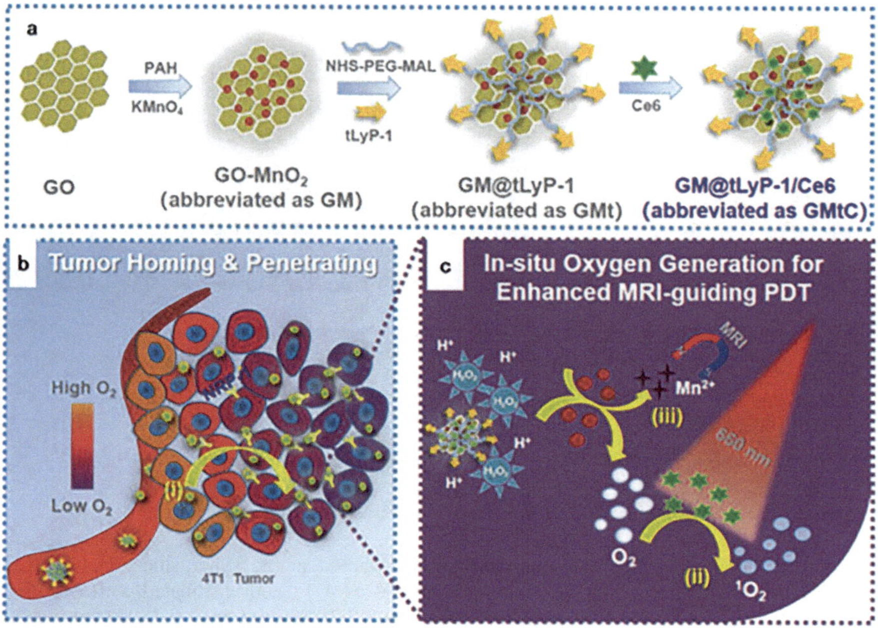

Fig. 4.10 (**a**) Schematic illustration of the assembly of GMtC and (**b**, **c**) its mediated highly efficient MRI-guiding PDT by the integration strategy of tumor homing-penetrating and in situ oxygen generation (reproduced from Wu et al. 2022a, b)

Tumor hypoxia impairs the effectiveness of photodynamic therapy (PDT) due to the critical role of local oxygen concentration in the production of cytotoxic singlet oxygen ($^1O_2$). This study presents multifunctional mesoporous nanoenzymes (NEs) derived from metal-organic frameworks (MOFs) designed to generate endogenous oxygen in situ under bioimaging guidance to enhance the efficacy of photoconductive radiotherapy (PDT) (Fig. 4.11). The mesoporous NE is fabricated by initially coating a manganese-based MOF with mesoporous silica, followed by a simple annealing process in ambient conditions. Subsequently, after removing the mesoporous silica shell and post-modifying with polydopamine and poly (ethylene glycol) to enhance biocompatibility, the resulting mesoporous NE is loaded with chlorin e6 (Ce6), a widely utilized photosensitizer in PDT, achieving a high loading capacity. Upon O_2 generation through the catalytic reaction between the catalytic amount of NE and endogenous H_2O_2, the hypoxic tumor microenvironment is alleviated. Consequently, Ce6-loaded NE functions as a H_2O_2-activated oxygen supplier, increasing the local O_2 concentration and significantly enhancing antitumor PDT efficacy both in vitro and in vivo. Additionally, the NE exhibits T2-weighted magnetic resonance imaging capabilities for in vivo tracking. T This study demonstrates the interesting biomedical applications of MOF-derived mesoporous NE as a multifunctional therapeutic agent in cancer treatment.

Photodynamic therapy (PDT) is a promising strategy for the treatment of tumors; however, the efficacy of PDT is often limited due to insufficient cytotoxicity against

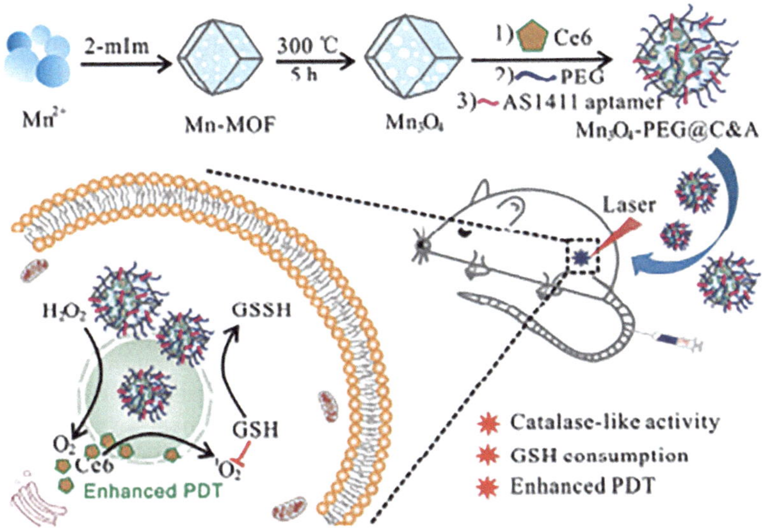

Fig. 4.11 A schematic of the preparation of Mn_3O_4-PEG@C&A nanoparticle for enhanced nucleus-targeted PDT. The nucleus-targeted Mn_3O_4 PEG@C&A particle can catalyze intracellular H_2O_2 to O_2 and consume intracellular GSH simultaneously, which leads to high PDT efficacy (reproduced from Zeng et al. 2020)

specific subcellular organelles and challenging properties of the tumor microenvironment, such as hypoxia and elevated glutathione levels. In this study, we developed a novel nanoenzyme derived from a metal-organic framework (MOF) with inherent catalase-like activity, capable of decomposing hydrogen peroxide into oxygen and simultaneously consuming glutathione to enhance PDT efficacy. The resulting Mn_3O_4 nanoparticles exhibit a larger pore size and greater surface area compared to native MOF particles, allowing for the loading of higher doses of photosensitizers. Upon modification with AS1411 aptamer and polyethylene glycol (PEG), the Mn_3O_4-PEG@C&A particles demonstrate superior stability and targeted delivery to the cell nucleus. Notably, in a mouse model, the Mn_3O_4-PEG@C&A particles effectively inhibited tumor growth with high efficacy and no observed biotoxicity. This research marks the first application of MOF-derived nanoparticles for nucleus-targeted PDT, potentially offering a novel approach for the design of functional nanoenzymes aimed at subcellular organelle-targeted tumor modulation.

4.3.5 The Combination of Nanozyme and Photothermal Therapy

In recent years, photothermal therapy (PTT) has emerged, leveraging the superior light absorption and high photothermal conversion efficiency of photothermal agents. These agents absorb light energy and efficiently transform it into heat energy, elevating the temperature of localized tumors. This process leads to the disruption of tumor cell membranes or protein denaturation, all while minimizing harm to surrounding healthy cells (Chen et al. 2009; Dariva et al. 2019). In recent years, photothermal therapy (PTT) has received widespread attention.

Photothermal therapy (PTT) represents a form of thermal therapy resulting from the conversion of near-infrared (NIR) light into heat through the action of photothermal agents. Its notable capabilities include deep penetration, high efficiency, and non-invasiveness, leading to widespread interest. This study involves the modification of glucose oxidase (GOx) onto nitrogen-doped carbon nanoparticles (NPs), creating a bionic nanozyme (NC@GOx NPs). This modification aims to enhance tumor PTT by promoting starvation therapy. GOx disrupts the supply of energy and nutrients through decomposition, inducing starvation treatment, thereby reducing the levels of adenosine triphosphate (ATP) and down-regulating heat shock proteins. These actions create a more favorable microenvironment for improving PTT efficacy. Additionally, the endogenous hydrogen peroxide (H_2O_2) generated can be transformed into hydroxyl radicals, which target cancer cells. Both in vitro and in vivo experiments have demonstrated that NC@GOx NPs effectively eliminate cancer cells and tumors (Fig. 4.12) (Wu et al. 2022a, b). NC nanoparticles (NPs) were initially synthesized via the template method, followed by the preparation of GoX-modified NC (NC@GOx) NPs through electrostatic interaction. Upon entering tumor cells, GOx catalyzes the conversion of glucose to gluconic acid and hydrogen peroxide (H_2O_2), thereby depleting energy and nutrient supplies and initiating starvation therapy. NC NPs are capable of generating heat and enhancing photothermal therapy (PTT) under near-infrared (NIR) laser irradiation. Consequently, NC@GOx NPs demonstrate significant synergistic effects encompassing starvation therapy, PTT, and chemodynamic therapy (CDT) against tumors. A biomimetic nanoenzyme was designed to concurrently promote starvation therapy, photothermal therapy, and chemodynamic therapy of tumors by synthesizing GOX-modified NC NPs. Once GOx enters tumor cells, it first decomposes glucose into H_2O_2, facilitating starvation treatment. Concurrently, the reduction in glucose levels decreases the ATP concentration, downregulates HSP70 expression, and augments the photothermal treatment efficacy of NC NPs. Additionally, the generated H_2O_2, along with endogenous H_2O_2, can be transformed into hydroxyl radicals (OH•) through the peroxidase activity of NCs, thereby enhancing the effectiveness of chemodynamic therapy. NC@GOx NPs effectively eliminate tumor cells through the combined impact of starvation therapy, photothermal therapy, and chemodynamic therapy.

This study reports the synthesis of semiconductor polymer nanozymes endowed with near-infrared (NIR) photothermal properties. To demonstrate its proof-of-concept application, the in situ and on-demand activation of these nanozymes was performed in live mice to enhance cancer therapy. The nanoenzyme comprises two crucial components: the semiconductor polymer Amphiphilic serves as a photothermal nanosensor, while bromelainase (Bro) functions as a temperature-sensitive enzyme (Fig. 4.13) (Li et al. 2018). Bro (Molecular Weight: 33,000) is a protease exhibiting optimal activity at approximately 45 °C. Consequently, under near-infrared (NIR) laser irradiation, semiconductor polymer nanoenzymes can effectively

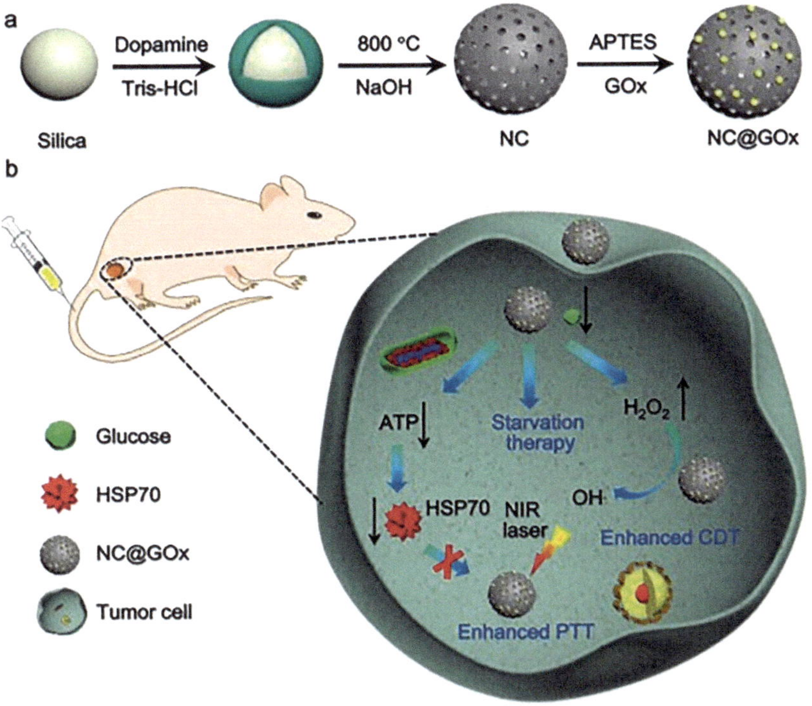

Fig. 4.12 (**a**) Synthetic method oNC@GOx NPs and (**b**) NC@GOx NPs for treatment. GOx has a catalytic effect after entering tumor cells glucose-induced starvation treatment reduced ATP level and produced H_2O_2, in which the decreased ATP level led to the downregulation of HSP expression and enhanced the performance of PTT, and at the same time, H_2O_2 was produced, which could catalyze the production of OH• to improve the performance of CDT (reproduced from Xu et al. 2022)

convert photothermal energy, leading to a localized temperature rise and consequently activating collagen digestion. Given that collagen is the predominant extracellular matrix (ECM) protein in tumors, iso-photothermally induced collagen digestion can augment nanoparticle accumulation within the tumor, thereby enhancing photothermal therapy (PTT). Upon systemic administration of PCB1-Bro to live mice, NIR light can remotely trigger in situ collagen digestion, thereby increasing nanoparticle accumulation in tumors. PCB1-Bro generated higher photothermal temperatures, resulting in superior PTT efficacy compared to its unbound precursor (PCB1). To our knowledge, this study presents the first instance of photothermally activated enzymatic activity of an organic nanoenzyme. Although the current nanoparticles are more suited for surface PTT due to the limited penetration depth of 808-nm laser, the treatment remains feasible through integration within the tumor combined with other light delivery methods. Beyond PTT, our approach can also enhance the drug delivery effects of other therapeutic modalities, such as photodynamic therapy, chemotherapy, and gene therapy. Non-invasive remote control of

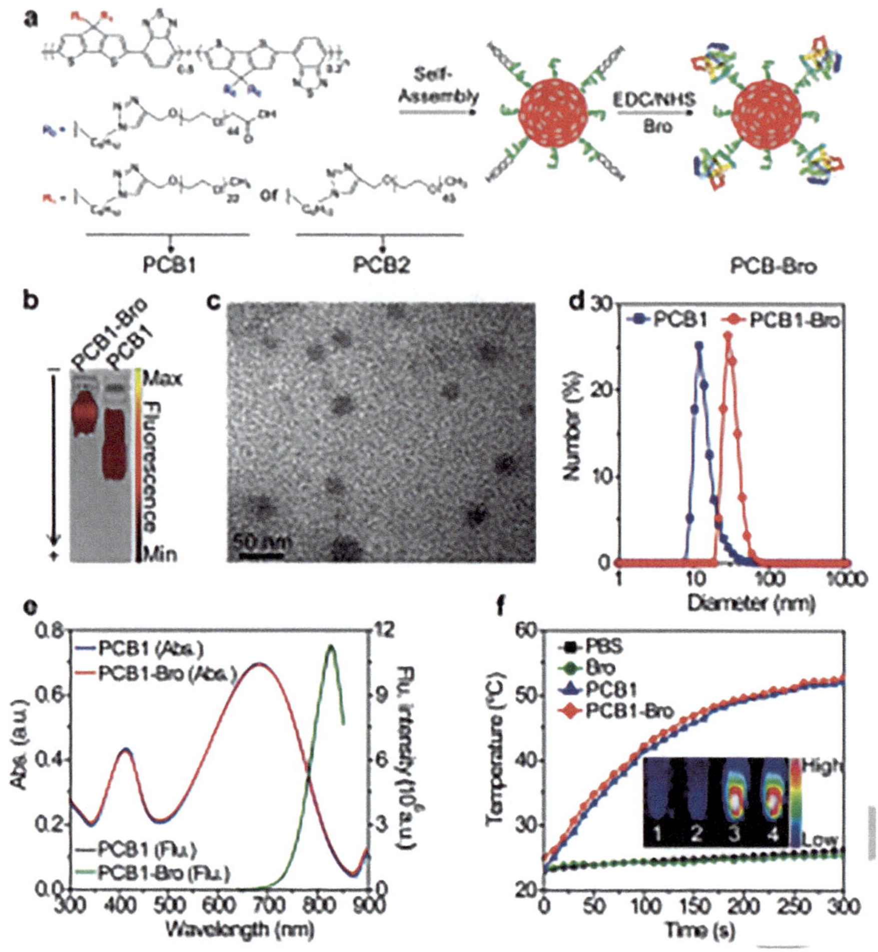

Fig. 4.13 (**a**) Chemical structures of PCB1 and PCB2 and schematic for the synthesis of PCB-Bro. (**b**) Agarose gel electrophoresis of PCB1 and PCB1-Bro. (**c**) Representative TEM image of PCB1-Bro. (**d**) Dynamic light scattering profiles of PCB1 and PCB1-Bro. (**e**) UV-vis absorption and fluorescence spectra of PCB1 and PCB1-Bro. (**f**) Temperature increase of PBS, Bro, PCB1 and PCB1-Bro as a function of laser irradiating time. Inserts are the thermal images of PBS (1), Bro (2), PCB1 (3), and PCB1-Bro (4) at their respective maximum temperatures (reproduced from Li et al. 2018)

enzyme activity in living systems has seldom been demonstrated and utilized therapeutically. Here, we have developed a photothermally active semiconductor polymer nanoenzyme designed to digest collagen, thereby enhancing cancer therapy. Under NIR light irradiation, the activity of nanoenzymes can be enhanced by 3.5 times to efficiently digest collagen in the tumor ECM, thereby increasing nanoparticle accumulation in the tumor and improving PTT.

Due to hypoxia in the tumor microenvironment, satisfactory therapeutic effect cannot be achieved. Consequently, it is imperative to create an appropriate nanoplatform to synergize photothermal therapy (PTT) with hypoxia reversal and innate immune activation. The integration of PTT with hypoxia reversal and stimulator of interferon genes (STING) activation could have significant implications for cancer treatment. As such, near-infrared (NIR)-responsive nanozymes (Pb-Mn/OVA NEs) were developed by incorporating manganese into OVA-templated Pb NPs (Fig. 4.14a, b) (Li et al. 2023). OVA was utilized not only as a template for assembling PB nanoparticles (NPs) but also as a model protein to examine the antigen cross-presentation capability of dendritic cells (DCs). Concurrently, manganese ions were incorporated into PB NPs via ion-exchange reactions. Upon exposure to 808-nm near-infrared (NIR) light, the prepared PB-Mn/OVA nano-envelopes (NEs) could induce immunogenic cell death (ICD) in tumor cells through localized hyperthermia and stimulate the release of numerous damage-associated molecular patterns (DAMPs). Additionally, the PB-Mn/OVA NEs, characterized by high catalytic activity, generate oxygen and alleviate the hypoxic tumor microenvironment. Furthermore, manganese ions released from the degradation of PB-Mn/OVA NEs activate the cyclic GMP-AMP synthase-stimulator of interferon genes (cGAS-STING) pathway, promoting DC maturation and antigen cross-presentation. Consequently, cytotoxic $CD8^+$T cells and memory T cells were significantly activated in the spleen and tumor-draining lymph nodes. Overall, this study presents a straightforward and versatile approach to developing PB NP-based nanoplatforms that integrate functional proteins and noble metals for combined tumor therapy. The PB-Mn/OVA NEs, with their high catalase activity, produce substantial amounts of oxygen, mitigating the hypoxic microenvironment and enhancing the therapeutic efficacy of photothermal therapy (PTT). Importantly, PB-Mn/OVA NEs can facilitate DC maturation and tumor-specific antigen presentation through the cGAS-STING pathway, thereby further activating cytotoxic T lymphocytes (CTL) and memory $CD8^+$ T cells for prolonged immune activation. Moreover, Mn^{2+} doping

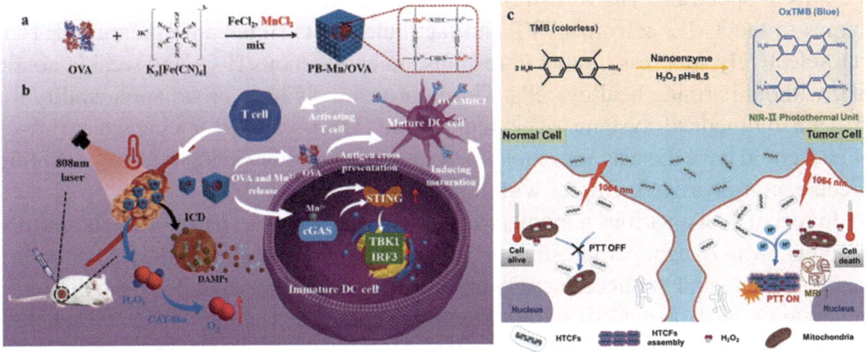

Fig. 4.14 Schematic illustration of synthesis process (**a**) and the antitumor effect of PB-Mn/OVA NEs (**b**). Illustration of HTCFs as an intelligent agent for MR imaging enhancement and photothermal therapy in NIR-II window (**c**) (reproduced from Chen et al. 2022)

confers T1-weighted magnetic resonance imaging (MRI) capabilities to PB-Mn/OVA NEs, enabling real-time in vivo tracking. This study offers a promising therapeutic strategy that combines STING agonists and hypoxia reversal to synergize PTT and antitumor immune activation.

Photothermal therapy (PTT) represents an innovative approach in tumor treatment, distinguished by its ability to selectively eradicate localized malignancies. In this modality, photosensitizers with high photothermal conversion efficiency concentrate within the tumor, transforming light energy into thermal energy upon laser exposure. Consequently, localized hyperthermia effectively destroys tumor cells, achieving in situ tumor ablation. Compared to conventional treatments such as surgery, chemotherapy, and radiotherapy, PTT offers distinct benefits including minimal invasiveness, reduced pain, and precise temporal and spatial control. Additionally, recent studies indicate that PTT can induce immunogenic cell death (ICD) in tumor cells, prompting the release of danger-associated molecular patterns (DAMPs) and tumor-associated antigens (TAA) from dendritic cells, which subsequently activate antitumor immune responses. However, eliciting a robust antitumor immune response through PTT remains challenging primarily due to hypoxia in the tumor microenvironment and suboptimal antigen recognition and presentation. Notably, Chen et al. have developed a TME-activated PTA HTCF predominantly composed of CeO_2-Fe_3O_4 hybrid magnetic nanozymes acting as peroxidase mediators and TMB serving as H_2O_2-activated prodrugs, featuring prolonged plasma residence time and targeted cellular delivery (Fig. 4.14c). Through peroxidase-mediated oxidation, HTCF achieves a substantial NIR-II absorption ranging from 1000 to 1200 nm with high photothermal properties, leading to a specific "on" response. In addition, this nanoenzymes-based platform has been cleverly designed to enhance magnetic resonance imaging (MRI) by exploiting the self-assembly of superparamagnetic htcf in tumor cells. To the best of our knowledge, this is the first nanoenzyme-based platform for enhanced MRI-guided photothermal therapy (Fig. 4.14c) (Chen et al. 2022). They have developed a therapeutic agent utilizing TME-activated nanoenzymes to enhance MRI-guided photothermal therapy within the NIR-II spectrum. The formation of TMB dimer, catalyzed by peroxidase nanozymes in the TME, activates PTT and facilitates MRI signal amplification. HTCF can selectively suppress tumors in response to endogenous H_2O_2 and weakly acidic pH, without harming healthy cells. This approach may have broad applicability for PTT in biomedical systems. The remarkable TME-responsive characteristics of HTCF could pave the way for innovative nanoenzyme-mediated platforms, enabling precise, non-invasive therapies within the NIR-II window.

In recent years, various nanomaterials, including catalase, perfluorocarbon nanodroplets, MnO_2 nanoparticles/nanosheets/nanoflowers, carbon nitrides, platinum nanoparticles, and semiconductor polymer nanoparticles, have been employed to generate O_2 or deplete GSH within the tumor microenvironment (TME). However, these approaches are hindered by challenges such as the instability of biological enzymes, rapid degradation of MnO_2, transient O_2 production or GSH depletion, and the necessity for external laser or ultrasound activation. These limitations impede effective regulation of tumor hypoxia and reduction. Additionally, this ad hoc strategy does not consistently and synchronously enhance the efficacy of ROS-based therapies. Consequently, a more sustainable approach is required to modulate the TME. To address this need, a biocompatible nanoplatform [$MnFe_2O_4$@metal-organic framework (MOF)] has been developed. This platform utilizes a porphyrin-based MOF coating as a photosensitizer and manganese ferrite nanoparticles ($MnFe_2O_4$) as a nanoenzyme (Fig. 4.14) (Yin et al. 2019). The synthesized $MnFe_2O_4$@MOF nanoplatform demonstrates catalase and glutathione peroxidase-like activities. Upon internalization into the tumor, this nanoplatform continuously catalyzes H_2O_2 to generate oxygen via the cyclic Fenton reaction, mitigating tumor hypoxia. Concurrently, $MnFe_2O_4$@MOF, in conjunction with the Fenton reaction, consistently consumes GSH in the presence of H_2O_2, thereby minimizing ROS consumption during photodynamic therapy (PDT), resulting in effective therapeutic outcomes both in vitro and in vivo. This approach provides sustained relief from tumor hypoxia, thereby enhancing the efficacy of PDT. Additionally, $MnFe_2O_4$@MOF, leveraging its glutathione peroxidase-like activity, depletes GSH, thereby reducing ROS consumption during laser irradiation, leading to improved PDT efficacy. Moreover, $MnFe_2O_4$@MOF is utilized in magnetic resonance imaging (MRI) for image-guided precision cancer therapy. Consequently, $MnFe_2O_4$@MOF emerges as a promising nanoplatform capable of continuously modulating the tumor microenvironment (TME) to enhance ROS-based therapies. Previous reports have rarely demonstrated nanomaterials capable of simultaneously producing O_2 and depleting GSH in such a sustained manner. Notably, our $MnFe_2O_4$@MOF effectively modulates the TME by concurrently supplying O_2 and consuming GSH in the presence of H_2O_2. Furthermore, $MnFe_2O_4$@MOF serves as an MRI contrast agent to accurately visualize tumors for image-guided precision cancer therapy, enabling selective irradiation of specific tumor regions without damaging adjacent healthy tissues or inducing toxic side effects. $MnFe_2O_4$@MOF possesses both catalase and glutathione peroxidase-like activities and functions as a H_2O_2-responsive nanostructure that continuously self-generates O_2 within the TME while diminishing GSH levels, thereby augmenting the antitumor therapeutic efficacy of PDT.

Hypoxia plays a crucial role within the tumor microenvironment (TME), significantly influencing tumor proliferation and metastasis, thereby posing a substantial challenge to effective cancer treatment. Integrating hypoxia regulation with imaging-based monitoring presents a promising approach that could potentially enhance therapeutic outcomes. In this context, a nanoenzyme comprising Prussian blue (PB) was synthesized as a metal-organic framework (MOF) to accommodate a second near-infrared (NIR-II) small molecule dye, IR1061 (Fig. 4.15a), which catalyzes hydrogen peroxide to oxygen and provides a photothermal conversion element for photoacoustic imaging (PAI) and photothermal therapy (PTT) (Fig. 4.15) (Xue et al. 2023). To enhance stability and biocompatibility, silica was employed as a coating for an integrated nanoplatform (SPI). SPI effectively mitigates the hypoxic characteristics of the tumor microenvironment (TME), thereby inhibiting tumor cell migration and down-regulating the expression of heat shock protein 70 (HSP70). This amplifies the near-infrared II (NIR-II) photothermal therapy (PTT) effect both in vitro and in vivo, guided by NIR-II photoacoustic imaging (PAI). Additionally,

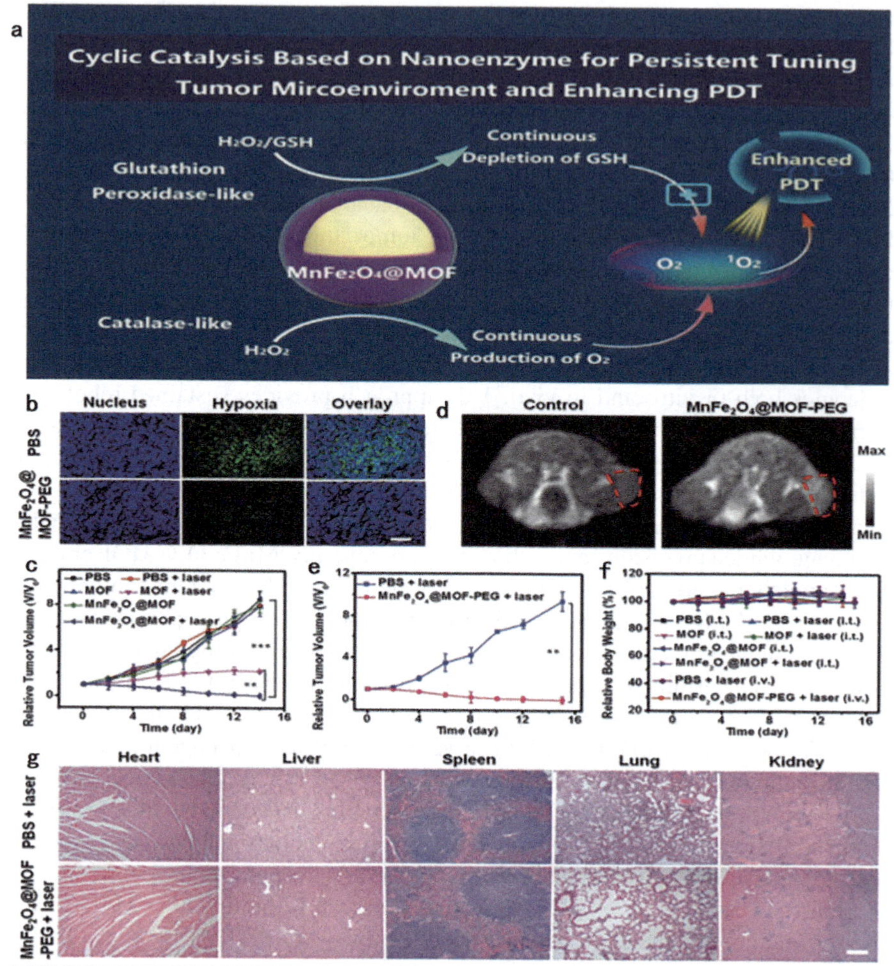

Fig. 4.15 (**a**) $MnFe_2O_4$@MOF Schematic representation of continuous provision of O_2 and depletion of GSH for effective enhancement of PDT. (**b**) HIF-1α staining of tumor sections treated with PBS or $MnFe_2O_4$@MOF-PEG. (**c**) Tumor growth curves of each group after laser irradiation. (**d**) In vivo T1-weighted MR Images of mice 24 h after administration. Injections of PBS or $MnFe_2O_4$@MOF-PEG were given. (**e**) Tumor growth curves after intravenous administration in each group. (**f**) Changes in body weight after the assigned treatment in each group. (**g**) H&E stained images of major organs 2 weeks after IV injection (reproduced from Yin et al. 2019)

label-free multispectral PAI facilitates real-time assessment of SPI as a potential tumor therapy. Clinical histological analysis confirms the therapeutic efficacy of this amplification. Consequently, SPI combined with PAI offers a novel approach for the diagnosis, treatment, and monitoring of tumors.

Photothermal therapy (PTT) is widely recognized as one of the most effective cancer treatments, utilizing laser near-infrared (NIR) light to convert into thermal energy, leading to the selective ablation of local tumor cells. However, it is important to note that the necrosis induced by high temperatures post-PTT can trigger an excessive inflammatory response and activate the self-protective mechanisms of the tumor, potentially resulting in metastasis and recurrence. A metal-organic framework (MOF) incorporating FDA-approved Prussian blue (PB) was synthesized to function as a nanoenzyme, mimicking the activities of peroxidase, catalase, and superoxide dismutase. Furthermore, a second near-infrared (NIR-II) small molecule dye, IR1061, was loaded onto the platform as a photothermal element (Fig. 4.16a, b). The NIR-II absorbance of IR1061 enabled an efficient high photothermal conversion for PAI, allowing precise and accurate in vivo assessment of SPI's behavior. SPI was observed to alleviate hypoxic TME both in vitro and in vivo, thereby inhibiting tumor cell migration and enhancing the PTT effect through HSP70 downregulation. PAI enables non-invasive detection of spatially resolved perfusion imaging of SPI and blood oxygen levels within the TME. Based on label-free PA imaging and histopathological analysis, SPI was found to inhibit tumor cell migration and significantly enhance the therapeutic effects of NIR-II PTT, including hypoxia inhibition and apoptosis-mediated anticancer activity. Moreover, SPI demonstrated good biosafety both in vitro and in vivo. These findings suggest that SPI holds significant potential for application in NIR-II PA image-guided tumor therapy.

4.3.6 The Combination of Nanozyme and Chemotherapy

Cancer is one of the leading causes of morbidity and mortality in the world today, and there is an urgent need to develop effective treatments. So far, traditional cancer treatment, including surgery, radiation, and chemotherapy, has faced various limitations and side effects. Chemo-kinetic therapy (CDT) is an evolving cancer treatment technology that kills tumor cells with nanoparticles containing transient metals (NPS) catalyzed by reaction. Fenton's effective response and good delivery of NPS nanoenzymes are key to successful cancer treatment. The opposite is true of the opposite is true of the opposite. We have developed a new type of nanoenzyme $MnFe_2O_4$@GFP for external and internal anti-neoplastic therapy (Fig. 4.17a). New $MnFe_2O_4$ nanoparticles containing two transient metal elements Fe and Mn were synthesized and used to strengthen Fenton's response. With the help of polymerization of dopamine, green fluorescent protein is adsorbed and covered with peg-strokes to obtain a common delivery of proteins with high biological compatibility. The abundance of H_2O_2 and glutathione (GSH) in oncological tissues provides a favorable microenvironment for the production of toxic free radicals at the source. Fe^{3+} and GSH cause an oxidative reductionist reaction that produces Fe^{2+}, which catalyzes H_2O_2 in •OH, while simultaneously consuming GSH antioxidants. By linking Fe^{3+} to another catalyst, the catalytic efficiency of nanoenzymes has increased significantly. In this way, nanoenzymes demonstrate an effective antineoplastonic ability both outside the body and inside. Thus, the multifunctional platform with nanofermentation demonstrates great prospects in antineoncological therapy by combining the catalyst Fe^{3+} and Mn^{2+} as well as the overall delivery of protein carriers.

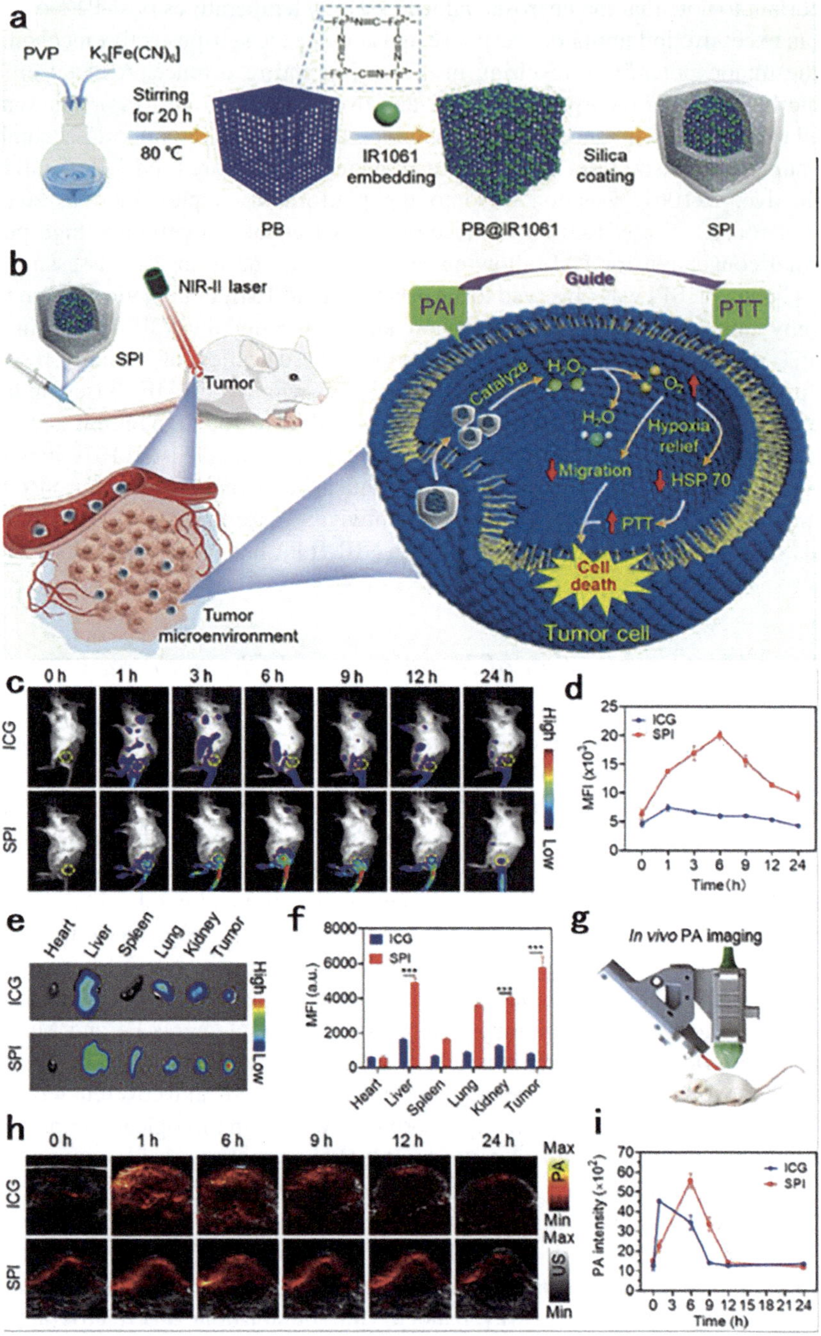
a
PVP
K3[Fe(CN)6]
Stirring for 20 h
80 °C
IR1061 embedding
Silica coating
PB
PB@IR1061
SPI
b
NIR-II laser
SPI
Tumor
Tumor microenvironment
Guide
PAI
PTT
Catalyze
Hypoxia relief
Migration
HSP 70
PTT
Cell death
Tumor cell
c
0 h
1 h
3 h
6 h
9 h
12 h
24 h
ICG
SPI
High
Low
d
ICG
SPI
MFI (x10³)
Time (h)
e
Heart
Liver
Spleen
Lung
Kidney
Tumor
f
MFI (a.u.)
g
In vivo PA imaging
h
Max
PA
Min
US
i
PA intensity (x10²)
Time (h)

Acoustic dynamic therapy (SDT) has recently been used as an alternative therapy. A photodynamic therapy (PDT) strategy has been proposed for cancer treatment. The SDT USES ultrasonic (US) activated acoustic sensitizers to generate an abundance of ROS that can damage biological molecules and additionally cause apoptosis or necrosis with non-radiotoxicity and high accuracy. Compared to previous optical mediated PDT, the SDT caused by us has a wide perpensibility of tissues and can be used for non-invasive treatment of internal or large tumors. Specific hypoxia in tumor microenvironment (TME) and glutatine hyperexpression (GSH)

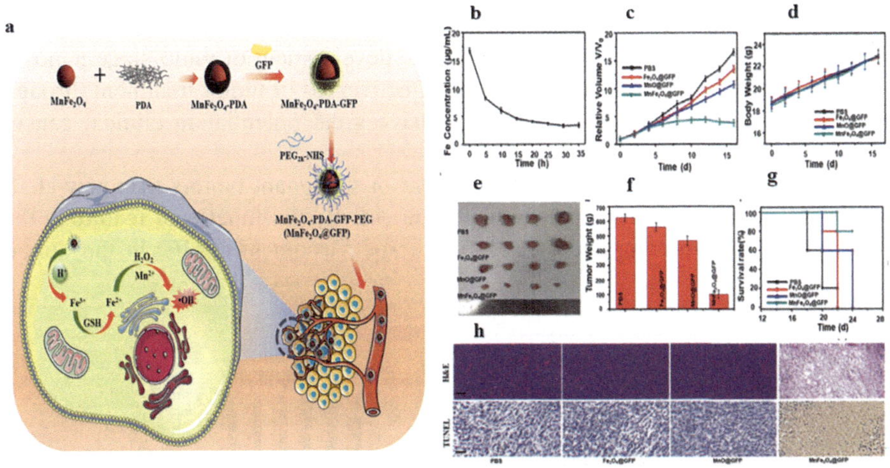

Fig. 4.17 (**a**) Schematic illustration of the preparation of $MnFe_2O_4$@GFP nanoenzymes and the mechanism of Fe/Mn dual-catalyzing Fenton reaction for tumor treatment. (**b**) Fe concentration of blood at different time points via ICP-MS. (**c**) Tumor growth recording, (**d**) body weight, (**g**) recording, (**e**) photographs of excised tumor tissues, (**f**) tumor weight, (**g**) comparison of all the groups, (**g**) survival rate in C6 cell after tail injection of PBS, Fe_3O_4@GFP MnO@GFP and $MnFe_2O_4$@GFP formulations. Data are present as the mean with a standard deviation ($n = 6$). (**h**) H&E and TUNEL analyzes of tumor sections from C6 cell various treatments. The scale bar represents 50 μm (reproduced from Li et al. 2023)

Fig. 4.16 Schematic representation of (**a**) SiO_2@PB-IR1061 (SPI) preparation and (**b**) mechanism of NIR-II photoacoustic (PA)-guided PTT to alleviate tumor microenvironment hypoxia amplification. (**c**) In vivo fluorescence images of subcutaneous 4T1-bearing mice at different time points after intravenous injection of free ICG and ICG-labeled SPI, respectively. Mean fluorescence intensity (MFI) quantification of the tumor region in (**c**, **d**). (**e**) Ex vivo fluorescence images of major organs and tumors excised from mice 24 h after injection of ICG and ICG-labeled SPI. (**f**) Quantitative biodistribution of ICG and ICG-labeled SPI in major organs and tumors. (**g**) In vivo PA imaging probe. (**h**, **i**) In vivo PA images and quantitative results of subcutaneous 4T1-bearing mice at different time points after intravenous injection of SPI and ICG, respectively (reproduced from Xue et al. 2023)

are bottlenecks affecting the effectiveness of intervention treatment with active oxygen (ROS). Here we report a new type of nanoenzyme Cu-$CuFe_2O_4$, which simultaneously reduces oxygen deficiency and GSH consumption and effectively improves chemical dynamics (CDT) acoustic dynamics (SDT) associated with ROS. This nanoenzyme has a catalytic activity similar to peroxidase and glutatine peroxidase and can produce oxygen (O_2) from hydrogen peroxide (H_2O_2) expressed in a continuous catalytic tumor at in vitro ultrasound, contributing to the development of monollinear oxygen and reaching SDT to facilitate hypoxia. At the same time, Cu-$CuFe_2O_4$NPs reacts with GSH, consuming GSH and releasing fenton ions Cu^+ and Fe^{2+}, which indirectly cut off the production of a large number of hydroxyl free radicals (•OH) in CDT. $CuFe_2O_4$-np has the ability to effectively kill MCF-7 cells and eliminate multicellular tumor spheres (MCTS) (Fig. 4.18a) (Gong et al. 2022). This study offers an effective strategy for the development of multi-stage nanosonary sensitizers TME, which increase the effectiveness of tumor treatment through continuous synchronization of TME and have great potential in clinical cancer therapy.

Accurate diagnosis and targeted treatment of malignant tumors face new challenges. Nanosets have a unique fermentation activity. Its functional feature is the presence of a multi-component catalytic microenvironment (TME) in the tumor.

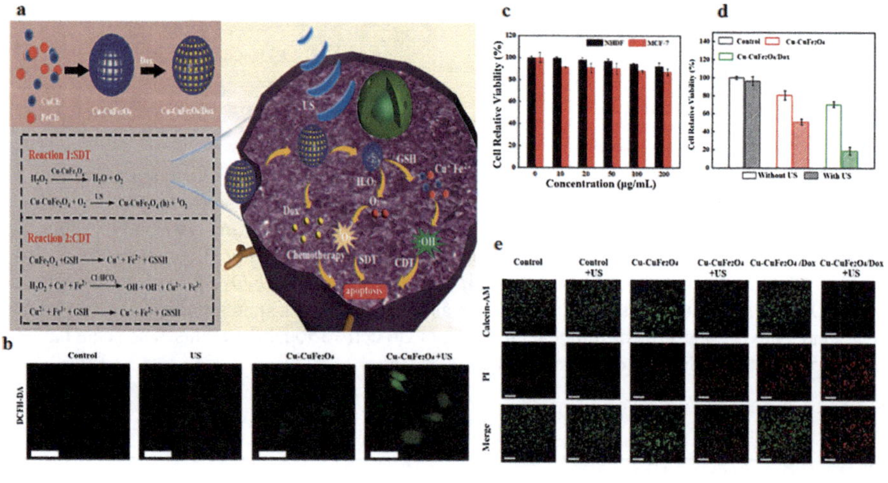

Fig. 4.18 (**a**) Fabrication procedure for Cu-$CuFe_2O_4$/Dox nanoplatform and its continuous GSH depletion and O_2 self-supply for enhancing CDT/SDT and CT against cancer. (**b**) Confocal laser scanning microscope (CLSM) analysis of ROS generation in MCF-7 cells received different treatments. Scale bar is 50 μm. (**c**) Cell viability of NHDF and MCF-7 cells incubated with Cu-$CuFe_2O_4$ NPs at different concentrations (10, 20, 50, 100, and 200 μg/mL). (**d**) In vitro cytotoxicity of MCF-7 cells incubated with Opti-MEM, Cu-$CuFe_2O_4$ NPs, and Cu-$CuFe_2O_4$/Dox NPs (200 μg/mL) with or without US irradiation for 10 min. (**e**) CLSM images of MCF-7 cells treated with Opti-MEM, Cu-$CuFe_2O_4$ NPs, and Cu-$CuFe_2O_4$/Dox NPs (200 μg/mL) with or without US irradiation for 10 min. Viable cells were stained green with Calcein-AM, and dead cells were stained red with PI. Scale bar is 200 μ (reproduced from Gong et al. 2022)

Even more exciting is that it can also be used as a photoacoustic imaging probe for NIR-II (PAI) and a cancer treatment agent. Over the past decade, a combination of nanotechnologies and biology has led to the development of various multifunctional nanoplatforms for biomedical research and tumor treatment. These new materials can catalyze the enzyme responses of relevant tumor active substances and follow dynamic similar to natural enzymes in physiological conditions. However, the high concentration of glutatione (GSH) in cancer cells can determine the intake of intracellular active oxygen (ROS), thereby reducing the effectiveness of nanase to support treatment (Fig. 4.19a) (Wang et al. 2024) to address these problems, researchers have developed nanomaterials with anti-H_2O_2 bifermentation or polyfermentation activity. Ox-POM@Cu developed and synthesized. Nanosets not only have a unique nanoenzyme function, but also have gsh-px and cat activity in TME. In the presence of GSH, it also demonstrated a high absorption and photothermal capacity in the ir-ii zone (Fig. 4.19b, d–f). The joint therapy NIRII PAI and CDT/PTT is used as a nucleotide acid that triggers the TME activation. Experimental proof Ox-POM@ Cunanase high accuracy in the treatment of tumors, OxPOM@Cu consumption of endogenous GSH reduces the resistance of cancer cells to intracellular oxidant

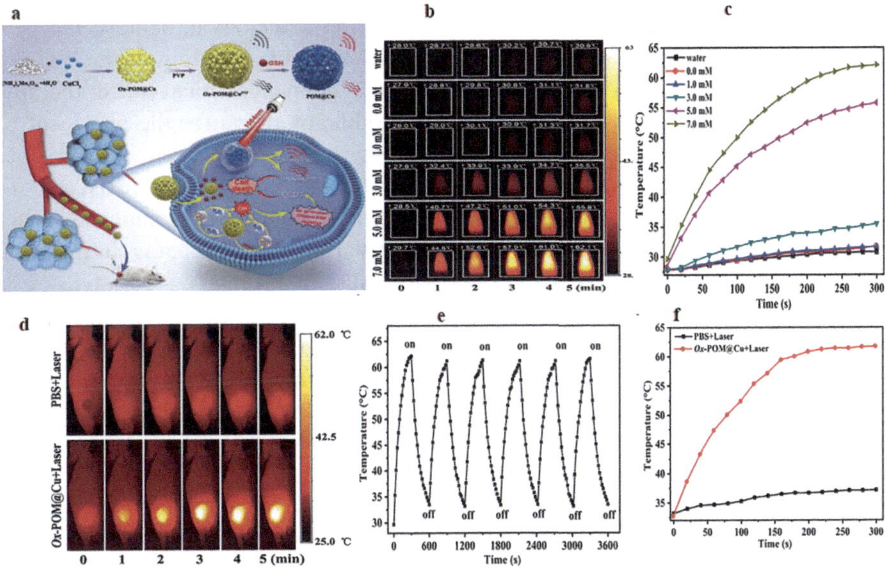

Fig. 4.19 (**a**) Diagram illustrating the preparation, PAI, and therapeutic mechanism of Ox-POM@ Cu nanoenzyme. (**b**) Thermal images of Ox-POM@Cu in the presence of GHS at different concentrations. (**c**) Temperature vs. time curves of Ox-POM@Cu incubated with GSH at various concentrations and irradiated with a 1064-nm laser for 5 min at 1.0 W cm^{-2}. (**d**) Heating curves of Ox-POM@Cu incubated with 7.0 mm GSH for conversion cycling under 1064-nm laser irradiation (heating for 5 min and cooling for 5 min, for six cycles). (**e**) IR-thermal images of mouse tumor tissue under 1064-nm laser irradiation after injecting with Ox-POM@Cu (20 mg/kg) or PBS. (**f**) Change in mouse tumor tissue temperature vs. laser irradiation time after injecting with Ox-POM@Cu or PBS (reproduced from Wang et al. 2020c)

stress and strengthens the action of CDT. The development of Ox-POM@Cunanase has good prospects for clinical application in the diagnosis and treatment of deep tissue cancer.

The application of nanomedicine in cancer treatment has been plagued by poor clinical efficacy. Researchers increasingly recognize that the failure of clinical nanomedical applications is caused by complex and unique biological characteristics of the tumor microenvironment (TME), consisting of an extracellular tumor matrix (ECM) associated with tumor fibroblastes, growth factors, blood system and immune cells. The reactionary nanogel structure of collagenase and precursor PTX (SP-NE) was successfully used for mechanical resimulation of ECM and strengthening chemotherapy in oncological stress. Collagenase is effectively released from SP-NE for digestion ECM, reducing the hardness of the tumor about 10 times, and the area of cell diffusion—4 times. This in turn greatly facilitates PTX chemotherapy in cancer cells, which reduces cell activity by 30% compared to free PTX. In mice with a Dutch tumor MCF-7 tumor inhibition at low doses SP-NE (equivalent PTX3 mg/kg) was 87.1%. To treat SP-NE, potential mechanisms have been proposed and tested to regulate the mechanical signal pathways of FAK-RhoA cells, strengthening the dynamics of F-actin assembly and facilitating mitotic division by means of the synthesis of fak perk 1/2 (Fig. 4.20a) (Zhong et al. 2019). This research approach enriches the library of TME modification strategies to achieve the best cancer treatment options. Our research broadens our understanding of the biology of TME mechanisms and can guide the development of nanomedical cancer therapy.

Cisplatin is one of the most effective chemotherapy drugs for the treatment of ovarian cancer at present, but more than 60% of patients with late ovarian cancer have relapses and subsequent metastatic tumors, mainly due to resistance to Cisplatin. Cysplatin resistance is more complex and difficult to overcome than other resistance and includes completely different enzymes and signal paths. In this context, it is essential to establish a comprehensive, sustainable, and cooperative development platform that will provide and even strengthen the treatment of this symptomatic cancer by addressing all of these key issues, such as platinum removal and detoxification, DNA recovery, and pathways of death. Platinum damaged DNA can be recovered using the DNA recovery enzyme. Removal of reparations cross-complementarity 1(ERCC1) is clinically more active in tissues of patients with platinum resistant ovarian cancer, which can make 5′ incisions on the DNA chain to remove pt-lesions. ATP7B creates multi-drug resistance by encouraging the export of nuclear drugs. In the worst case, the path of death of tumor cells becomes inactive, leading to the escape of death. Traditional strategies for the treatment of platinum resistance include preplatinum, radiotherapy, inhibitors, phototherapy, and gene therapy. These methods can temporarily ease resistance to platinum by aiming at one or two of the above-mentioned factors or ways, but unfortunately they cannot completely overcome resistance to platinum, because the complex microenvironment of the tumor can quickly restore these stability mechanisms. MoS_2 piezoelectric NFs, a biodegradable composer containing platinum and CF_3SO_2Na, is a new type of acoustic sensitizer that offers an American trigger treatment strategy to restore the tumor microenvironment, activate cytopathia, and process iron

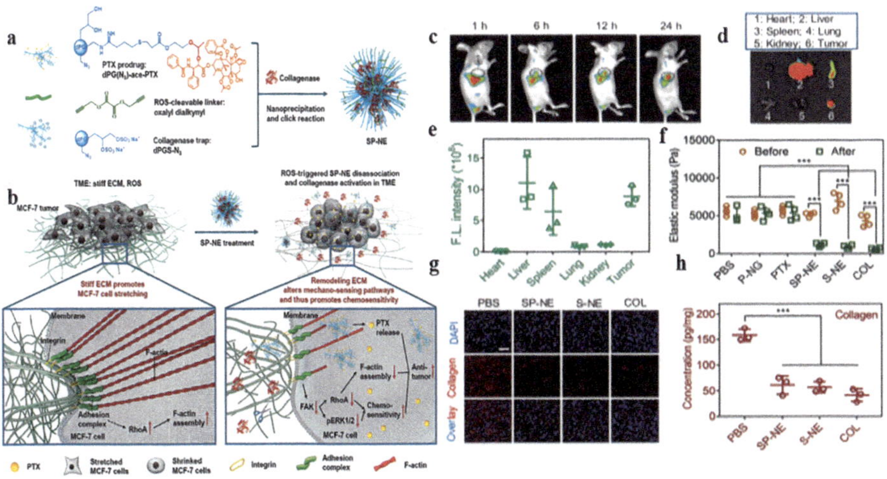

Fig. 4.20 (**a**) components and formation of SP-NE, and (**b**) a proposed mechanism of SP-NE-modulated mechanical remodeling of ECM to enhance MCF-7 cells' susceptibility toward chemotherapy. (**c**) In vivo fluorescence images of MCF-7 tumor-bearing mice receiving DIR-loaded SP-NG at different time points. (**d**) Ex vivo fluorescence images of the major organs or tissues treated with DIR-loaded SP-NG at 24 h. (**e**) Quantification of fluorescence intensity of the organs or tissues according to ex vivo fluorescence images (n = 3). (**f**) Elastic modulus (G′) of MCF-7 tumors before and after different treatments at a collagenase concentration of 0.05 mg/ml at 37 °C for 10 h measured by rheology at a strain of 0.1% and a frequency of 1 Hz (n = 4). ***$p < 0.001$, two-way ANOVA, Bonferroni's post-hoc analysis. (**g**) CLSM immunofluorescence images of collagen expression in MCF-7 tumors treated by different groups. Blue signals, nucleus; red signals, collagen (n = 6). Scale bar: 100 μm. (**h**) ELISA analysis of collagen levels in MCF-7 tumors treated by different groups (n = 3). ***p < 0.001, one-way ANOVA, Tukey's post-hoc analysis (reproduced from Zhong et al. 2023)

deposition. In the study of complete and comprehensive reversal of cisplatin resistance, we found for the first time that the introduction of •CF_3 induced downregulation of DNA repair enzymes can further ensure cisplatin induced DNA damage (Fig. 4.21a) (Ge et al. 2022) stay HA@MoCF_3Pt. The piezoelectric nanosystem demonstrated above may provide a promising new direction for the future treatment of cisplatin-resistant tumors.

Much research has been done on cancer over the past few decades, but in recent years, cancer remains one of the leading causes of morbidity and mortality worldwide, especially in the treatment of solid tumors, 2 where the extracellular matrix (ECM) of tumor tissue is usually denser than that of healthy tissue due to the presence of more collagen. Tumor ECM is mainly composed of collagen, 2, 4-hyaluronic acid, 5,6 and proteoglycan. 7,8 Due to the combination of various components and cells to form a dense and rigid three-dimensional macromolecular network 9,10, most nanoparticles can only be distributed around blood vessels and tissues, It is the most difficult to inject in the tumor environment, because it will inhale tumors with severe hypoxia, so it will reduce the efficacy of chemotherapy on tumors.

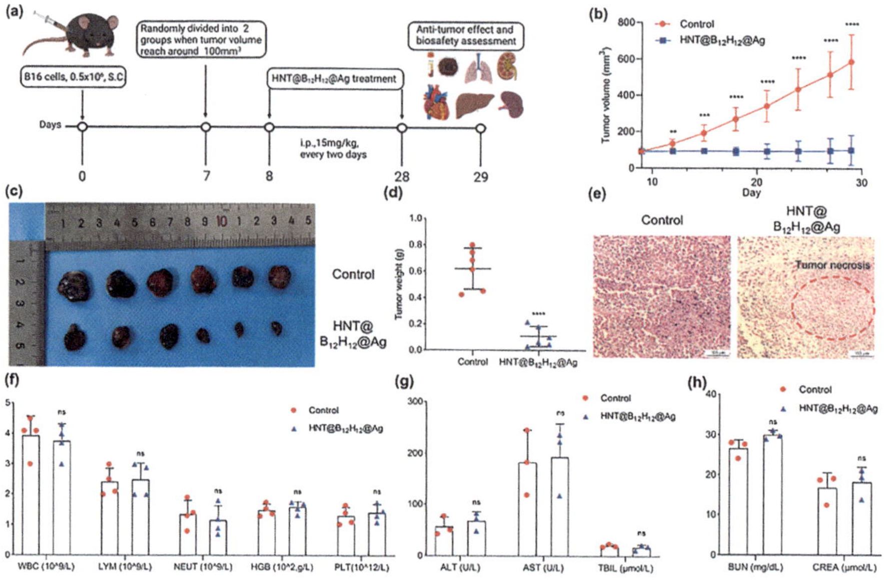

Fig. 4.23 HNT@$B_{12}H_{12}$@Ag nanoenzyme inhibits melanoma progression in vivo. (**a**) Animal experiment workflow. (**b**) Time-dependent tumor volume growth curves ($n = 6$). (**c**) Photograph of the harvested tumors. (**d**) Tumor weight ($n = 6$). (**e**) HE staining of tumors. (**f**) Hematological parameters of blood cell count (WBC, LYM, NEUT, HGB, and PLT) in mice ($n = 4$). (**g**) Test of liver function indices (ALT, AST and TBIL) in mice ($n = 3$). (**h**) Test of renal function indices (BUN and CREA) in mice ($n = 3$). The results are presented as mean ± standard deviation (SD). *$P < 0.05$, **$P < 0.01$, ***$P < 0.001$, and ****$P < 0.0001$, two-sided Student's t test (reproduced from Dong et al. 2023)

have limited nanometer (Fig. 4.23). Dong et al. (2023) propose a new method of silver-nanomaterials synthesis, increasing the likelihood of its clinical application.

4.3.7 *The Combination of Nanozyme and Radiotherapy*

Radiation therapy (RT) has been one of the most prevalent methods for cancer treatment, utilizing targeted irradiation as an antitumor approach. Various forms of in vitro accelerator therapy and in vivo radioisotope therapy can be categorized into external beam radiotherapy (EBRT) and internal radioisotope therapy (RIT). Both EBRT and RIT employ high-energy rays, such as X-rays or electron beams, particle beams like proton beams, or radioactive isotopes to either directly or indirectly damage nuclear DNA through ionizing radiation, ultimately leading to tumor cell apoptosis. Jiao successfully developed a decahedral PD-FA nanozyme and utilized it to enhance the efficacy of RIT with 125I seeds (Fig. 4.24) (Jiao et al. 2021). PD-FA effectively targeted tumor regions, catalyzing the conversion of H_2O_2 to

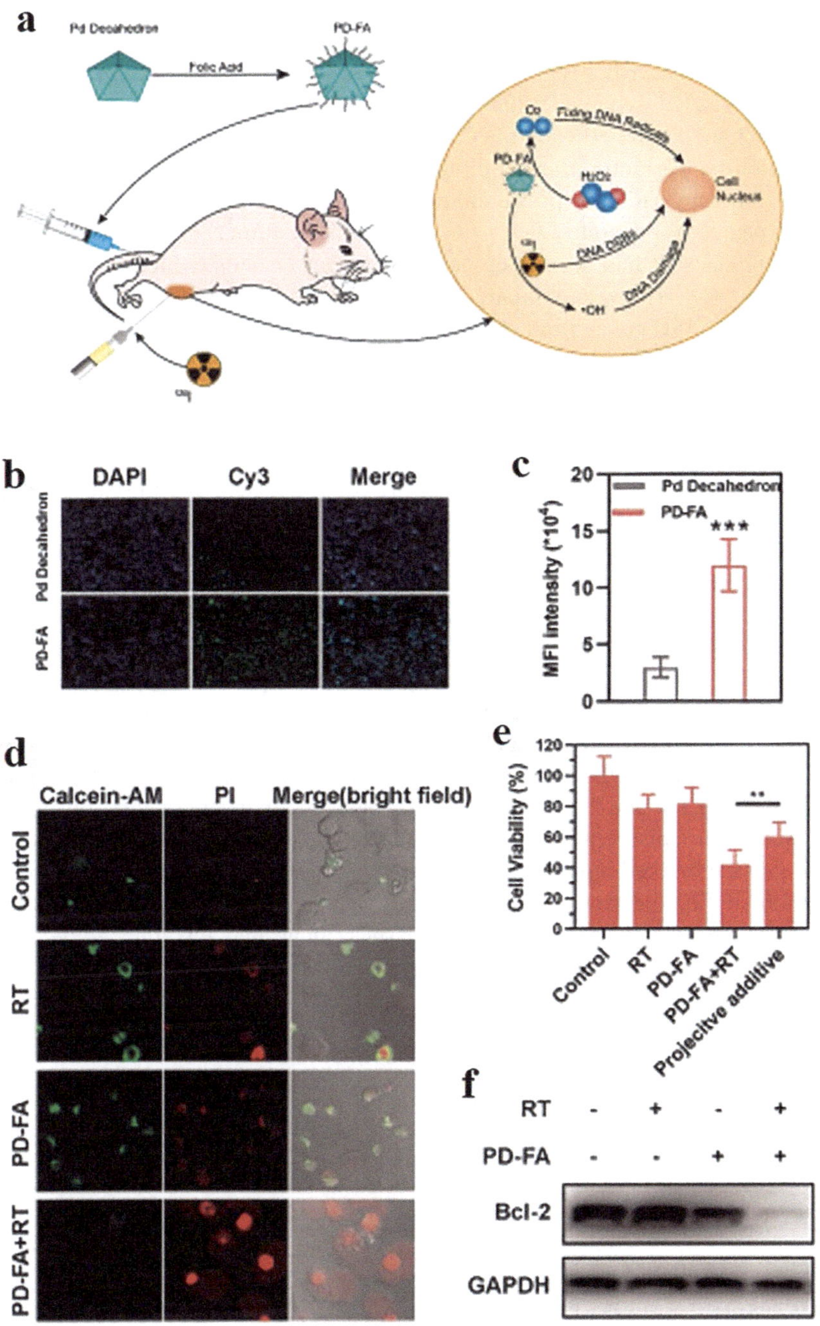

Fig. 4.24 (**a**) The synthesis of PD-FA and its application for in vivo radiosensitization Assessment of the in vitro diosensitization ability of PD-FA in vitro. (**b**) CLSM images of cells and (**c**) corresponding quantitative analyzes of image fluorescence. (**d**) CLSM images of live and dead cells following the indicated treatments and (**e**) results of CCK-8 assay. (**f**) Western blotting results for cells in the indicated treatment groups (reproduced from Wu et al. 2021)

oxygen and thereby alleviating intratumoral hypoxia. Additionally, PD-FA enhanced the energy deposition of 125I seeds within tumor areas, promoting the generation of DNA radicals that, when stabilized by oxygen, lead to irreversible cellular damage and apoptosis. Consequently, this PD-FA nanozyme demonstrates significant potential for enhancing the efficacy of 125I seed radiotherapy and shows promising biocompatibility, warranting further clinical investigation (Jiao et al. 2021).

Radiotherapy (RT), a prevalent modality in cancer treatment, utilizes high-energy beams to generate DNA radicals within the tumor, effectively eliminating the malignancy. It has been documented that cancer cells exhibit fourfold higher glutathione (GSH) levels compared to normal cells, indicating a negative correlation between GSH concentration and the efficacy of RT. Recently, various nanoplatforms have been developed to diminish GSH levels. To this end, a core-shell nanoenzyme composed of $Fe_3O_4@MnO_2$ was engineered and employed in conjunction with GOX therapy (Fig. 4.25) (Lyu et al. 2020). Fe_3O_4 nanoparticles, extensively studied, can initiate the Fenton reaction, producing substantial quantities of reactive oxygen species (ROS) under acidic conditions and targeting tumor regions through magnetic fields (MF). The excessive intracellular generation of hydrogen peroxide (H_2O_2), along with its production via glucose oxidase (GOX) in the presence of glucose, is catalyzed into large amounts of oxygen, mitigating hypoxia and immobilizing DNA free radicals. Crucially, the MnO_2 shell oxidizes glutathione (GSH) to glutathione disulfide (GSSG), thereby inhibiting double-strand break (DSB) self-repair following radiotherapy (RT). During this process, manganese in its high-valent state (Mn^{4+}) is reduced to lower-valent free Mn^{2+} ions, functioning as a high T1-weighted contrast agent. The Fe_3O_4 core provides superior contrast for T2-MRI, and the $Fe_3O_4@MnO_2$ composite demonstrates significant enhancement in dual-modality T1- and T2-weighted MRI. Given that manganese and iron are essential trace elements required daily, $Fe_3O_4@MnO_2$ exhibits significant biosafety. Owing to the marked improvement in hypoxia relief and repair inhibition, the combination of $Fe_3O_4@MnO_2$ and GOX is advantageous for radiotherapy (RT) and minimizes the dose to adjacent normal tissues. Smart nanoenzymes hold considerable promise for enhancing radiotherapy efficacy and MRI performance.

Fig. 4.25 (**a**) Preparation of $Fe_3O_4@MnO_2$ NPs and mechanism of combination therapy. (**b**) The production of oxygen treated by H_2O or $Fe_3O_4@MnO_2$ NPs (0.03 mg L^{-1}) in H_2O_2 (30 mm) ($n = 3$). (**c**) The production of oxygen treated by H_2O, glucose, GOX + glucose or GOX + glucose+NPs. (**d**) The GSH content of solution treated with Fe_3O_4 or $Fe_3O_4@MnO_2$ under various concentrations. (**e**) The GSH content of solution treated with Fe_3O_4 (100 μg mL^{-1}) or $Fe_3O_4@MnO_2$ (100 μg mL^{-1}) over time (reproduced from Yu et al. 2020)

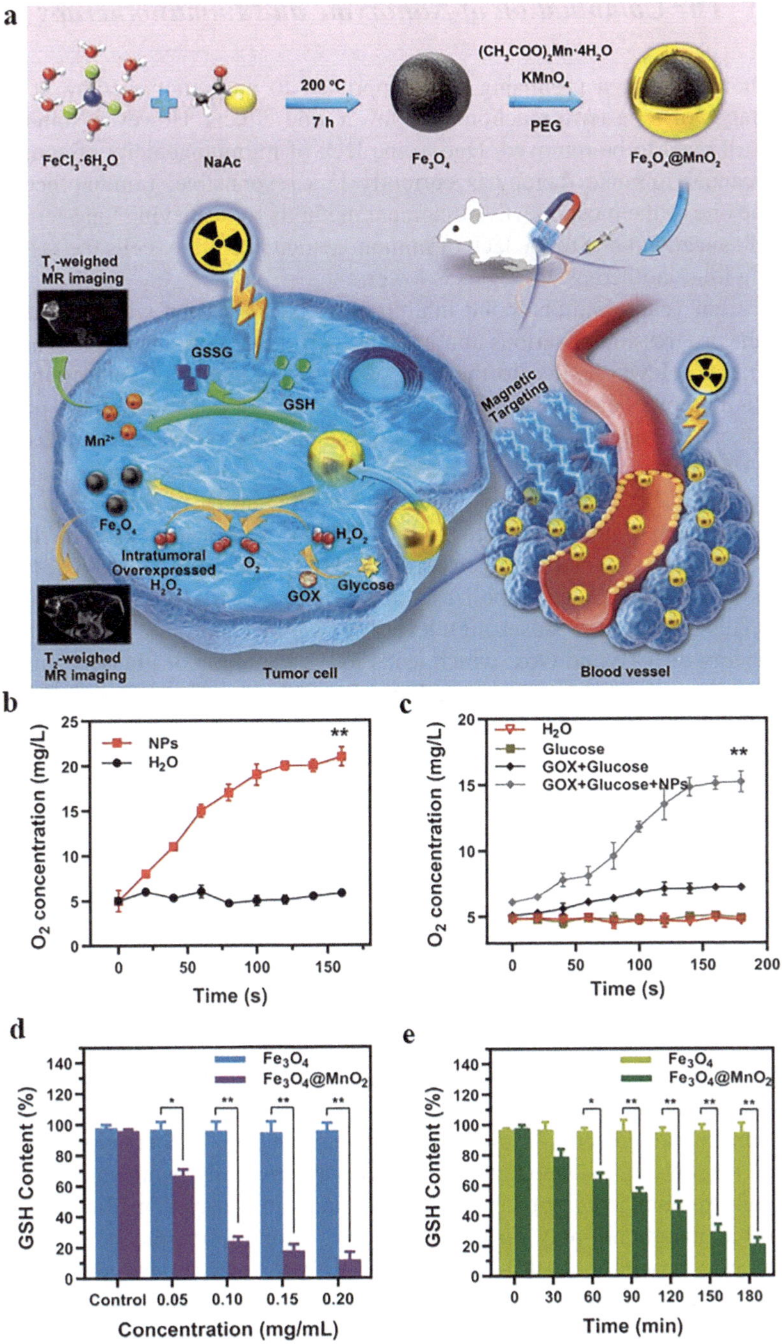
a
FeCl3·6H2O
NaAc
200 °C
7 h
Fe3O4
(CH3COO)2Mn·4H2O
KMnO4
PEG
Fe3O4@MnO2
T1-weighed MR imaging
GSSG
GSH
Mn2+
Fe3O4
Intratumoral Overexpressed H2O2
O2
H2O2
GOX
Glycose
T2-weighed MR imaging
Tumor cell
Magnetic Targeting
Blood vessel
b
O2 concentration (mg/L)
NPs
H2O
Time (s)
c
O2 concentration (mg/L)
H2O
Glucose
GOX+Glucose
GOX+Glucose+NPs
Time (s)
d
GSH Content (%)
Fe3O4
Fe3O4@MnO2
Control 0.05 0.10 0.15 0.20
Concentration (mg/mL)
e
GSH Content (%)
Fe3O4
Fe3O4@MnO2
Time (min)

4.3.8 The Combination of Nanozyme and Immunotherapy

Immunotherapy is a promising basic approach to antitumor therapy. It has the advantages of long-term reactions and lower side effects. However, some restrictions still need to be removed. Due to the lack of immunogenicity of some tumor cells, cancer immune therapy is currently less responsible. Immunotherapy has become one of the most effective treatment methods for achieving long-term remission of cancer, but due to the lack of immunogenicity of tumor cells the response to existing immunotherapy strategies is lower.

In recent years, immune point inhibitors (ICIs) have been widely used to suppress the progression of various tumors. It has advantages of long reaction and small side effects. However, due to the lack of immunogenicity of some tumor cells, immunotherapy of cancer showed a lower response level. To solve these problems, an acid-specific nanoreactor was designed to induce immunogenicity, causing iron atrophy in tumor cells. The nanoreactor glucose oxide (GOx) and mono-atomic nanoenzymes (SAE) complexes show high activity similar to peroxidase (POD)—in the microenvironment of acid tumors (TME). This acid sensitivity has been defined as an endogenous hydrogen peroxide (H_2O_2) converted to cytotoxic hydroxyl free radicals (•OH). Gox metabolizes glucose in tumor cells, producing glucose and H_2O_2. This nanoreactor causes high levels of oxidative stress in tumor cells due to the synergy of SAE and Gox, which leads to the depletion of glutatine and subsequent iron apoptosis. Nanoreactor-induced ferroptosis promotes targeted immunogenic cell death (ICD) and high expression aggregation of T lymphocyte infiltration in tumor tissue (Fig. 4.26). Based on the idea of launching an iron-deficiency mechanism for bladder cancer cells, an acid-specific nanoreactor was developed to increase the immunotherapy effect of bladder cancer, which offers new ways of immunotherapy of bladder cancer.

General cancer treatment strategies, such as surgery, chemotherapy, and radiotherapy, are largely unqualified due to serious side effects and significant drug resistance. Immunotherapy is now considered to be one of the most effective methods of cancer treatment, as it can stop the growth and metastasis of the tumor, causing a constant and broad immune response. To avoid immune observation, the immunogenicity of tumor cells remains low. The use of anti-cancer drugs that induce cell death is another major cancer treatment strategy. Immunotherapy is one of the most effective treatments for achieving long-term remission of cancer. However, due to the poor immunogenicity of tumor cells, the current immunotherapy strategy has a limited response rate. Nanomedical strategies are currently being studied to maintain high immunogenicity of tumor cells by inducing cascade mediated iron. The PdMoP nanoparta not only causes the release of iron in the primary immunogenic tumor cells through its multifermentation simulation activity, but also accelerates the transition from Mo(IV) to Mo(VI), which increases glitatine depletion (GSH), inactivates glitatine peroxide 4 (GPX4), and leads to excessive production of free radicals. It promotes p53 expression and reduces SLC7A11, thereby contributing to effective iron descent and cell apoptosis (Fig. 4.27).

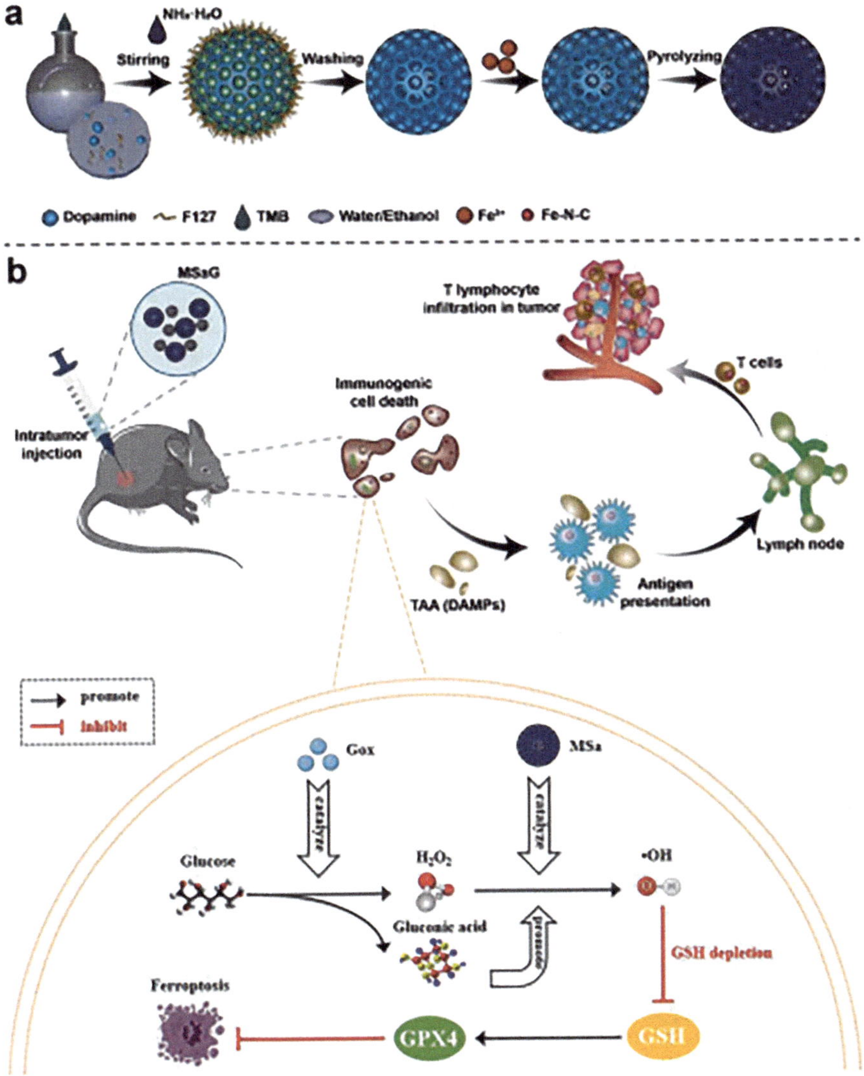

Fig. 4.26 Schematic Illustration of the working mechanisms of MSaG promoting ferroptosis in tumor cells. (**a**) Preparation of the SAE (MSa). (**b**) The nanoreactor integrates GOx with MSa, which exhibits high POD-like activity in the TME. This specific acid sensitivity transforms endogenous H_2O_2 into cytotoxic •OH. GOx enhances the POD-like SAE activity in the nanoreactor by metabolizing glucose in tumor cells, producing gluconic acid and H_2O_2. This nanoreactor induces high levels of oxidative stress within tumor cells through the synergistic action of SAE and GOx, leading to depletion of GSH and subsequently triggering ferroptosis. The resulting nanoreactor-induced ferroptosis leads to ICD and significantly recruits T lymphocyte infiltration in tumor tissues (reproduced from Liu et al. 2024)

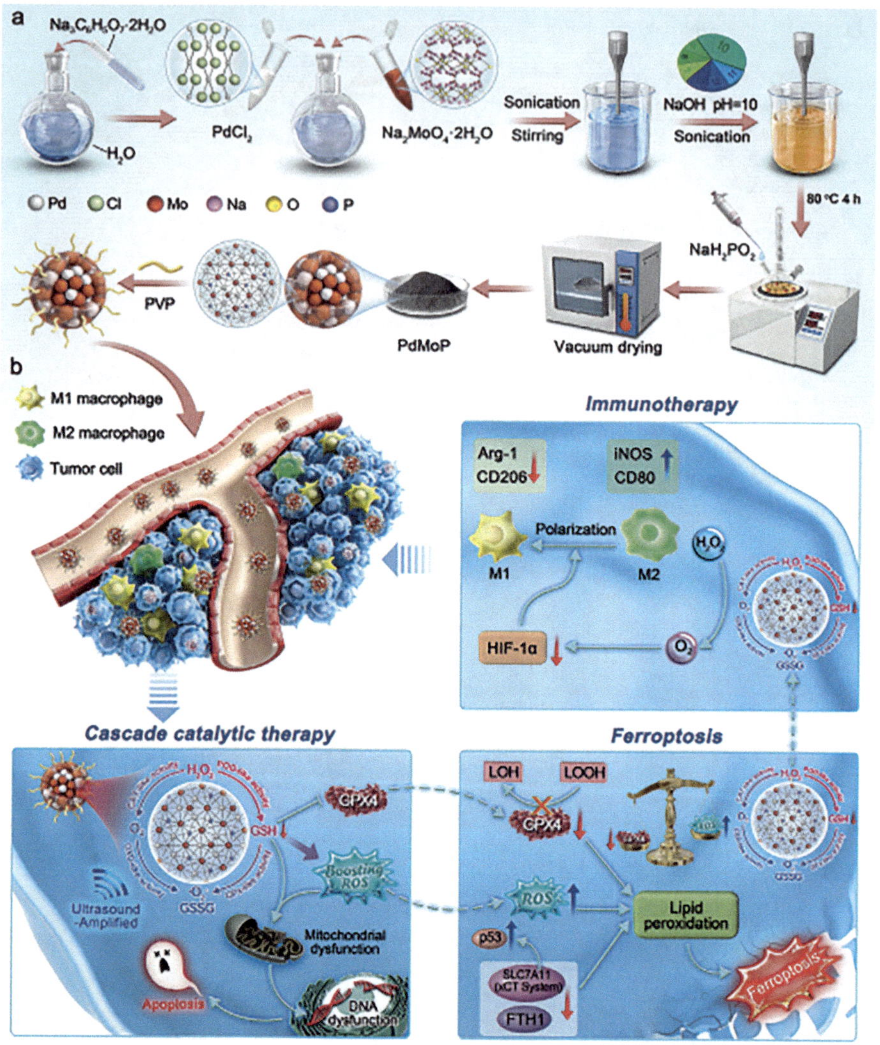

Fig. 4.27 Schematic illustration of the (**a**) synthetic procedure and (**b**) therapeutic mechanism of PdMoP nanoenzyme. The engineered PdMoP nanoenzyme was constructed by the solvothermal reduction and hydration sonication method and PVP loading. The intrinsic therapeutic mechanism enabled by PdMoP nanoenzyme includes cascade catalytic therapy, ferroptosis, and immunotherapy. PdMoP nanoenzymes with multienzyme properties were purposefully developed to trigger the production of ROS and decrease of GPX4, which ultimately include the apoptosis and ferroptosis of tumor cells, accompanied by DNA dysfunction, xCT, and FTH1 downregulation, HIF-1 reduction and reprogram immunosuppressive TME (reproduced from Song et al. 2024)

Immunotherapy is a treatment that artificially strengthens or suppresses the immune function of the body in response to weakened or overactive immune status in order to achieve the goal of treating the disease. There are many immunotherapy methods that are suitable for the treatment of various diseases. Oncological immunotherapy is designed to activate the human immune system and destroy cancer cells and tumor tissues with autoimmune function. A large number of intelligent reaction nanomaterials containing polyvalent metal elements, as excellent TME regulators and therapeutic agents, have a high catalytic activity against H_2O_2 and are of widespread concern. CAR-NK cells modified by the MnOX nanoenzyme were built to expand immunotherapy of solid tumors and to facilitate immunosuppressive microenvironment. Synthetic nanoenzymes MnOX of dithioglicoyanacid (DMSA) have excellent water dispersion, can produce oxygen by catalyzing H_2O_2, and are recovered by GSH to Mn^{2+} in TME, which can enhance the T1-weighted MRI effect. Nanasa MnOX-DMSA communicates with the CD56 antibody (MnOX-CD56). CARNK cells are constructed, which are steadily expressed in fibroblastic activated protein (FAP) car. Based on the connection of antibodies to cells, the MnOX-CD56 nanoenzyme can serve as a backpack for CARNK cells. The target capacity of CARNK cells contributes to the efficient delivery of MnOX NPs, and TME modification caused by the mnox nanoenzyme improves the ability of CAR-NK batteries to kill. CAR NK engineering cells modified by the MnOX-CD56 nanoensemble showed iterative reinforcement in the treatment of cancer (Fig. 4.28) (Duan et al. 2024).

4.3.9 The Combination of Nanozyme and Multimodal Therapy Strategy

Although significant progress has been made in cancer treatment, the current cancer therapies (surgery, chemotherapy, radiotherapy, and many more) still facing challenges, such as poor specificity and serious side effects. Thus, developing multimodal therapy strategy for the special characteristics of the tumor, such as lower pH, hypoxia, and excess hydrogen peroxide (H_2O_2), is an important strategy. Recently, new tumor therapies including PTT, PDT, SDT, ST, and immunotherapy have been developed vigorously. These therapies can overcome the disadvantages of traditional surgery and chemotherapy, such as systemic drug resistance and severe side effects. However, the effect of monotherapy is often unsatisfactory. Therefore, it is of great significance to fabricate nanocomposite materials for multimodal combined therapy.

Since conventional cancer therapies can no longer meet the current demand for precision and personalized cancer treatment, it is crucial to develop new therapeutic modalities and investigate new combined antitumor mechanisms. CDT, as the most

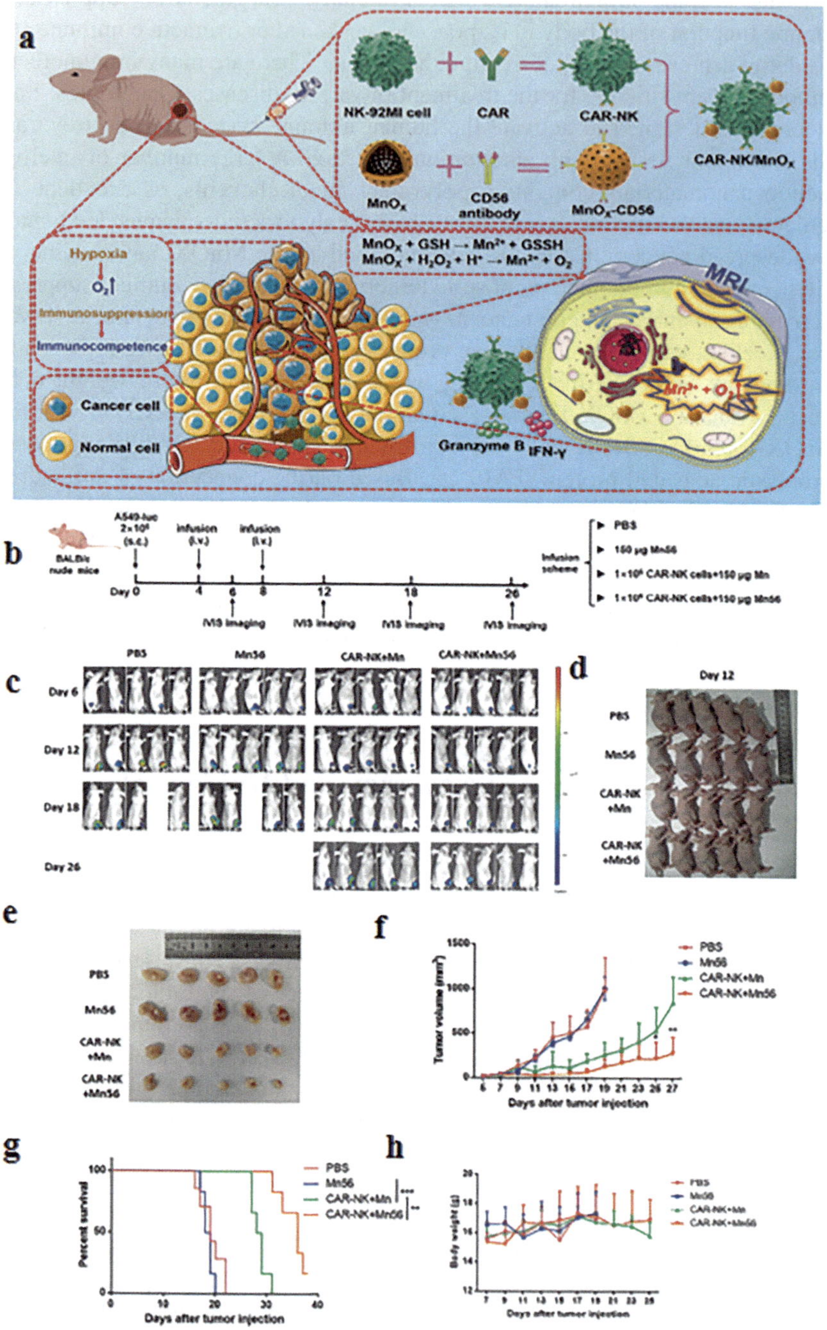
a
NK-92MI cell
CAR
CAR-NK
CAR-NK/MnO$_x$
MnO$_x$
CD56 antibody
MnO$_x$-CD56
MnO$_x$ + GSH → Mn^{2+} + GSSH
MnO$_x$ + H$_2$O$_2$ + H$^+$ → Mn^{2+} + O$_2$
Hypoxia
O$_2$↑
Immunosuppression
Immunocompetence
Cancer cell
Normal cell
MRI
Mn^{2+} + O$_2$↑
Granzyme B
IFN-γ
b
c
PBS
Mn56
CAR-NK+Mn
CAR-NK+Mn56
Day 6
Day 12
Day 18
Day 26
d
Day 12
PBS
Mn56
CAR-NK +Mn
CAR-NK +Mn56
e
PBS
Mn56
CAR-NK +Mn
CAR-NK +Mn56
f
Tumor volume (mm^3)
Days after tumor injection
PBS
Mn56
CAR-NK+Mn
CAR-NK+Mn56
g
Percent survival
Days after tumor injection
PBS
Mn56
CAR-NK+Mn
CAR-NK+Mn56
h
Body weight (g)

well-known ROS-based anti-cancer strategy, has attracted much attention. The use of novel non-metallic peroxidase mimetic nanoenzymes for CDT to treat cancer is an active and challenging field. PTT is an important phototherapeutic method that converts light energy into heat energy to kill cancer cells. Thus, the design and fabrication of nanoenzymes for CDT-PTT is an advantageous strategy. Li reported an endogenous H_2S-activated CuMOF (HKUST-1) nanoenzyme, which synergistically mediate H_2S-activated NIR PTT and CDT in the effective treatment of colon cancer (Li et al. 2020). This "turn-on" strategy for in situ generation of therapeutic agents triggered by endogenous biomarkers could largely simplify the composition of nanomedicines and avoid the introduction of cargoes, thus offering great promise for the precise treatment of colon cancer. The biomarker-triggered "turn-on" strategy employed in this work demonstrates a viable approach to designing and preparing smart therapeutic systems that can largely improve antitumor efficacy and diagnostic accuracy (Fig. 4.29a). Besides, Tan et al. synthesized a novel hollow pH-sensitive CeO_2 nanoenzyme after a cavity is loaded with indocyanine green (ICG), as well as with surface modification of tumor targeting peptides, Arg-Gly-Asp (denoted as $HCeO_2$@ICG-RGD), to successfully target tumor cells via $\alpha v\beta 3$ recognition (Tan et al. 2022). Importantly, in comparison with single chemodynamic therapy, a large amount of reactive oxygen species in cytoplasm were induced by enhanced chemodynamic therapy with PTT (Fig. 4.29b). In addition, the combination of photothermal and chemodynamic therapies effectively kills tumor cells, successfully building a synergistic therapy. This is mainly due to the precise delivery of ICG and the release of $HCeO_2$ after decomposition in the cytoplasm, in which the thermal effect is effectively generated under 808-nm laser irradiation. This work paved a fascinating way to obtain self-reinforced PTT + CDT with high specificity based on CeO_2-based nanoplatform, which will have a potential value in cancerous therapy. In fact, exploiting new non-metal-based peroxidase-mimic nanoenzymes for CDT in cancer treatment is an active and challenging field. Wu et al. found that activated carbon nanoparticles (denoted as ANs) fabricated from coconut shell have

Fig. 4.28 (**a**) Schematic illustration of the mechanism by which MnOX nanoenzyme-armed CAR-NK cells enhance solid tumor immunotherapy with immunosuppressive microenvironment alleviation. CAR-NK cells armed with Mn56 NPs exhibited superior antitumor cytotoxicity in vivo. (**b**) Experimental design. (**c**) Bioluminescent imaging (BLI) of solid tumor xenografts over time in the indicated groups ($n = 5$ mice per group). (**d**) Images of tumor-bearing mice. (**e**) The tumors were isolated at day12. (**f**) Tumor size was measured once every 2 days. The tumor volume was calculated and compared individually. All values are presented as the mean ± SEM. *$P < 0.05$ and **$P < 0.01$ by One-way ANOVA. (**g**) Survival curve of tumor-bearing mice, and statistical analysis was performed using the log-rank (Mantel–Cox) test. **$P < 0.01$ and ***$P < 0.001$. (**h**) Body weight per mouse was normalized against their starting weight before the cell infusion (reproduced from Duan et al. 2024)

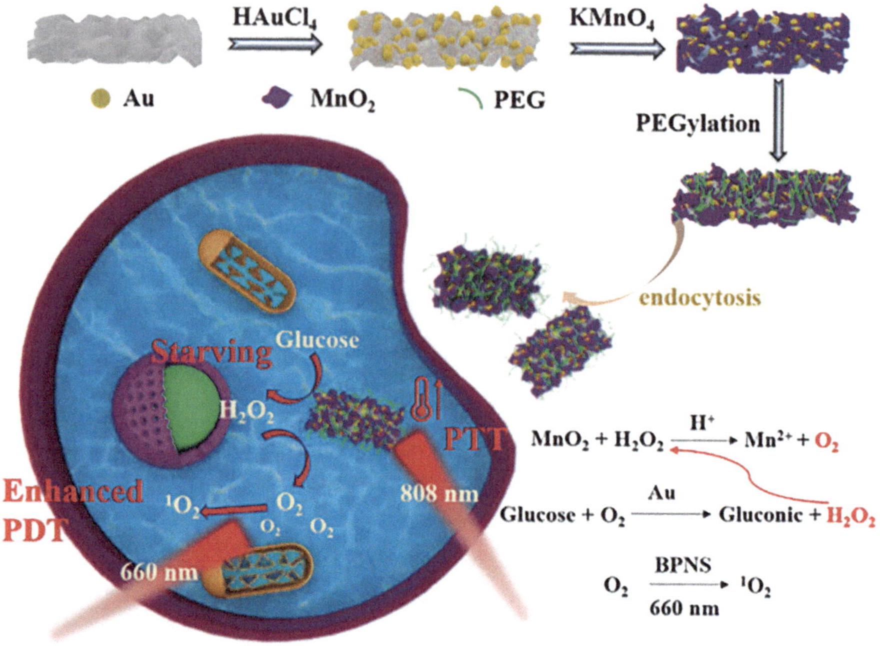

Fig. 4.31 Schematic diagram of the construction of AMGP and its application in the combination therapy of tumors (reproduced from Du 2023)

Photothermal and photodynamic therapies are non-invasive and highly specific and have recently been extensively studied. Photothermal therapy uses a photothermal reagent to convert light energy into heat energy to ablate tumor cells at high temperatures. Photodynamic therapy uses laser light to activate photosensitisers, which interact with oxygen in the tumor tissue to produce cytotoxic reactive oxygen species (ROS). However, the typical hypoxic characteristics of the tumor microenvironment severely limit the efficacy of photodynamic therapy; therefore, alleviation of tumor hypoxia is the key to photodynamic therapy. Recently, starvation therapy has received increasing attention due to its promising clinical applications. The depletion or impaired intake of nutrients in the tumor will affect the stable supply of energy to the tumor cells, leading to tumor starvation (Du et al. 2023). Two-dimensional black phosphorus nanosheets (BPNSs) have been used as photothermal reagents and photosensitisers due to their unique physical properties as well as nanocarriers for in situ reductive deposition of gold nanoparticles and fragmented MnO_2 coatings. Finally, polyethylene glycol was introduced to obtain the nanocomposite BP@Au@MnO_2-PEG (i.e., AMGP) (Fig. 4.31). In the tumor microenvironment, MnO_2 can catalyze endogenous H_2O_2 to produce oxygen, improving the effect of ST and PDT. H_2O_2 generated via the catalytic oxidation of glucose by Au nanoparticles can continue to be used as the substrate catalyzed by MnO_2. Meanwhile, black phosphorus nanosheets have excellent photothermal properties for photothermal treatment. It exerted excellent antitumor effects in ST/PDT/PTT combination therapy with good biosafety. This work provides a new perspective for the development of BP-based multifunctional nanocomposites for tumor therapy.

4.3.10 Others

Despite the significant progress that has been made, cancer mortality rates are still on the rise, leading to the disease becoming one of the leading causes of death today. There is an urgent need to develop minimally or non-invasive therapeutic strategies to reduce the severe side effects and suffering associated with conventional options such as conventional chemotherapy, radiotherapy, and surgical resection. Over the past decade, emerging externally triggered treatment strategies, such as photochemotherapy and PTT, have been extensively explored for cancer treatment. However, these approaches are often intrinsically limited by the depth of penetration of light into tissues, resulting in ineffective treatment of deep-seated tumors.

In addition, ultrasound (US) is widely used for cancer diagnosis and treatment. In contrast to light-triggered photosensitisers for local radiotherapy, ultrasound can trigger acoustic sensitisers to produce reactive oxygen species (ROS) for cancer treatment. This is called sonodynamic therapy (SDT). Zhu Piao et al. report on augmenting the SDT efficacy based on catalytic nanomedicine, which takes the efficient catalytic features of nanoenzymes to modulate the tumor TME (Zhu et al. 2018). The multifunctional nanosonosensitizers were developed by the integration of MnOx component with biocompatible/biodegradable hollow mesoporous organosilica nanoparticles (HMONs), then conjugated with protoporphyrin (PpIX, as the sonosensitizer) and cyclic arginine-glycine-aspartic pentapeptide c (RGDyC) (as the targeting peptide) (Fig. 4.32). The MnOx component worked as the inorganic nanoenzyme for converting the tumor-overexpressed hydrogen peroxide (H_2O_2) molecules into oxygen and enhancing the tumor oxygen level, which has been demonstrated to facilitate the SDT-induced reactive oxygen species production and enhance the SDT efficacy subsequently. This work not only improves the efficacy of SDT to a great extent by modulating the specificity of TME but also provides a highly effective therapeutic modality for fighting cancer through US irradiation without the tissue penetration limitations of conventional light-triggered therapies such as PDT and PTT.

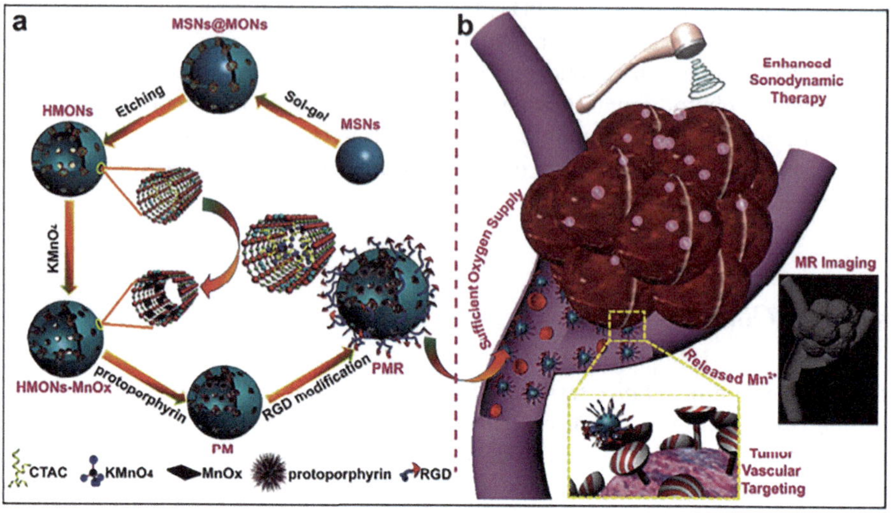

Fig. 4.32 Schematic illustration of the construction of PMR nanosonosensitizers and their MRI-guided and catalytic oxygen generation-enhanced SDT against cancer. (**a**) The detailed steps of the fabrication of PMR nanosonosensitizers. MnOx were in situ introduced into the hollow interior and mesopores of HMONs via a simple redox reaction, followed by PpIX anchoring into the hollow interior and mesopores of HMONs-MnOx. The targeting peptide RGD (Arg-Gly-Asp) was further modified onto the surface to construct the targeting nanosonosensitizers. (**b**) Schematic illustration of targeted accumulation of PMR nanosonosensitizers into tumor for in vivo tumor microenvironment-responsive MR imaging and nanoenzyme-enhanced SDT (reproduced from Zhu et al. 2018)

4.4 Conclusions and Perspectives

Malignant tumors and cancers still pose a serious threat to human life and health. To address this issue, scientists have been committed to develop more accurate and practical diagnostic methods as well as more efficient treatment strategies. In recent years, nanozymes have received widespread attention due to their excellent performance. And a lot of effective work in the integration of nanozyme tumor diagnosis and treatment have been reported. Although significant progress has been made, there are still some problems that limit the further development and application of nanozymes.

Firstly, compared with natural enzymes, the catalytic activity and selectivity of nanozymes are still relatively low. Actually, the vast majority of nanozymes cannot reach the level of natural enzymes in vivo. Thus, it is necessary to develop nanozymes with higher catalytic efficiency and selectivity; secondly, compared to natural enzymes, nanozymes simulate fewer types of catalytic activities. Therefore, it is necessary to prepare more types of nanozymes to better simulate the activities of many natural enzymes; in addition, long-term use of nanozymes may pose potential toxicity issues, which also hinder their clinical translation. Therefore, it is necessary to explore nanozymes with good biocompatibility and biosafety to accelerate their clinical application.

In summary, the research progress of nanozymes is rapid, but there are still many unknowns waiting to be explored. How to achieve the clinical translation of nanozymes is one of the key issues. We believe that nanozymes will play a greater role and play a more important role in medical treatment in the future.

References

Ali SS, Hardt JI, Quick KL, Sook Kim-Han J, Erlanger BF, Huang T-t, Epstein CJ, Dugan LL (2004) A biologically effective fullerene (C60) derivative with superoxide dismutase mimetic properties. Free Radic Biol Med 37(8):1191–1202

Attar F, Shahpar MG, Rasti B, Sharifi M, Saboury AA, Rezayat SM, Falahati M (2019) Nanozymes with intrinsic peroxidase-like activities. J Mol Liq 278:130–144

Bornscheuer UT, Huisman GW, Kazlauskas RJ, Lutz S, Moore JC, Robins K (2012) Engineering the third wave of biocatalysis. Nature 485(7397):185–194

Breaker RR (1997) DNA enzymes. Nat Biotechnol 15(5):427–431

Cai S, Qi C, Li Y, Han Q, Yang R, Wang C (2016) PtCo bimetallic nanoparticles with high oxidase-like catalytic activity and their applications for magnetic-enhanced colorimetric biosensing. J Mater Chem B 4(10):1869–1877

Cai R, Liu JB, Wu XC (2021) Research progress of noble metal-based nanozymes. Chem J Chin Univ 42(4):1188–1201

Cao C, Zou H, Yang N, Li H, Cai Y, Song X, Shao J, Chen P, Mou X, Wang W, Dong X (2021) $Fe_3O_4/Ag/Bi_2MoO_6$ photoactivatable nanozyme for self-replenishing and sustainable cascaded nanocatalytic cancer therapy. Adv Mater 33(52)

Chang M, Wang Z, Dong C, Zhou R, Chen L, Huang H, Feng W, Wang Z, Wang Y, Chen Y (2022) Ultrasound-amplified enzyodynamic tumor therapy by perovskite nanoenzyme-enabled cell pyroptosis and cascade catalysis. Adv Mater 35(7)

Chang Z, Liu C, Guo L, Shu B, Liang H, Ding J, Zhang X, Sun Q (2023) Single-dye NIR-II chemiluminescence system for H_2O_2 imaging. Chem Commun 59(41):6171–6174

Chen Z, Zhang L, Sun Y, Hu J, Wang D (2009) 980-nm laser-driven photovoltaic cells based on rare-earth up-converting phosphors for biomedical applications. Adv Funct Mater 19(23):3815–3820

Chen Y, Liu Y, Kuang P, Guo C, Zan J, Xie C, Yin C, Fan Q (2022) Tumor microenvironment activated nanoenzyme-based agents for enhanced MRI-guided photothermal therapy in the NIR-II window. Chem Commun 58(16):2742–2745

Cheng H, Lin S, Muhammad F, Lin Y-W, Wei H (2016) Rationally modulate the oxidase-like activity of nanoceria for self-regulated bioassays. ACS Sens 1(11):1336–1343

Dariva CG, Coelho JFJ, Serra AC (2019) Near infrared light-triggered nanoparticles using singlet oxygen photocleavage for drug delivery systems. J Control Release 294:337–354

Ding Y, Williams NH, Hunter CA (2019) A synthetic vesicle-to-vesicle communication system. J Am Chem Soc 141(44):17847–17853

Dong A, Huang S, Qian Z, Xu S, Yuan W, Wang B (2023) A pH-responsive supramolecular hydrogel encapsulating a CuMnS nanoenzyme catalyst for synergistic photothermal–photodynamic–chemodynamic therapy of tumours. J Mater Chem B 11(45):10883–10895

Du W, Chen WN, Wang J, Zhang H, Song L, Hu Y, Ma X (2023) A dual-nanozyme-loaded black phosphorus multifunctional therapeutic platform for combined photothermal/photodynamic/starvation cancer therapy. J Mater Chem B 11(5185-5194)

et al. 2022). The discovery of fullerenes began with theoretical predictions, and ultimately in 1985, C60 molecules were successfully prepared under laboratory conditions by British chemist Dr. Harold Walter Croto and American scientist Richard Smalley. C60 is the most common and extensively studied member of the fullerene family. It is composed of 60 carbon atoms connected by 20 hexagonal rings and 12 pentagonal rings, forming a cage like structure with high symmetry, similar to the shape of a football, and is therefore also known as "football alkene." Fullerene molecules are extremely small, with C60 having a diameter of only 0.7 nm. Fullerenes have the properties of electron deficient compounds, which can easily react with nucleophilic reagents and undergo both nucleophilic and electrophilic addition reactions.

Fullerenes and their derivatives, especially water-soluble fullerene nanomaterials modified with multifunctional groups, have good biocompatibility and stability (Li et al. 2023a; Illescas and Martin 2006). These materials can serve as imaging probes to achieve precise localization and diagnosis of tumor tissue through specific imaging techniques such as fluorescence imaging, magnetic resonance imaging, and many more. The fluorescence characteristics of fullerene make it have strong fluorescence signals in the near-infrared region, which can be detected in complex biological environments and provide real-time images of tumor tissues. Fullerene nanomaterials can also serve as core components of biosensors for detecting tumor markers. By specifically identifying and binding to tumor markers, fullerene nanomaterials can achieve early screening and diagnosis of tumors.

Fullerenes and their derivatives have a large specific surface area and rich π-electron structure, which can efficiently load drug molecules (Zhu et al. 2018; Liu et al. 2020b). These fullerene nanomaterials loaded with drugs can enter the interior of tumor cells through targeted delivery systems, achieving precise drug release. The cage like structure of fullerenes also provides the ability to protect drug molecules, preventing degradation or inactivation of drugs during delivery. Some fullerene derivatives can generate a large number of biologically toxic reactive oxygen species (such as singlet oxygen, superoxide anions, and many more) under specific wavelength light conditions. These reactive oxygen species can damage the cell membrane and DNA structure of tumor cells, leading to apoptosis and necrosis of tumor cells (Li et al. 2023c; Cao et al. 2022). Therefore, fullerene nanomaterials can be used as photosensitizers in photodynamic therapy for the treatment of tumors. Fullerene nanomaterials can also inhibit tumor growth and spread by activating the body's immune system (Huo et al. 2022). For example, fullerene nanomaterials can serve as immune adjuvants to enhance the immunogenicity of vaccines. Alternatively, as a carrier of immune cells, immune cells can be directly delivered into tumor tissues for immunotherapy. Fullerene nanomaterials can also be combined with other therapeutic methods, such as chemotherapy, radiotherapy, and many more, to form a combined treatment plan. Through the targeted delivery and drug controlled release functions of fullerene nanomaterials, multiple killing mechanisms against tumor cells can be achieved, improving therapeutic efficacy and reducing side effects. Bai et al. developed a hypoxia-sensitive nanotherapeutic system

(FTCD-SRGD) based on fullerene (C70) and anoxic activating chemical prodrug tirapazamine (TPZ) to multimodal therapy of deep hypoxic tumors (Li et al. 2024).

In recent years, with the continuous development of nanotechnology and biomedicine, the application research of fullerene in the field of tumor diagnosis and treatment has also been deepening. Researchers have explored the potential applications of fullerenes in tumor diagnosis, treatment, and prognosis monitoring through various methods and means and have made a series of important progress. These studies not only provide new ideas and methods for the clinical treatment of tumors but also lay a solid foundation for the further development and application of fullerene nanomaterials. Fullerenes have broad application prospects and important research value in the field of tumor diagnosis and treatment. With the continuous advancement of technology and in-depth research, it is believed that fullerene will play a more important role in early diagnosis, precise treatment, and prognosis monitoring of tumors.

5.2.2 Carbon Nanotubes

Carbon nanotubes are a one-dimensional quantum material with a special structure, mainly composed of hexagonal carbon atoms arranged in several to dozens of coaxial circular tubes, with a fixed distance between layers, about 0.34 nm (Anzar et al. 2020; Zare et al. 2021). Its radial size is in the nanometer range, its axial size is in the micrometer range, and both ends of the tube are basically sealed. According to the number of layers on the tube wall, carbon nanotubes can be divided into single-walled carbon nanotubes (SWCNTs) and multi-walled carbon nanotubes (MWCNTs). Single-walled carbon nanotubes are composed of a single layer of graphene sheets, while multi-walled carbon nanotubes contain multiple layers of graphene sheets, shaped like coaxial cables. Since its first discovery by Japanese electron microscope scientist Iijima in 1991 during the preparation of C60 by arc method, carbon nanotubes have attracted the attention of many scientists in the fields of materials, physics, electronics, chemistry, and so on due to their unique structure and excellent mechanical, electrical, and chemical properties. There are various methods for preparing carbon nanotubes, including arc discharge method, chemical vapor deposition method, laser evaporation method, and many more. Each method has its unique advantages and limitations, and choosing the appropriate preparation method is crucial for obtaining high-quality carbon nanotubes.

The application of carbon nanotubes (CNTs) in the field of tumor diagnosis and treatment is becoming increasingly widespread. Their unique physical, chemical properties, and biocompatibility make them powerful tools for tumor diagnosis and treatment (Kumar et al. 2017). Magnetic nanoparticles such as superparamagnetic iron oxide (SPIO) can bind with carbon nanotubes to form contrast agents with enhanced MRI signals. This complex can significantly improve the imaging contrast of tumor tissue, which is helpful for early detection of tumor lesions. Carbon nanotubes have strong fluorescence signals in the near-infrared region and can be detected in complex biological environments. This makes carbon nanotubes an ideal material for near-infrared imaging, which can be used for in vivo imaging and real-time monitoring of tumor tissues. By modifying carbon nanotubes to give them a radioactive signal on their surface, PET imaging can be achieved. PET technology can display functional changes in the human body, making it particularly suitable for detecting subclinical lesions before morphological changes and evaluating treatment efficacy. Carbon nanotubes can serve as the core component of biosensors for detecting tumor markers or monitoring changes in the tumor microenvironment. This type of sensor has high sensitivity and specificity, which can provide important information for early diagnosis of tumors.

Carbon nanotubes have a large specific surface area and good drug loading capacity, which can carry peptides, proteins, nucleic acids, and small molecule drugs into tumor cells (Yan et al. 2017). By covalent modification or non-covalent forces, small molecules targeting tumors and therapeutic drugs can be connected to carbon nanotubes to construct a nano-drug delivery system targeting tumors. Research has shown that carbon nanotubes (PEG CNTs) covalently modified with polyethylene glycol (PEG) can significantly increase the half-life of drugs in the body, thereby enhancing drug efficacy and reducing side effects. Some carbon nanotubes with photothermal properties, such as gold carbon nanotubes, can generate local high temperatures under laser irradiation, thereby killing tumor cells. This photothermal therapy has the advantages of non-invasiveness, high selectivity, and minimal side effects. The photothermal effect can also promote the rapid release of drugs on the surface of carbon nanotubes, further enhancing the effectiveness of chemotherapy. Researchers have also developed a technique for mechanical nanosurgery using magnetic field controlled iron containing carbon nanotubes (mCNTs). By controlling the magnetic field, mCNT generates mechanical motion and applies mechanical stimulation to tumor cells, leading to tumor cell death. This technology provides a new means of treating drug-resistant tumors. Bai et al. developed bifunctional nanocomposite based on phosphor and carbon nanotubes for tumor ablation in a photothermal fiber system with temperature feedback (Fig. 5.1) (Su et al. 2022).

The application of carbon nanotubes in the field of tumor diagnosis and treatment has demonstrated its enormous potential and value. With the continuous deepening of research and advances in technology, it is believed that carbon nanotubes will play a more important role in early diagnosis, precise treatment, and prognosis monitoring of tumors. At the same time, attention should also be paid to the safety of carbon nanotubes in vivo to ensure their safety and effectiveness in clinical applications.

5.2.3 Graphene

Graphene is a two-dimensional carbon nanomaterial composed of carbon atoms in sp^2 hybridized orbitals, forming a hexagonal honeycomb lattice (Reina et al. 2017; Riley and Narayan 2021). It is an allotrope of carbon. The lattice of graphene is a

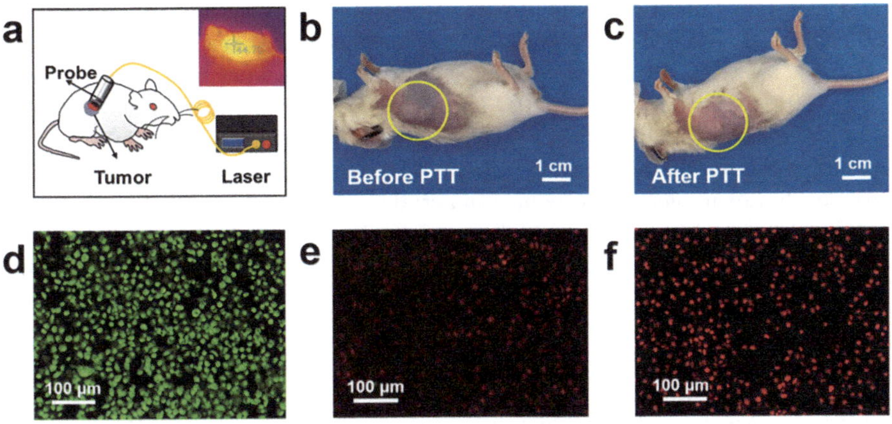

Fig. 5.1 Bi-functional nanocomposite based on phosphor and carbon nanotubes for tumor ablation in a photothermal fiber system (reproduced from Su et al. 2022) Copyright © 2022, Elsevier

hexagonal structure surrounded by six carbon atoms, forming a single-layer sheet-like structure and a two-dimensional material with only one carbon atom thickness. Its thickness is approximately 0.335 nm, making it the thinnest known nanomaterial to date. Graphene has excellent optical, electrical, and mechanical properties. Graphene has always been considered a hypothetical structure and cannot exist stably on its own. Until 2004, physicists Andre Heim and Konstantin Novoselov of the University of Manchester, UK, successfully separated graphene from graphite in their experiments. This discovery confirmed the existence of graphene alone and made them jointly win the 2010 Nobel Prize in Physics. There are various methods for preparing graphene, including mechanical exfoliation, chemical oxidation, crystal epitaxial growth, chemical vapor deposition (CVD), organic synthesis, carbon nanotube exfoliation, and so on. Among them, the mechanical exfoliation method can prepare high-quality graphene, but with low yield and high cost; the chemical vapor deposition method can make breakthroughs in the large-scale preparation of high-quality graphene, but the cost is also relatively high; the oxidation–reduction method has become a commonly used method for preparing graphene due to its low cost and easy implementation.

The application of graphene in tumor diagnosis and treatment is gradually demonstrating its enormous potential and value (Alagarsamy et al. 2021). Although the application of graphene in direct tumor diagnosis is still in the research stage, its unique physical and chemical properties provide the possibility for developing new diagnostic tools. For example, graphene and its derivatives can be functionalized to bind with specific biomolecules, forming targeted and sensitive biosensors. These sensors can detect tumor markers or monitor changes in the tumor microenvironment, providing important information for early screening and diagnosis of tumors. Berger et al. developed functional graphene for peritumoral brain microenvironment modulation therapy in glioblastoma (Chin et al. 2023).

The application of graphene in tumor treatment has made significant progress. Graphene and its derivatives have a large specific surface area and excellent loading capacity, which can efficiently carry anticancer drugs into tumor cells (Younis et al. 2021; Martelli et al. 2020). Through specific modification and targeting strategies, graphene drug carriers can achieve precise drug delivery, improve therapeutic efficacy, and reduce side effects (Szczepaniak et al. 2021; Wang et al. 2020b). For example, graphene nanomaterials can be combined with various anticancer drugs to form stable nanocomposites. These complexes remain stable in the bloodstream and release drugs through specific triggering mechanisms (such as pH changes, light exposure, and many more) when they reach tumor tissue, achieving targeted therapy. Graphene has excellent photothermal conversion performance and can generate local high temperatures under near-infrared light irradiation, thereby killing tumor cells. This photothermal therapy has the advantages of non-invasiveness, high selectivity, and minimal side effects. Research has shown that graphene heating film can quickly and uniformly transmit far-infrared light waves to tumor tissues, enhance the penetration and tissue absorption ability of far-infrared through resonance effect, effectively inhibit cancer cell growth and metastasis, and promote cancer cell apoptosis. Graphene and its derivatives can also inhibit the growth and spread of tumors by activating the body's immune system. For example, small graphene particles can activate immune cells, induce cytokine release, and regulate immune responses. By surface modification of graphene, its immunogenicity can be further enhanced and its immunotoxicity reduced, making it more suitable for cancer immunotherapy. Graphene can also be combined with other treatment methods to form a combination therapy plan. For example, the combination therapy of drug chemotherapy and graphene far-infrared has been proven to have a significant synergistic enhancement effect. Targeted delivery and photothermal effects of graphene can improve the efficacy of chemotherapy drugs and reduce their side effects. Non-invasive treatment methods such as graphene photothermal therapy can reduce patients' pain and risk. Through functional modification and targeting strategies, graphene drug carriers can achieve precise drug delivery and improve therapeutic efficacy. Graphene materials have good biocompatibility and degradability, which can reduce side effects and risks during the treatment process. Wang et al. developed new features of edge-selectively hydroxylated graphene nanosheets as NIR-II photothermal agent and sonothermal agent for tumor therapy (Zhang et al. 2024).

With the continuous deepening of research on graphene and the continuous advancement of technology, the application prospects of graphene in tumor diagnosis and treatment will be even broader. In the future, graphene is expected to play a more important role in early diagnosis, precise treatment, and prognosis monitoring of tumors, providing safer and more effective treatment options for cancer patients. Meanwhile, the large-scale production and cost control of graphene are also important directions for future research.

5.2.4 Carbon Quantum Dots

Carbon quantum dots (CQDs), also known as carbon dots or carbon nanodots, are a class of zero-dimensional carbon nanomaterials with significant fluorescence properties (Azam et al. 2021; Rosenkrans et al. 2020). Carbon quantum dots are composed of ultrafine, dispersed, quasi spherical carbon nanoparticles with a size below 10 nm. It is a new type of carbon nanofunctional material that once again demonstrates the diversity and potential of carbon materials, following carbon nanotubes, nanodiamonds, and graphene (Desmond et al. 2021). Carbon quantum dots have excellent optical properties, including photoluminescence and ultraviolet absorption ability (Qi et al. 2021). There is a strong absorption peak in the ultraviolet region and a long tail in the visible region. Carbon quantum dots have the characteristic of photoluminescence, which means that under illumination, they emit bright fluorescence and have good optical stability. Carbon quantum dots also possess upconversion photoluminescence (UCPL) properties, which enable them to absorb low-energy light and emit high-energy light. This has promising applications in fields such as biological sciences and energy technology. Compared to traditional metal quantum dot materials such as lead and cadmium, carbon quantum dots have higher biocompatibility and lower cytotoxicity, with minimal environmental harm. Many hydrophilic functional groups such as carboxyl groups on the surface of carbon quantum dots have excellent solubility in water, further enhancing their biocompatibility (Iannazzo et al. 2021; Liang et al. 2021). Carbon quantum dots are mainly composed of elements such as carbon, hydrogen, and oxygen, with a relatively high content of carbon. Its structure is usually composed of amorphous and crystalline carbon nuclei, with different oxygen-containing functional groups on the surface, such as hydroxyl and carboxyl groups (Devi et al. 2022). The luminescence properties of carbon quantum dots are size and excitation wavelength dependent, and the luminescence is stable without photobleaching phenomenon. There are various methods for preparing carbon quantum dots, including arc discharge method, laser ablation method, electrochemical synthesis method, chemical oxidation method, combustion method, hydrothermal synthesis method, microwave synthesis method, template method, and many more. These methods each have their own advantages and disadvantages, but they can effectively prepare carbon quantum dots. Since its first discovery in 2004, significant progress has been made in the study of carbon quantum dots. However, more efficient methods for preparing high-quality carbon quantum dots with uniform particle size and morphology still need to be developed, and the fluorescence quantum yield of carbon quantum dots also needs to be improved. In addition, research on the luminescence mechanism of carbon quantum dots and the influence of surface functional groups on their properties is constantly deepening.

The application of carbon quantum dots in tumor diagnosis and treatment is gradually expanding, and it is showing its unique advantages and potential (Chung et al. 2021; Rahimi et al. 2022). Carbon quantum dots are widely used in the field of biological imaging due to their excellent fluorescence properties and good biocompatibility. In tumor diagnosis, carbon quantum dots can serve as fluorescent markers for labeling and imaging tumor cells, improving the sensitivity and accuracy of tumor detection. Carbon quantum dots have strong fluorescence stability, are not easily bleached, and can maintain stable fluorescence signals for a long time, which is beneficial for continuous and dynamic monitoring of tumors. Compared with traditional organic fluorescent dyes, the fluorescence emission spectrum of carbon quantum dots is narrow, the excitation spectrum is wide, and the fluorescence intensity is higher, which makes it have higher resolution and clearer image quality in tumor imaging. Carbon quantum dots can be surface modified to connect specific targeting molecules (such as antibodies, peptides, and many more), achieving specific recognition and binding to tumor cells. This targeting property enables carbon quantum dots to have higher accuracy and specificity in tumor diagnosis. For example, some studies have successfully combined carbon quantum dots with tumor specific antigens or receptors for early screening and diagnosis of tumors. Zhou et al. developed acidity activatable graphene quantum-dot-based nanotransformers for long-period tumor imaging and repeated photodynamic therapy (Yang et al. 2023).

Carbon quantum dots, due to their large specific surface area and excellent loading capacity, can serve as drug carriers to efficiently deliver anticancer drugs into tumor cells (Ge et al. 2014; Sinha et al. 2023; Kuo et al. 2023). This nano-drug delivery system can significantly improve the bioavailability and targeting of drugs, reducing the toxic side effects of drugs on normal tissues. Carbon quantum dots can also control drug release through specific triggering mechanisms such as pH changes, light exposure, and so on, achieving intelligent drug release and on-demand therapy. Carbon quantum dots have excellent photothermal conversion performance and can generate local high temperatures under near-infrared light irradiation, thereby killing tumor cells (Liu et al. 2020c). This photothermal therapy has the advantages of non-invasiveness, high selectivity, and minimal side effects. Research has shown that by adjusting the size and surface modification of carbon quantum dots, their photothermal conversion efficiency can be optimized and the effectiveness of photothermal therapy can be improved. Carbon quantum dots also have potential applications in photodynamic therapy. Photodynamic therapy is a method that uses a combination of photosensitizers and light to kill tumor cells. Carbon quantum dots can serve as carriers for photosensitizers, delivering photosensitizers to the interior of tumor cells and generating reactive oxygen species such as singlet oxygen under illumination, disrupting the biological membrane and DNA structure of tumor cells, thereby achieving therapeutic goals. Carbon quantum dots can also be combined with other therapeutic methods to form a combined treatment plan. For example, carbon quantum dots can be used as drug carriers in combination with chemotherapy drugs to achieve synergistic effects of chemotherapy and photothermal therapy; alternatively, carbon quantum dots can be combined with immunotherapy to enhance anti-tumor effects by activating the body's immune system. Han et al. developed the graphene quantum dots gated nanoplatform for photothermal-enhanced synergetic tumor therapy (Fig. 5.2) (Wang et al. 2024a, b).

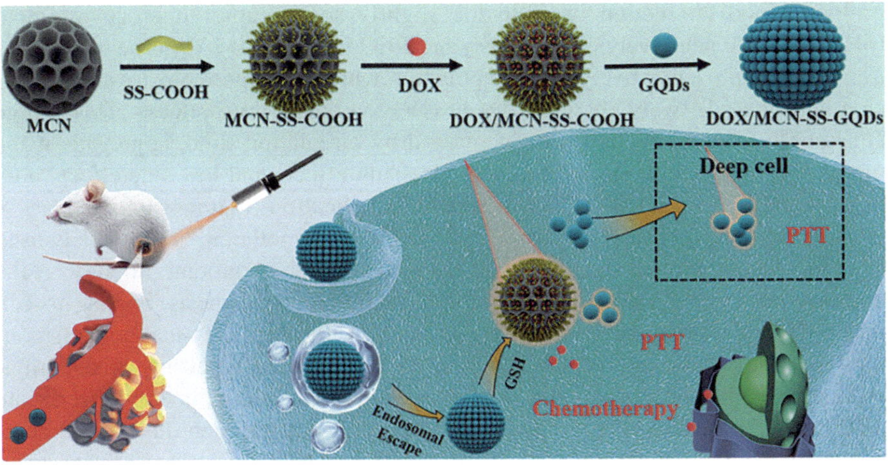

Fig. 5.2 Acidity activatable graphene quantum-dot-based nanotransformers for long-period tumor imaging and repeated photodynamic therapy (reproduced from Wang et al. 2024a, b) Copyright © 2024, MDPI

Due to the presence of functional groups such as $-NH_2$, –OH, and –COOH on the surface of carbon dots, they have become a novel nanocarrier platform that can be covalently coupled or molecularly interacted (such as electrostatic interactions and π-π stacking) with anticancer drugs [such as doxorubicin (DOX), camptothecin (CPT), cisplatin (cis-Pt)], photosensitizers (such as Ce6, porphyrin, phthalocyanine, and so on), and anticancer genes loaded on the surface of carbon dots (Yan et al. 2020; Yang et al. 2023; Jiang et al. 2021). Compared with other nanomaterials, carbon dots as carriers have the following advantages: (1) Carbon dots have good water solubility and biocompatibility, which can improve the stability and dispersibility of anticancer drugs or photosensitizers in aqueous solutions, effectively avoiding aggregation and quenching of photosensitizers to produce reactive oxygen species; (2) Carbon dots can be used to construct fluorescence resonance energy transfer (FRET) systems with other photosensitizers, utilizing the wide absorption wavelength range of carbon dots to broaden the excitation light range and improve the photodynamic therapy effect; (3) By enhancing the permeability and retention (EPR) effect of nanomaterials in vivo, photosensitizers can be more effectively enriched in tumor tissue sites, improving their killing effect on cancer cells.

Carbon dots are regarded as an excellent carrier due to their excellent water solubility, low toxicity, and good biocompatibility. Zhao used citric acid as a carbon source and ethylenediamine as a passivator to prepare aminated carbon dots. CDs Pt (IV) were synthesized by acylation reaction with cisplatin prodrug, and then self-assembled with PEG-(PAH/DMMA) (dimethyl malee) by electrostatic interaction to form the composite CDs Pt (IV) @ PEG-(PAH] DMMA). PEG-(PAH/DMMA), as an invisible layer, can prevent drug carriers from being quickly cleared by the immune system, resist protein adsorption caused by electrostatic repulsion, and thus prolong blood circulation time. In the slightly acidic tumor microenvironment (pH = 6.8), the negatively charged polymer PEG-(PAH/DMMA) in the complex is transformed into a positively charged polymer, which undergoes electrostatic repulsion with the positively charged CDs Pt (IV) to control drug release. This charge convertible CDs carrier helps to increase drug circulation time, aggregate at the tumor site, and exhibit strong tumor suppression effects and low side effects both in vitro and in vivo. It has important potential applications in tumor treatment.

Doxorubicin is a broad-spectrum anthracycline chemotherapy drug widely used in clinical treatment. It can penetrate the cell nuclear membrane directly to the nucleus, induce DNA damage, and inhibit nucleic acid synthesis, leading to cell death. DOX can be loaded onto the surface of CDs through chemical bonding coupling or electrostatic interactions. Dong and Choi et al. added evenly mixed ethylenediamine and phosphoric acid to glucose, and after high-temperature reaction, dialysis, and freeze-drying, prepared novel phosphorus nitrogen dual doped composite multifunctional hollow carbon dot PNHCDs. The central cavity structure and surface phosphate groups give PNHCDs a high loading capacity. Through electrostatic interactions, doxorubicin is loaded into PNHCDs and assembled into a composite PNHCDs DOX, with a loading rate of up to 35.4%. This nanosystem has fluorescence tunability, and intracellular fluorescence signals can be observed in the blue, green, and red channels of confocal microscopy. Under acidic conditions, PNHCDs release DOX at a faster rate than in neutral microenvironments, effectively inhibiting the proliferation of tumor cells. In vitro and in vivo experiments have shown that PNHCDs DOX can more effectively inhibit tumor growth and reduce the toxic side effects of anticancer drugs than free DOX (Feng et al. 2016). Yang synthesized NLS-CDs using PEG2000, citric acid, and ethylenediamine as carbon sources. After amidation, the surface of the NLS-CDs was enriched with amino functional groups, which were then modified with DOX to assemble into a nanocomposite DOX-CD. Through cell imaging, MTT (thiazole blue) experiments, and flow cytometry experiments, it has been demonstrated that compared to free DOX, DOX-CD can be more efficiently absorbed by tumor cells and induce cell apoptosis. The tumor treatment results of mice showed that compared with the control group (PBS and DOX groups), DOX-CD had a stronger inhibitory effect on tumors in mice (Gong et al. 2016).

Paclitaxel is a hydrophobic anticancer drug with significant toxic side effects such as allergic reactions, nausea, vomiting, and many more. It is commonly loaded into nanocomposites and widely used in cancer treatment. Zhang et al. prepared nanocomposites using magnetic Fe_3O_4 as the core Fe_3O_4@CDs @ $mSiO_2$@PTX @ $MSiO_2$, in which CDs and PTX are embedded inside the shell and covered with mesoporous silica on the surface. Among them, CDs and Fe_3O_4 are used for single/two-photon cell imaging and magnetic resonance imaging, respectively. Due to the supramolecular gate stacking and mesoporous double shell structure between the conjugated ring of PTX molecule and the aromatic ring of CDs, this nanocomposite material has a high drug loading capacity. CDs in nanocomposites can absorb near-infrared light and convert it into thermal energy, thereby breaking the chemical bond between PTX and $mSiO_2$ and releasing PTX, effectively killing cancer cells. In vivo and in vitro experiments have shown that the nanocomposite releases anticancer drugs under near-infrared light activation, with advantages such as good tumor inhibition and low side effects (Yang et al. 2016).

Carbon quantum dots have good biocompatibility and low toxicity and are less harmful to the body. It has excellent fluorescence and photothermal conversion properties, suitable for fields such as biological imaging and photothermal therapy. The surface is easy to modify and functionalize, enabling targeted delivery and intelligent release of drugs. At present, the preparation process of carbon quantum dots is not mature enough and needs further optimization to improve its yield and quality. The long-term safety and stability of carbon quantum dots in vivo still need further research and verification. More personalized and precise treatment plans need to be developed for different types of tumors and individual differences in patients. In summary, carbon quantum dots have broad application prospects and important research value in tumor diagnosis and treatment. With the continuous optimization of preparation processes and in-depth research, it is believed that carbon quantum dots will play a more important role in tumor diagnosis and treatment.

5.3 Diagnosis Applications

Carbon nanomaterials have shown many advantages in early diagnosis of tumors, mainly due to their unique physical, chemical, and biological properties. Carbon nanomaterials, such as carbon nanotubes, carbon quantum dots, and many more, have excellent optical, electrical, and magnetic properties. These performances enable them to serve as efficient signal amplifiers, significantly improving the sensitivity and specificity of tumor biomarker detection. By modifying or functionalizing carbon nanomaterials, they can be combined with specific tumor markers to achieve early and accurate detection of tumors. Carbon nanomaterials can be combined with various imaging techniques, such as fluorescence imaging, photoacoustic imaging, and magnetic resonance imaging (MRI), to achieve multimodal imaging. This multimodal imaging technology can provide more comprehensive and accurate tumor information, which helps doctors to diagnose and stage tumors more accurately. Carbon nanomaterials have advantages in early diagnosis of tumors, such as high sensitivity and specificity, multifunctionality, good biocompatibility and stability, targeting, and real-time monitoring and dynamic evaluation. These advantages make carbon nanomaterials have broad application prospects and important research value in the field of early diagnosis of tumors.

5.3.1 Fluorescence Imaging

Carbon nanomaterials, such as carbon quantum dots, carbon nanotubes, and many more, have excellent fluorescence properties, good biocompatibility, and low toxicity, which make them show great potential in the biomedical field, especially in tumor fluorescence imaging (Chen et al. 2019; Lesani et al. 2021). Carbon quantum

specific recognition of two types of nucleic acids. When imaged using super-resolution microscopy, the 3D structures of chromosomes and nucleoli were reconstructed. Finally, in situ imaging of DNA and RNA was achieved in live nematodes, and green and red fluorescence signals were observed at 488-nm and 543-nm excitation, respectively, indicating that the cationic cQD probe has the ability to simultaneously track DNA and RNA in real time (Han et al. 2019).

In 2020, Yang Dayong's research group synthesized chiral nanocatalysts L-CDs and D-CDs derived from cysteine using cysteine as a carbon source. It can simulate topoisomerase to regulate the topological rearrangement of supercoiled DNA and has enantioselectivity. D-CDs can better embed into double helix DNA than L-CDs. The inserted CDs can catalyze the generation of hydroxyl radicals to cleave the phosphate backbone on the double helix single strand of DNA, leading to topological rearrangement of supercoiled DNA. After treatment with L-CDs, the rearrangement rate of pDNA was less than 55.0%, while after treatment with D-CDs, the rearrangement rate of pDNA was higher than 91.3%. And the molecular dynamics simulation results indicate that compared with L. cysteine, D-cysteine has a stronger affinity for DNA, which may be due to the chiral selectivity of CDs toward pDNA (Li et al. 2020).

In 2023, Wang has developed iron doped CDs (Fe CDs) for the first time, which exhibit high quantum rates (QY = 1.27%) between 900 and 1200 nm. They can be used as effective probes for NIR-II biological imaging in vivo, with a penetration depth of 4 mm, reflecting real-time pH changes. Fe CDs probes can non-invasively monitor changes in gastric pH during digestion in mice (Ci et al. 2023).

With the continuous development and progress of nanotechnology, the application prospects of carbon nanomaterials in tumor fluorescence imaging will be even broader. In the future, we can look forward to the emergence of more new carbon nanomaterials and their innovative applications in tumor diagnosis and treatment. At the same time, it is necessary to strengthen the safety assessment and standardization research of carbon nanomaterials to ensure their safety and effectiveness in clinical applications. In summary, carbon nanomaterials play an important role in tumor fluorescence imaging, and their unique physical and chemical properties provide strong support for early diagnosis and precise treatment of tumors. With the continuous deepening of research and the continuous advancement of technology, we have reason to believe that carbon nanomaterials will show broader application prospects in the field of tumor diagnosis and treatment.

5.3.2 Photoacoustic Imaging

The application of carbon nanomaterials in tumor photoacoustic imaging is becoming increasingly widespread, and their unique physical and chemical properties provide strong support for early diagnosis and precise treatment of tumors (Lee et al. 2018). Carbon nanomaterials, such as graphene, carbon nanotubes (including single-walled carbon nanotubes and multi-walled carbon nanotubes), carbon quantum dots, and mesoporous carbon nanospheres, have good biocompatibility, high photothermal conversion efficiency, and strong absorption characteristics in the near-infrared region (Zhao et al. 2023). These characteristics make carbon nanomaterials an ideal contrast agent for tumor photoacoustic imaging. Photoacoustic imaging is a hybrid mode biological/medical imaging method based on the photoacoustic effect. In photoacoustic imaging, pulsed laser illuminates the imaging site, and the absorbed light energy is converted into thermal energy, causing thermal expansion of nearby tissues and generating broadband ultrasound waves (Guo et al. 2020). This type of ultrasonic wave can be detected by an ultrasonic transducer and form an image. Graphene and carbon nanotubes have been widely studied and used as contrast agents for photoacoustic imaging due to their excellent optical properties and ease of preparation. They can absorb near-infrared light and convert it into heat energy, thereby generating strong photoacoustic signals. Carbon quantum dots also have near-infrared light absorption ability, good biocompatibility, and low toxicity, therefore they have potential application value in photoacoustic imaging. In recent years, mesoporous carbon nanospheres have attracted attention due to their strong photothermal conversion ability and rich mesoporous structure. Research has shown that mesoporous carbon nanospheres exhibit excellent performance in photoacoustic imaging and can produce high contrast imaging effects. Researchers have successfully applied carbon nanomaterials to photoacoustic imaging of tumors. After delivering carbon nanomaterials into the body through intravenous injection and other methods, they can target and enrich at the tumor site, and generate strong photoacoustic signals under pulsed laser irradiation, thereby achieving high-resolution imaging of tumors. For example, studies have shown that mesoporous carbon nanospheres generate significantly higher photoacoustic signal intensity in agar mimics under near-infrared laser irradiation than other control photothermal materials (such as graphene, single-walled carbon nanotubes, and gold nanorods), and also exhibit excellent photoacoustic imaging ability in mice. Zhao et al. developed nucleus-targeted carbon dots as peroxidase nanozyme for photoacoustic imaging and phototherapy of tumor (Fig. 5.5) (Liu et al. 2024).

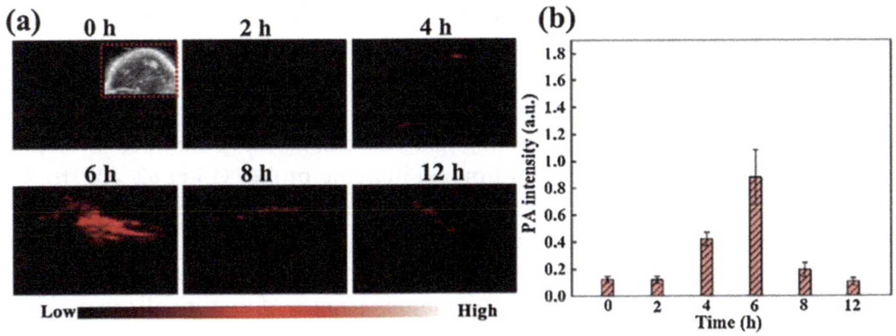

Fig. 5.5 In vivo PAI of mice injected with CDs (reproduced from Liu et al. 2024) Copyright © 2024, Elsevier

With the continuous development and progress of nanotechnology, the application of carbon nanomaterials in tumor photoacoustic imaging will become more extensive and in-depth. In the future, we can look forward to the emergence of more new carbon nanomaterials and their innovative applications in tumor diagnosis and treatment. At the same time, it is necessary to strengthen the safety assessment and standardization research of carbon nanomaterials to ensure their safety and effectiveness in clinical applications. In summary, carbon nanomaterials have broad application prospects and important research value as contrast agents in tumor photoacoustic imaging.

5.3.3 Magnetic Resonance Imaging

The application of carbon nanomaterials in tumor magnetic resonance imaging (MRI) is receiving increasing attention, and their unique physical and chemical properties provide strong support for early diagnosis and precise treatment of tumors (Tomczyk et al. 2020; Lu et al. 2019a, b). Carbon nanomaterials, such as carbon nanotubes (including single-walled carbon nanotubes and multi-walled carbon nanotubes), graphene, and so on, have high specific surface area, good biocompatibility, and modifiability. These characteristics enable carbon nanomaterials to serve as carriers, loading magnetic nanoparticles, fluorescent nanoparticles, and targeting biomolecules, thereby preparing magnetic resonance imaging contrast agents with multiple functions (Zhou et al. 2019; Li et al. 2023b). Magnetic carbon nanocomposites, such as magnetic multi-walled carbon nanotubes (MAGCNTs), are prepared by modifying magnetic nanoparticles (such as superparamagnetic iron oxide SPIO) onto carbon nanotubes to achieve excellent magnetic properties. These materials can significantly shorten the proton spin transverse relaxation time (T2) and effective transverse relaxation time (T2*) in MRI, thereby enhancing the contrast between healthy tissue and diseased tissue, improving the sensitivity and diagnostic accuracy of MRI. For example, there are studies that further load folate targeted biomolecules onto functionalized magnetic multi walled carbon nanotubes through amide reactions to prepare targeted magnetic resonance contrast agents. This contrast agent exhibits excellent performance in specific magnetic resonance imaging at the cellular level, specifically labeling tumor cells and enhancing the MRI signal intensity of tumor tissue. Carbon nanomaterials can also be combined with other imaging techniques to prepare multimodal imaging probes (Li et al. 2021b). For example, by modifying fluorescent nanoparticles onto magnetic carbon nanotubes, nano-molecular probes with both optical and magnetic resonance functions can be prepared. This probe can not only provide high-definition anatomical images in MRI, but also monitor the dynamic changes of tumor cells in real time through optical imaging, providing more comprehensive information for the diagnosis and treatment of tumors. In addition to serving as imaging contrast agents, carbon nanomaterials can also be used as drug delivery systems to load anti-tumor drugs onto carbon nanotubes, achieving targeted delivery and controlled release of drugs. This drug delivery system can prolong the retention time of drugs in tumors, slow down tumor growth, and reduce adverse reactions of drugs to normal tissues. Zhao et al. present the synthesis, characterization, and application of gadolinium and ytterbium co-doped carbon dots (Yb/Gd-CDs) for triple-modal fluorescence (FL)/magnetic resonance (MR)/computed tomography (CT) imaging (Fig. 5.6) (Shu et al. 2024).

In recent years, with the continuous development and progress of nanotechnology, significant progress has been made in the application of carbon nanomaterials in tumor magnetic resonance imaging. For example, research has successfully applied magnetic multi-walled carbon nanotubes to MRI of mouse tumor models, achieving high-resolution imaging and precise localization of tumor tissues. In addition, carbon nanomaterials have been combined with other therapeutic methods such as photothermal therapy and photodynamic therapy, providing new ideas and methods for the comprehensive treatment of tumors. With the continuous deepening of research on carbon nanomaterials and the continuous innovation of technology, we can expect the emergence of more new types of carbon nanomaterials and their innovative applications in tumor magnetic resonance imaging. At the same time, it is necessary to strengthen the safety assessment and standardization research of carbon nanomaterials to ensure their safety and effectiveness in clinical applications.

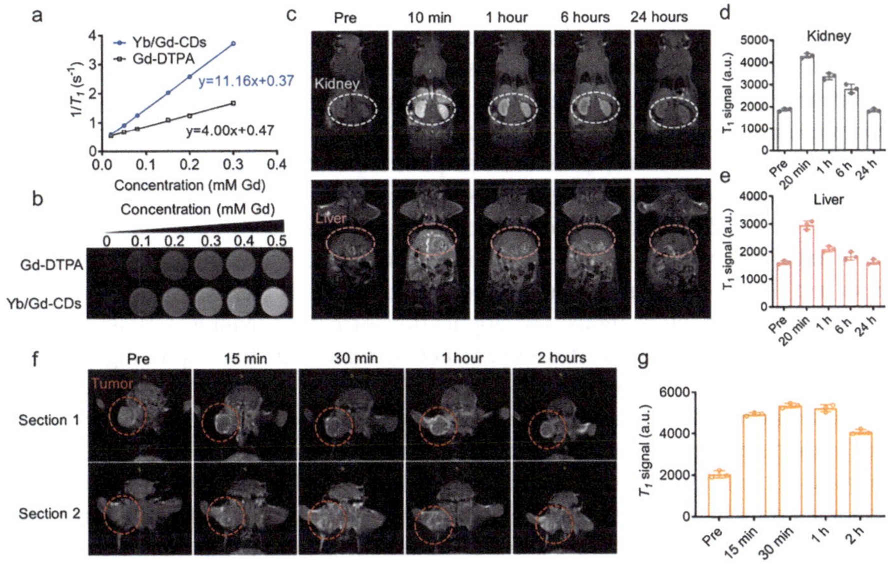

Fig. 5.6 In vitro and in vivo MR imaging of Yb/Gd-CDs and Gd-DTPA (reproduced from Shu et al. 2024) Copyright © 2024, Elsevier

tunable fluorescence emission and efficient light absorption ability. This enables them to serve as ideal photosensitizers, generating sufficient ROS to kill tumor cells under specific wavelengths of light irradiation. Carbon nanomaterials have good biocompatibility, can exist stably in vivo, and exert therapeutic effects, reducing the risk of immune reactions, and other adverse reactions caused by the materials themselves. The surface of carbon nanomaterials can be chemically modified to bind with specific ligands, achieving specific recognition of tumor tissues. This can not only improve the targeting of PDT, but also reduce damage to healthy tissues. Carbon nanomaterials can not only serve as photosensitizers for PDT, but also be combined with other therapeutic methods such as photothermal therapy (PTT), chemotherapy, and so on, to achieve multimodal synergistic therapy. This collaborative therapy can further improve the effectiveness of tumor treatment.

Carbon dots, as a new type of fluorescent nanomaterial, have excellent optical properties, extremely small size, and good biocompatibility (Li et al. 2021a, b). Partial carbon dots have the potential to act as photosensitizers to induce reactive oxygen species in tumor photodynamic therapy, enabling them to play an important role in the biomedical field. Through chemical modification, functional groups or molecules such as ferrocene can be introduced onto the surface of carbon nanomaterials. These functionalized carbon nanomaterials not only improve the tumor targeting properties of the materials, but also enhance the therapeutic efficacy of PDT (Wei et al. 2024). For example, ferrocene carbon dot crosslinked nanoparticles (Fc-CD NPs) can achieve enhanced tumor accumulation and specific amplification in synergistic therapy (including PDT and PTT). Carbon nanomaterials also have good imaging properties, such as photoacoustic imaging (PAI). Through multimodal imaging technology, the distribution of carbon nanomaterials in tumor tissue and the therapeutic effect of PDT can be monitored in real time, providing strong support for precision medicine.

Carbon dots are used as small molecule photosensitizer carriers for photodynamic therapy due to their good biocompatibility, low toxicity, and good stability, effectively improving the therapeutic efficiency of photosensitizers. Chen used PEG2000N as a passivator to synthesize carbon dots with rich amino groups on the surface. Through amidation and coupling with Ce6, a FRET process was constructed between the carbon dots and Ce6 to enhance the photodynamic ability of Ce6. When irradiated with 671-nm laser, the CDs-Ce6 nanosystem exhibits stronger red fluorescence signal and higher ROS generation efficiency compared to free Ce6, while avoiding the dark toxicity of Ce6 and improving the stability of the photosensitizer. In addition, the CDs. Ce6 system can be more effectively enriched at the tumor site through EPR effect, achieving efficient fluorescence imaging and photodynamic therapy (Huang et al. 2012). Hahn et al. used the same method to amidate carbon dots and couple them with Ce6, followed by modification with sodium hyaluronate to prepare a nanocarrier system Cdot-Ce6-HA. The hyaluronic acid contained in Cdot-Ce6-HA can recognize overexpressed hyaluronic acid receptors in melanoma cells and perform photodynamic therapy. In vitro experiments using melanoma cells have shown that the Cdot-Ce6-HA nanosystem has a much higher ability to generate reactive oxygen species and kill tumor cells after exposure to light than free Ce6 and Cdot-Ce6. In vivo experiments on mice also showed that subcutaneous administration of Cdot-Ce6-HA can effectively deliver to tumor tissues, and compared to the non-illuminated area, the growth of tumor tissues in the illuminated area is significantly inhibited (Beack et al. 2015). Camptothecin (CPT) drug was grafted onto the surface of carbon dots through thiol linkers (TL), and CPT was effectively released under acidic environment and light, achieving image-guided photodynamic and chemotherapeutic with on demand drug release (Fig. 5.7) (Getachew et al. 2023).

Due to the fact that the nucleus is the main site of biological genetic information and the regulatory center of cellular genetics and metabolism, damage to the nucleus can easily lead to cell apoptosis or direct necrosis. Therefore, photosensitizers targeting the cell nucleus are receiving increasing attention from researchers. Wu used meta-phenylenediamine and L-cysteine as carbon sources to synthesize nucleic acid targeted carbon dots CDs via hydrothermal method in alkaline environment. After amidation, they were connected to protoporphyrin (PplX) to prepare CD-PplX. During cell imaging, CDs can quickly target the nucleolus, and their fluorescence signal intensity remains constant for 24 h. The intracellular nucleic acid digestion experiment and fluorescence titration response experiment showed that compared to commercial nucleic acid dyes SYTO@RNA Select™ and Hoechst 33342, CDs preferentially bind to RNA and their dynamic changes can be tracked in real-time for a long time. After binding to PpIX, CD-PpIX still maintains its ability to target nucleic acids. Upon exposure to light, PpIX directly generates singlet oxygen in the nucleus, disrupting the genetic information carrier within the cell and

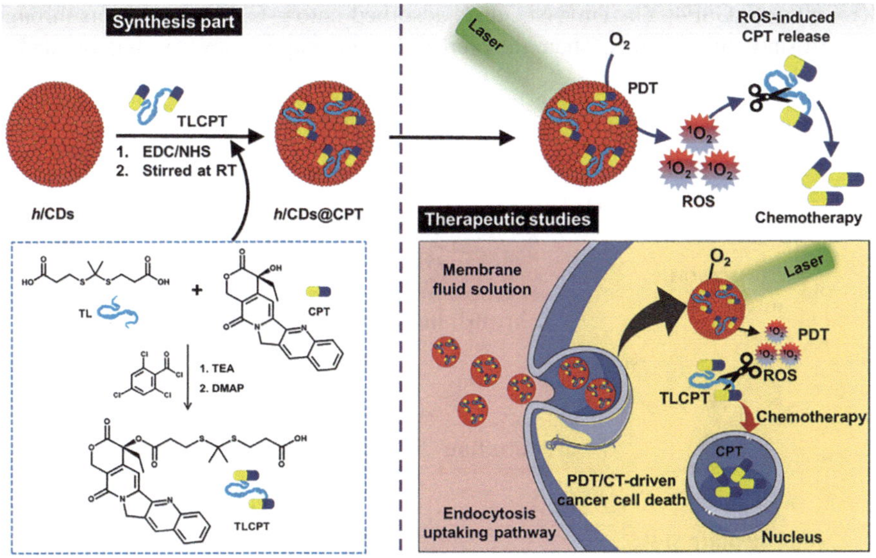

Fig. 5.7 Schematic illustrations of synthesis of h/CDs@CPT for PDT and chemotherapy of cancer (reproduced from Getachew et al. 2023) Copyright © 2023, Elsevier

causing apoptosis or necrosis. In vitro and in vivo experiments have confirmed that CD-PplX has stronger killing ability on tumor cells compared to pure PplX (Hua et al. 2018). Ru-CDs have strong red light emission and efficient generation of reactive oxygen species (ROS), which can achieve DNA photolysis and thus achieve photodynamic therapy for tumors (Fig. 5.8) (Yue et al. 2020).

The malignant growth of tumor cells leads to an increase in intracellular oxygen consumption, making hypoxia an important feature of the tumor microenvironment ($pO_2 < 2.5$ mmHg). During photodynamic therapy, intracellular hypoxia is exacerbated, and the decrease in intracellular oxygen levels limits the efficiency of photodynamic therapy. This vicious cycle can lead to irreversible tumor metastasis or drug resistance. Therefore, the development of nano-systems for hypoxic environments is particularly important. Zhang used citric acid and urea as carbon sources to prepare carbon dots using a hydrothermal method, assembled with C3N4 as a carrier, and used PEG as a connecting arm. PplX and RGD (Arg-Gly-Asp), a peptide chain targeting tumor cells, were used to form an amphiphilic polymer PplX-PEG-RGD, which was then self-assembled with C3N4 through the stacking effect of π-π to prepare a nano polymer PCCN (Fig. 1.19a). When irradiated with a 630-nm laser, C3N4 in PCCN can effectively decompose water to produce oxygen, increase the concentration of oxygen in tumor cells, and promote the production of singlet oxygen by PplX during illumination to kill tumor cells. And the killing ability of 4 T1 cells was basically the same when tested at normal O_2 concentration or 1% O_2 concentration through in vitro experiments. Finally, the tumor was injected into the tail vein of mice, and the RGD peptide chain actively targeted and EPR effect aggregated at the tumor site, effectively inhibiting tumor growth after exposure to light (Zheng et al. 2016). Yang's upconversion nanoparticles with core–shell structure (UCNPs, NaGdF4: Yb, Tm@NaGdF4) adsorbed onto g-C3N4 nanosheets through electrostatic interactions. Then, the prepared nanocomposites were encapsulated in

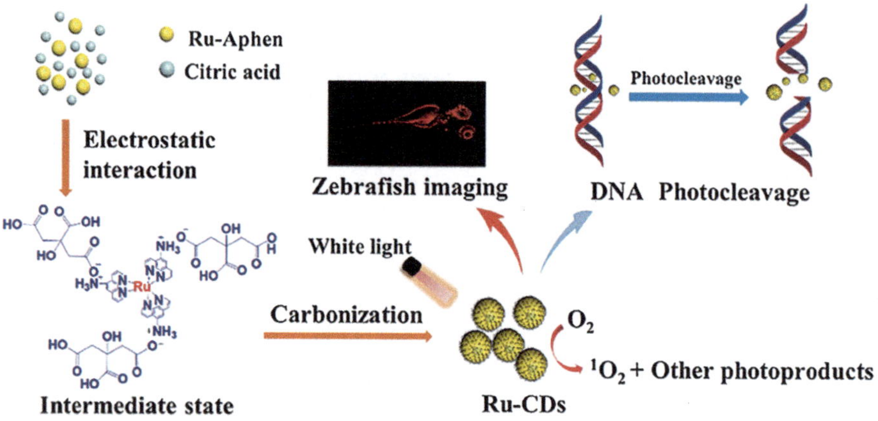

Fig. 5.8 Schematic diagram of preparation and DNA photocleavage of Ru-CDs (reproduced from Yue et al. 2020) Copyright © 2020, ACS

situ with carbon dots within a metal organic framework (ZIF-8) to achieve a hybrid nanosystem UCNPs-g-C3N4-CDs@ZIF-8 with dual photosensitizers. Among them, the upconversion nanoparticles convert near-infrared light into high-energy ultraviolet light for the excitation of g-C3N4 and CDs. On the one hand, it improves the hypoxia situation in tumor cells, and on the other hand, it increases the rate of reactive oxygen species production. Moreover, it can store O_2 or H_2O in the ZIF-8 shell, further regulating the oxygen content in real tumors and enhancing the ability to inhibit tumor growth (Yang et al. 2017).

By modifying photosensitizers on the surface of carbon dots using various methods for photodynamic therapy, such as covalent coupling, electrostatic interactions, and π–π stacking, as well as activating photosensitizers through methods such as FRET and direct irradiation, tumor cells can be effectively damaged. With the development of carbon dots, researchers have found that carbon dots also have photosensitizing ability, which can generate reactive oxygen species for killing tumor cells under light without the need for cumbersome steps to modify small molecule photosensitizers. Markovic and Trajkovic's research group prepared graphene carbon quantum dots GQD by electrochemical oxidation method. Under 1.0 W/cm^2 and 470-nm blue light irradiation, GQD produces singlet oxygen, kills glioma cells, and induces autophagy. Through TEM observation and Western blot experiments, it was confirmed that autophagic vesicles, LC3-I and LC3-II transformation, and p62 degradation accelerate glioma cell death in cells (Markovin et al. 2012). Bi Hong's research group used acrylic acid and $Cu(NO_3)_2$ as raw materials to polymerize Cu complexed polyacrylic acid in situ, and then obtained Cu doped carbon dots Cu-CDs through high-temperature cracking. This carbon dot has a high fluorescence quantum yield (24.4%), good dispersibility, low toxicity, and a singlet oxygen quantum yield of 36.0% under white light irradiation (400–700 nm, 40.0 mW/cm^2). It can be used for fluorescence imaging of tumor spheres in HeLa cells and neuroblasts and can effectively inhibit the growth of tumor spheres after illumination (Wang et al. 2019).

The efficiency of generating reactive oxygen species is an important parameter for evaluating the performance of photosensitizers. Due to the high activity and short lifespan of ROS, the stronger the ability of photosensitizers to produce reactive oxygen species, the more favorable they are for causing damage to tumor cells. Wang Pengfei's research group synthesized carbon dot GQDs using polythiophene derivatives as a carbon source via hydrothermal method. GQDs have a wide absorption wavelength in the visible light region and a strong emission peak at 680 nm. When using white light irradiation, the yield of singlet oxygen in GQDs is 1.3, which is much higher than that of commercial photosensitizers and has enhanced photodynamic therapy ability. The author believed that the high quantum yield of singlet oxygen generated by GQDs is due to the polymorphic sensitization mechanism. By adjusting the energy level between S1 and T1, the energy gap ΔE_{ST} (energy gap between S1 and T1) can be made larger than the energy required to sensitize oxygen to produce singlet oxygen. There are two ways in which this multi-sensitization mechanism generates singlet oxygen: one is the traditional pathway from T1 and O_2 sensitization to S0, and the other is the transition from S1 to T1, which generates singlet oxygen, and then generates singlet oxygen again from T1 to S0. As a result, the singlet oxygen quantum yield of GQDs (1.3) is higher than the theoretical maximum value (1.0). Song prepared carbon dots (R-CDs) using methylene blue (MB) and phosphate. R-CDs have a high singlet oxygen generation ability (0.91) due to the extended triplet lifetime, achieving enhanced photodynamic therapy (Fig. 5.9) (Xu et al. 2021a, b).

With the in-depth study of carbon dots, it has been found that carbon sources with special properties are beneficial for preparing carbon dots with special properties, that is, carbon dots can obtain some specific properties of carbon sources, such as functional groups, targeting, and photodynamic or photothermal properties. Porphyrin carbon dot TPP CDs were prepared using monohydroxyphenyl triphenylporphyrin (TPP) and chitosan as raw materials. Compared to TPP, TPP CDs have better water solubility, can quickly enter cells, and can be localized in lysosomes. They have a higher counterstaining coefficient with commercial lysosome trackers and can also generate singlet oxygen. In vitro cell experiments and mouse tumor model experiments have confirmed that TPP CDs have strong killing ability against tumor cells under 625-nm irradiation. Shuai prepared carbon dot DPP CDs using pyrrolidinedione and chitosan as carbon sources. Compared to the small molecule photosensitizer DPP, this carbon dot has good biocompatibility and dispersibility, can be localized to lysosomes, and can generate singlet oxygen under light, effectively inhibiting tumor proliferation (He et al. 2018; Li et al. 2017). Peng used the same strategy to synthesize carbon dot MB CDs using methylene blue as the carbon source and hydrothermal method. Compared with methylene blue, MB CDs have good biocompatibility and low toxicity (Figs. 5.10 and 5.11). MB CDs inherit the excellent photosensitizing properties of methylene blue, with a triplet lifetime of up to 11.4 μs and a singlet oxygen quantum yield of 24.0%. More importantly, during

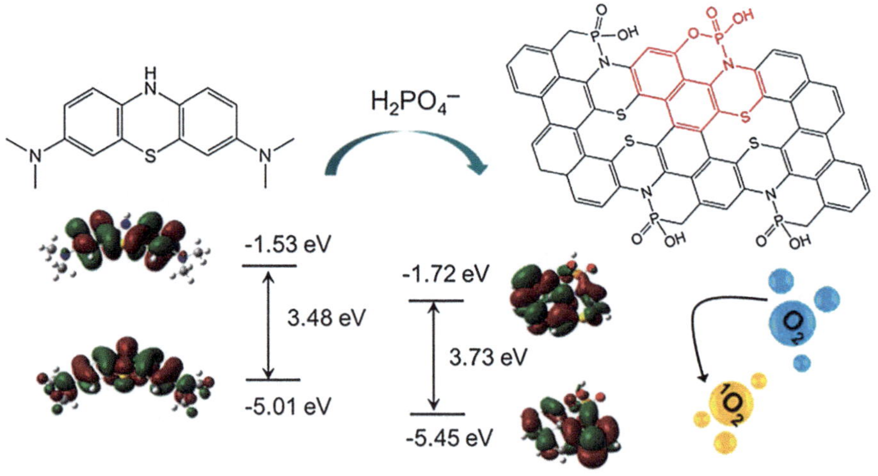

Fig. 5.9 Schematic diagram of preparation of Ru-CDs for photodynamic therapy (reproduced from Xu et al. 2021a, b) Copyright © 2021, ACS

the photodynamic therapy of tumors, the disadvantage of methylene blue being easily reduced by reducing atmosphere, aggregation, and DNA interference, which leads to the loss of photodynamic performance, is avoided. Both in vitro and in vivo experiments have strong killing ability against tumor cells (Xu et al. 2021). Ge prepared Mn doped carbon dot Mn CDs using manganese (II) phthalocyanine as a carbon source, and then self-assembled into Mn CDs NPs by PEG modified distearoyl phosphatidylethanolamine. This nanosystem exhibits near-infrared absorption and emission, with a singlet oxygen quantum yield of 12.0%. Due to the efficient catalysis of H_2O_2 to generate oxygen by Mn doping, the oxygen concentration in the tumor microenvironment is increased, overcoming the hypoxic microenvironment of tumor tissue and the rapid consumption of oxygen in photodynamic therapy, thereby improving photodynamic efficiency. Meanwhile, Mn^{2+} is also a contrast agent that enables Mn CDs NPs to possess magnetic resonance imaging capabilities, enabling precise photodynamic therapy under multi-mode guidance. In vitro and in vivo experiments have confirmed that Mn CDs NP is both an excellent photosensitizer and a good contrast agent, with good inhibitory effects on tumor growth (Xu et al. 2021a).

In 2021, Lan prepared red fluorescent N, S-doped carbon dots (N, S-CDs) using polythiophene derivatives, with a singlet oxygen quantum yield of about 8%, mainly enriched in lysosomes and mitochondria of tumor cells, achieving photodynamic therapy (Fig. 5.10) (Zhao et al. 2021).

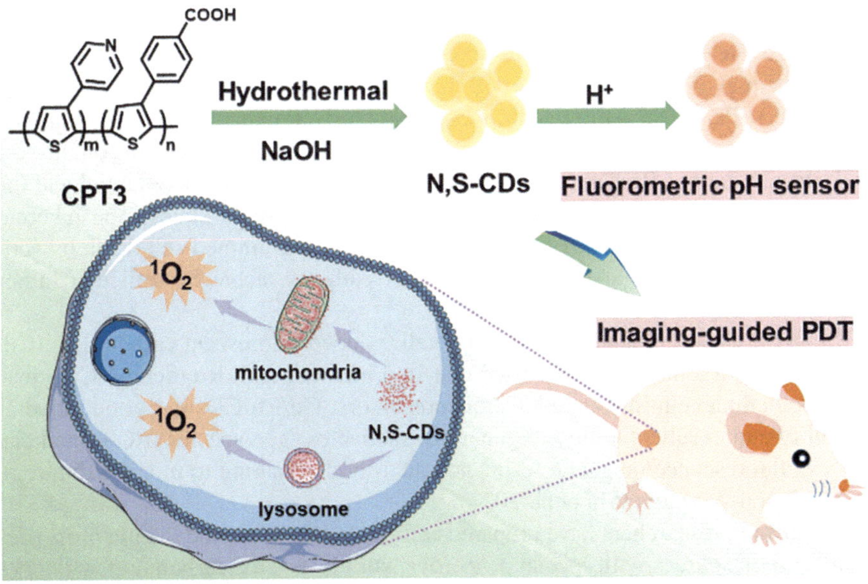

Fig. 5.10 Illustration of preparing N,S-CDs and its application in imaging-guided PDT (reproduced from Zhao et al. 2021) Copyright © 2021, ACS

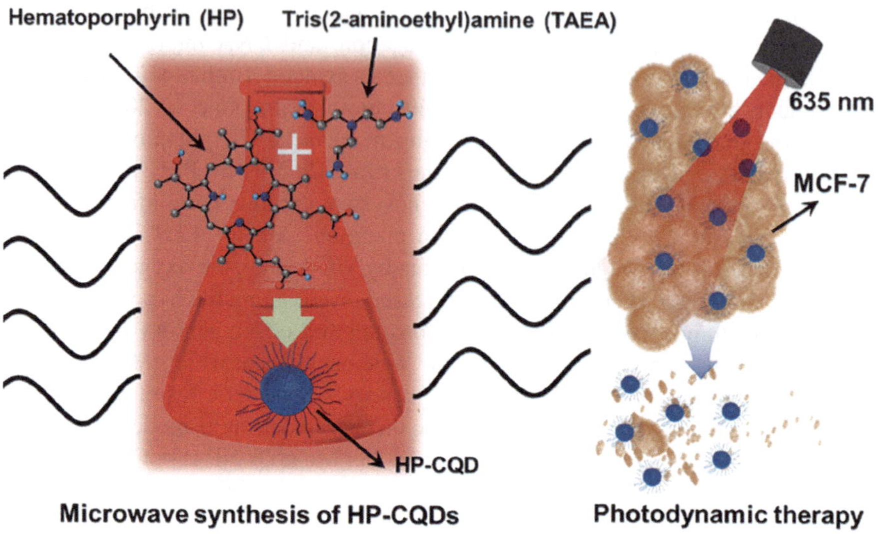

Fig. 5.11 Illustration of preparing HP-CQDs for PDT (reproduced from Murali et al. 2022) Copyright © 2022, ACS

In 2022, In used Hematoporphyrin (HP) photosensitizer encapsulated carbon quantum dots (HP CQDs) as precursors and prepared them through a one-step microwave reaction, avoiding complex processes. Compared to HP, HP-CQDs exhibited very high phototoxicity and low dark toxicity toward MCF-7 cells (Fig. 5.11) (Murali et al. 2022).

In 2023, Ge prepared three types of red emission carbon dots RCDs, which could generate equal amounts of superoxide anions (type I PDT) and tunable singlet oxygen (type II PDT). Structural and optical characterization, as well as DFT calculations, indicate that the formation of novel I/II type ROS mainly depends on the size and surface state of RCDs, which determine their identical redox potential and the gradually narrowing energy gap between singlet and triplet states. The inherent mitochondrial targeting ability of RCD can induce programmed cell death by activating the mitochondrial mediated apoptosis pathway, achieving efficient tumor eradication (Fig. 5.12) (Zhang et al. 2023).

In 2024, Cheng prepared Gd/Ru bimetallic doped fluorescent carbon dots (Gd/Ru CDs) using a microwave-assisted one-step method, with Ru (dcbpy)$_3$Cl$_2$, citric acid, polyethyleneimine, and GdCl3 as precursors. Gd/Ru CDs has good biocompatibility and excellent ability to generate reactive oxygen species (ROS), and can achieve fluorescence/magnetic resonance dual-mode imaging to monitor the treatment of mouse tumors (Shi et al. 2024).

At present, researchers have prepared carbon dots with photodynamic properties using carbon sources with special properties, but most of the carbon dots with photodynamic properties do not have targeting ability, resulting in limited killing ability against cancer cells. Research has shown that photodynamic carbon dots with

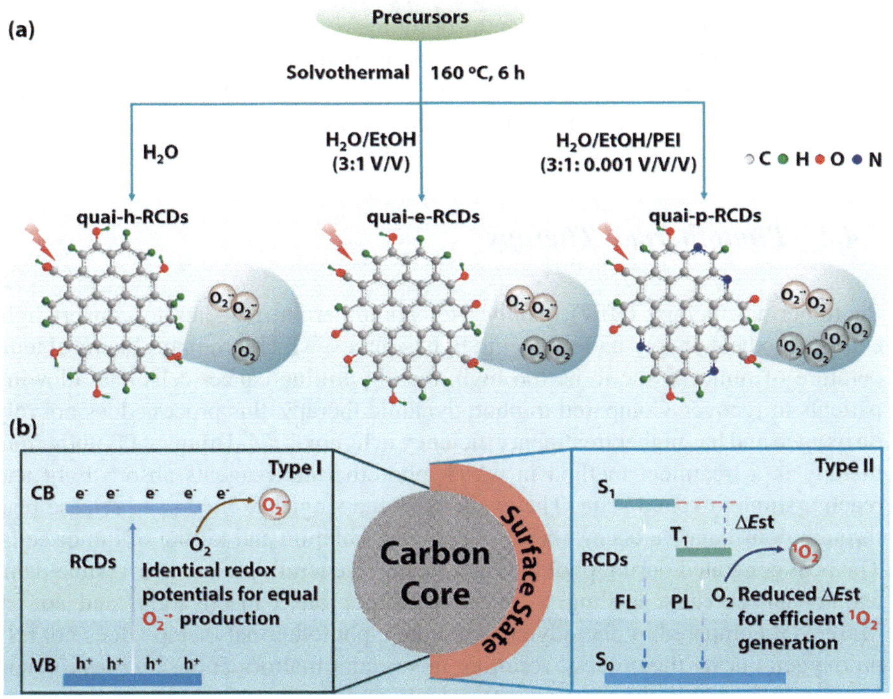

Fig. 5.12 Illustration of RCDs as NanoPSs with tunable ROS generation (reproduced from Zhang et al. 2023) Copyright © 2023, Elsevier

targeted properties have better therapeutic effects than carbon dots without targeted properties. In 2020, Gomes et al. synthesized two types of carbon dots using curcumin as the carbon source, one containing folate fragments as the carbon dot CDCF, and the other being the bare carbon dot CDc (Fig. 5.11). Through cell imaging experiments, it has been confirmed that CDCF can target the nucleus of cells, while CDc does not have targeting ability. Flow cytometry and MTT experiments have also shown that CDCF with targeting ability has significantly higher lethality toward H413 cells than CDc (Nasrin et al. 2020).

In 2020, Wei prepared carbon dots C-dots using citric acid and ethylenediamine as carbon sources (Figs. 5.11 and 5.12). The singlet oxygen quantum yield of C-dots is 4.8%. Compared with other carbon dots with photodynamic properties, their ability to produce reactive oxygen species is lower. However, this carbon dot can be located in the nucleolus and bind to nucleic acids, enhancing its photodynamic therapy ability. MTT assay and subcutaneous tumor model experiment showed that its targeting ability enhanced the inhibitory effect of C-dots on tumor cells (Pang et al. 2020).

Nanomaterials have demonstrated significant advantages and potential applications in photodynamic therapy for tumors. With the continuous advancement of technology and research, the application of carbon nanomaterials in PDT will become more extensive and in-depth, bringing new breakthroughs and developments to the field of tumor therapy. However, it should be noted that the application of carbon nanomaterials in PDT is still in the research stage, and their safety and effectiveness need further verification and improvement.

5.4.2 Photothermal Therapy

Photothermal therapy (PTT) mainly uses photothermal reagents to convert light energy into heat energy under specific light sources, which promotes the local temperature of tumor tissue to be too high, thereby killing cancer cells and allowing patients to recover. Compared to photodynamic therapy, this process does not rely on oxygen and has higher treatment efficiency in hypoxic solid tumors. Photothermal therapy is a treatment method in which photothermal reagents absorb light and reach a singlet excited state. The excitons in the singlet excited state release heat through non-radiative decay, thereby achieving ablation and killing of tumor cells. The heat generated during photothermal therapy generally causes irreversible damage to cancer cells, leading to tissue edema, protein inactivation, and so on. Moreover, compared to photodynamic therapy, photothermal therapy does not rely on oxygen during the process, resulting in vascular malformations and insufficient blood supply within the tumor tissue, which leads to much lower heat dissipation capacity within the tumor tissue compared to normal tissue. When the normal tissue temperature rises to around 41 °C, the temperature inside the tumor tissue can reach over 48 °C. It is now widely believed that there are two main aspects that affect the efficacy of photothermal therapy: the level of temperature increase and the duration of treatment. At that time, when the temperature was between 41 and 48 °C, proteins would aggregate and require more than 60 min of irradiation to cause irreversible damage to cancer cells. But at this temperature, it can enhance the effectiveness of radiotherapy and chemotherapy treatment. When the temperature exceeds 48 °C, irreversible coagulation necrosis of cells will occur within a few minutes (Pang et al. 2020; Jaque et al. 2014). However, in high temperatures, excessive heat can not only kill cancer cells, but also cause damage to normal tissues near the tumor. Although there are still shortcomings in the clinical application of photothermal therapy, its prospects in tumor treatment are beyond doubt.

The application of carbon nanomaterials in tumor photothermal therapy (PTT) is becoming increasingly widespread, and their unique physical and chemical properties provide new avenues for tumor treatment. Photothermal therapy utilizes materials with high photothermal conversion efficiency to convert light energy into heat energy under external light sources (usually near-infrared light), thereby generating local high temperatures within tumor tissue and inducing apoptosis or necrosis of tumor cells (Han et al. 2022; Zhao et al. 2021). This treatment method has the advantages of strong penetration, good efficacy, and minimal toxic side effects. Carbon nanomaterials, such as carbon nanotubes and graphene, have excellent photothermal conversion properties and can efficiently convert absorbed near-infrared light into thermal energy, thereby generating high temperatures within tumor tissues. Carbon nanomaterials have good biocompatibility, can exist stably in vivo, and exert therapeutic effects, reducing the risk of immune reactions and other adverse reactions caused by the materials themselves (Wang et al. 2021a; Lu et al. 2019a, b). The surface of carbon nanomaterials can be chemically modified to bind with specific ligands, achieving specific recognition of tumor tissues. This can not only improve the targeting of PTT, but also reduce damage to healthy tissues. Carbon nanomaterials can not only serve as photothermal conversion agents for PTT, but also be combined with other therapeutic methods such as chemotherapy, photodynamic therapy, and so on, to achieve multimodal synergistic therapy. This collaborative therapy can further improve the effectiveness of tumor treatment. Functionalized carbon nanotubes have been widely used in PTT due to their unique one-dimensional structure and excellent photothermal conversion performance. Through surface modification, carbon nanotubes can achieve specific enrichment and efficient photothermal conversion within tumor tissues. Graphene and its derivatives, such as graphene oxide (GO), have shown great potential in PTT due to their excellent photothermal conversion performance and good biocompatibility. GO can load chemotherapy drugs or photosensitizers through π-π stacking, electrostatic adsorption, and other methods to achieve synergistic treatment of chemotherapy and PTT. Hollow mesoporous carbon nanospheres have a large specific surface area and porosity and can be loaded with a large amount of chemotherapy drugs or photosensitizers. Meanwhile, its unique hollow structure facilitates multiple reflections and scattering of light within the sphere, improving the efficiency of photothermal conversion. Therefore, hollow mesoporous carbon nanospheres have also shown promising application prospects in PTT. Some carbon nanomaterials also have good imaging performance, such as photoacoustic imaging (PAI), computed tomography (CT), and so on. By combining imaging technology, real-time monitoring of the distribution of carbon nanomaterials within tumor tissue and the therapeutic effect of PTT can be achieved, providing strong support for precision medicine.

Carbon dots in PTT can not only serve as carriers for photothermal reagents, but also act as photothermal reagents themselves to achieve photothermal therapy guided by fluorescence imaging. Wang Pengfei's research group used polythiophene propionic acid synthesized within the group as a carbon source for the hydrothermal synthesis of carbon dots C-dots. This carbon dot has a wide spectral band of absorption (400–800 nm) and red light emission (640 nm). Under the irradiation of a 671 nm laser (2.0 W/cm^2), its photothermal conversion efficiency reaches 38.5%, and it also has the performance of dual-mode imaging of fluorescence and photoacoustic. MTT assay and live/dead cell staining assay showed that C-dots could effectively kill cancer cells after incubation with HeLa and HBE cells, and the temperature rapidly increased after 10 min of light exposure. C-dots are injected into tumor mice through the tail vein and aggregate at the tumor site through EPR effect. When using laser irradiation, the temperature of tumor tissue can reach 58.4 °C. It has a good inhibitory effect on tumor growth during the treatment process (Ge et al. 2015). Subsequently, in order to improve its photothermal conversion efficiency, Wang and Zhang collaborated to prepare S and Se doped carbon dots S/Se-CDs using polythiophene and diphenyl ethylene as carbon sources under alkaline conditions. This carbon dot has a large two-photon absorption cross section (~3×10^4 GM) at 880 nm, which can be used to guide photothermal therapy through two-photon imaging. When irradiated with a 635 nm laser, the photothermal conversion efficiency of S/Se-CDs can reach 58.2%, which is much higher than the photothermal efficiency of C-dots. MTT experiments have shown that only 40.0 μg/mL carbon dots have a high killing effect on HeLa cells after light irradiation, with a cell survival rate of less than 15%. The mouse tumor model experiment also proved that S/Se CDs can effectively inhibit tumor growth after photothermal therapy (Lan et al. 2017).

By targeting important substances such as major organelles, proteins, or nucleic acids of tumor cells, rapid damage to these important substances can be caused, leading to apoptosis or necrosis of tumor cells, thereby improving the efficiency of photothermal therapy. Lin dispersed citric acid in formamide solution and synthesized near-infrared carbon dot R-CDs through a microwave reactor, with a maximum emission wavelength of 640 nm and a fluorescence quantum yield of 22.9%. When irradiated with a 671-nm laser (2.5 W/cm^2), the photothermal conversion efficiency can reach 43.9%. During cell imaging, R-CDs are mainly located in the nucleolus of the nucleus and have excellent photothermal killing ability against MCF-7 cells. This carbon dot can also serve as a drug loading platform, where probes or drugs can be assembled and quickly delivered into cells for use in biological imaging and therapy (Sun et al. 2016). Wu used ortho phenylenediamine and L-cysteine as carbon sources to prepare sulfur and nitrogen co-doped carbon dots, which were then mixed with copper ions and surface modified with HS-PEG and HS-PEG-RITC (Rhodamine B isothiocyanate) to prepare copper/carbon dot cross-linked nanosheets CuCD NSs. The nanosheets have a particle size of approximately 20–30 nm, are uniformly dispersed, and have good biocompatibility. When irradiated with an 808-nm laser (2.0 W/cm^2), its photothermal conversion efficiency is 41.3% and it has good photostability. When CuCD NSs initially enter the cell, they are mainly localized in the mitochondria, and gradually transfer from the mitochondria to the nucleus after light exposure, leading to cell apoptosis after 10 minutes. Subsequently, the tumors in the mice were treated, and the tumor tissues showed good enrichment ability. After exposure to light, there was a significant inhibitory effect on tumor growth. The experimental results demonstrate that CuCD NSs transfer from cytoplasm to nucleus when triggered by light, greatly improving the efficiency of PTT, which will contribute to the development of novel light controlled drug delivery nanoplatforms (Bao et al. 2018).

Based on the above carbon dots, high-power lasers are required for photothermal therapy, which may damage surrounding tissues and cause serious side effects. Therefore, carbon dots with high photothermal conversion efficiency are needed for tumor treatment. Pan used 1,3,6-trinitropyrene (TNP) and branched polyethyleneimine (BEPI) as carbon sources to synthesize nitrogen and oxygen co-doped carbon dot N-O-CDs via a one-step hydrothermal method. Due to the doping of a large amount of pyrrole N and graphite N, N-O-CDs exhibit strong absorption in the

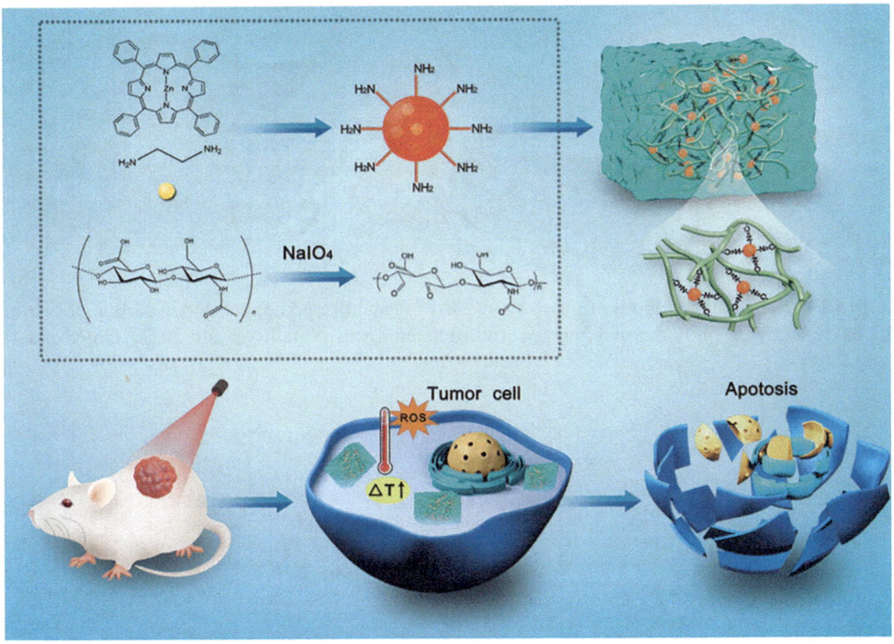

Fig. 5.13 Schematic diagram of the synthesis process of CD@Hydrogel and its application in cancer treatment (reproduced from Yue et al. 2022) Copyright © 2022, ACS

near-infrared region. When irradiated with a low-power (0.8 W/cm^2, 808 nm) laser, the temperature can rise to 55.4 °C and the photothermal conversion efficiency can reach 38.3%. MTT assay confirmed that N-O-CDs have strong cytotoxicity against cancer cells. At the same time, during photothermal therapy in tumor bearing mice, it can effectively inhibit tumor growth and cause almost no damage to normal tissues (Geng et al. 2018).

In 2022, Mei prepared injectable hydrogel by the Schiff base reaction between HA-CHO and carbon dots (CDs), achieving PTT and PTT simultaneously. In vitro and in vivo experiments, CDs@hydrogel with good biosafety showed high anti-tumor effect after 660-nm laser irradiation (Fig. 5.13) (Yue et al. 2022). Khosroushahi prepares copper-doped CDs (Cu doped CDs) using Alcea extract as the carbon source and $CuSO_4$ as the dopant. The photothermal conversion efficiency of Cu-CDs is 39.3%, and it can induce cell thermal death under 800-nm near-infrared laser (Fig. 5.14) (Najaflu et al. 2022). Zheng prepared AS1411 ligand coupled gadolinium-doped CD (AS1411-Gd-CD) using solvothermal method for targeted fluorescence / MR imaging-guided photothermal therapy, achieving significant tumor inhibition effect owing to the outstanding photothermal effect (Fig. 5.15) (Jiao et al. 2022).

In 2023, Petralia prepared carbon-based polymer dots (CPDs-PNM) by a one-pot thermal process from poly(N-isopropylacrylamide) (PNIPAM) without using organic solvent and additional reagents. The CPDs-PNM exhibited high photothermal conversion efficiency and photothermal-controlled drug release, achieving

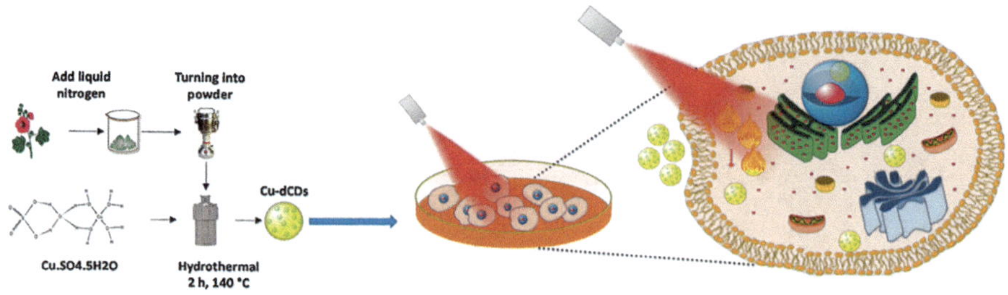

Fig. 5.14 Schematic diagram of green synthesis of near-infrared copper-doped carbon dots from alcea for cancer photothermal therapy (reproduced from Najaflu et al. 2022) Copyright © 2022, ACS

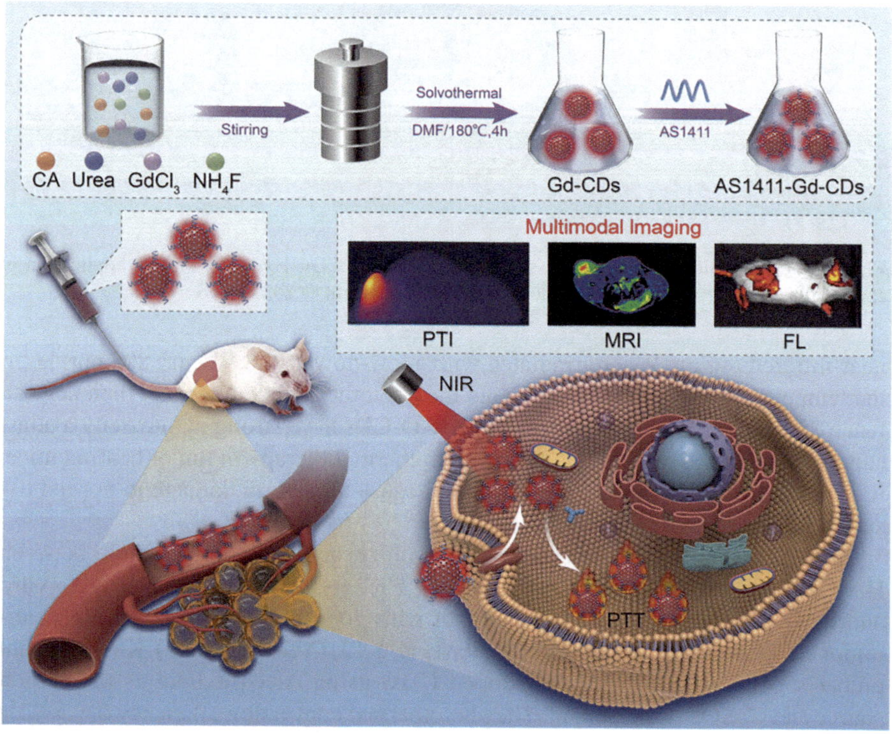

Fig. 5.15 Schematic illustration of preparation of AS1411-Gd-CDs and FL/MR-guided photothermal therapy of tumor (reproduced from Jiao et al. 2022) Copyright © 2022, ACS

combined chemo-photothermal cancer therapy (Fig. 5.16) (Consoli et al. 2023). In 2023, Cavallaro prepared CDs-Gd using citric acid, GdCl3, and urea as raw materials via solvothermal decomposition method. CDs-Gd exhibited contrast-enhancing properties in T1-weighted magnetic resonance imaging (MRI) experiments, achieving multimodal image-guided photothermal therapy (Fig. 5.17) (Mauro et al. 2023).

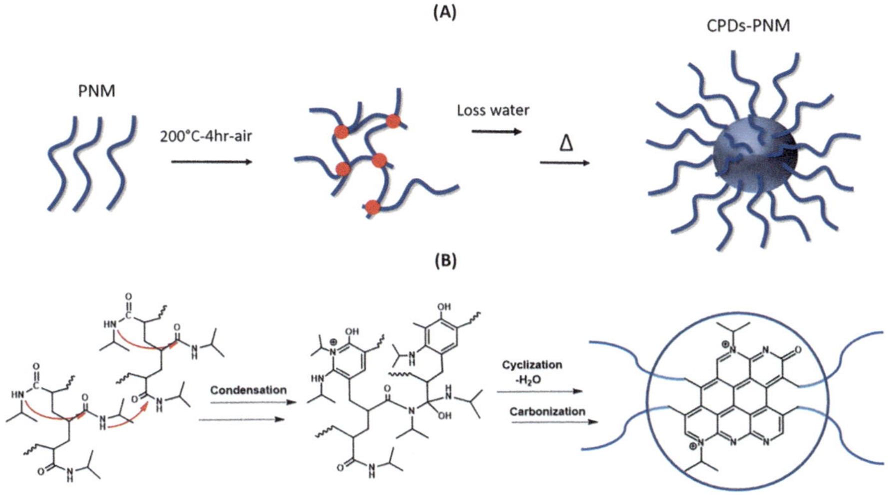

Fig. 5.16 Schematic representation of the CPDs/PNM formation process and proposed mechanism (reproduced from Consoli et al. 2023) Copyright © 2023, ACS

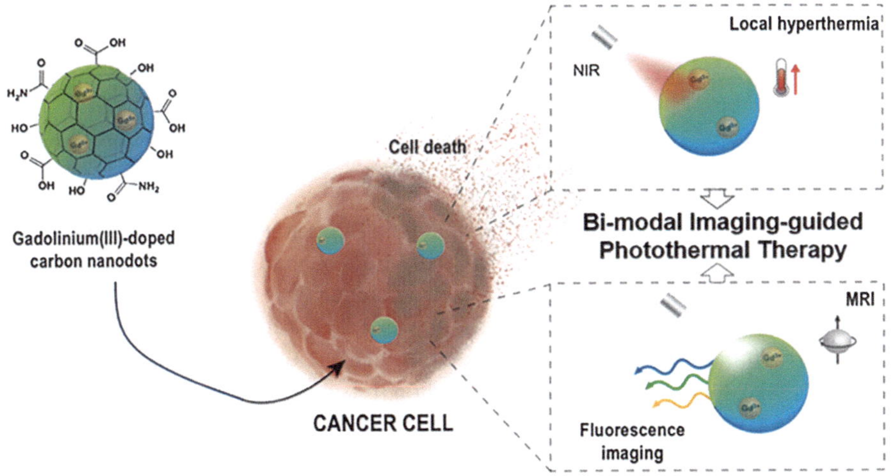

Fig. 5.17 Schematic representation of gadolinium-doped carbon nanodots as potential anticancer tools for multimodal image-guided photothermal therapy and tumor monitoring (reproduced from Mauro et al. 2023) Copyright © 2023, ACS

Zhou synthesized a chemotherapeutic agent, PDA-Ce-CDs, which combines the photothermal conversion ability and the generation of hydroxyl radicals (•OH), enabling synergistic enhancement of anti-tumor effects (Fig. 5.18) (Sheng et al. 2023).

The absorption and emission wavelengths of most carbon dots are mainly concentrated in the blue or green region, while carbon dots with near-infrared second

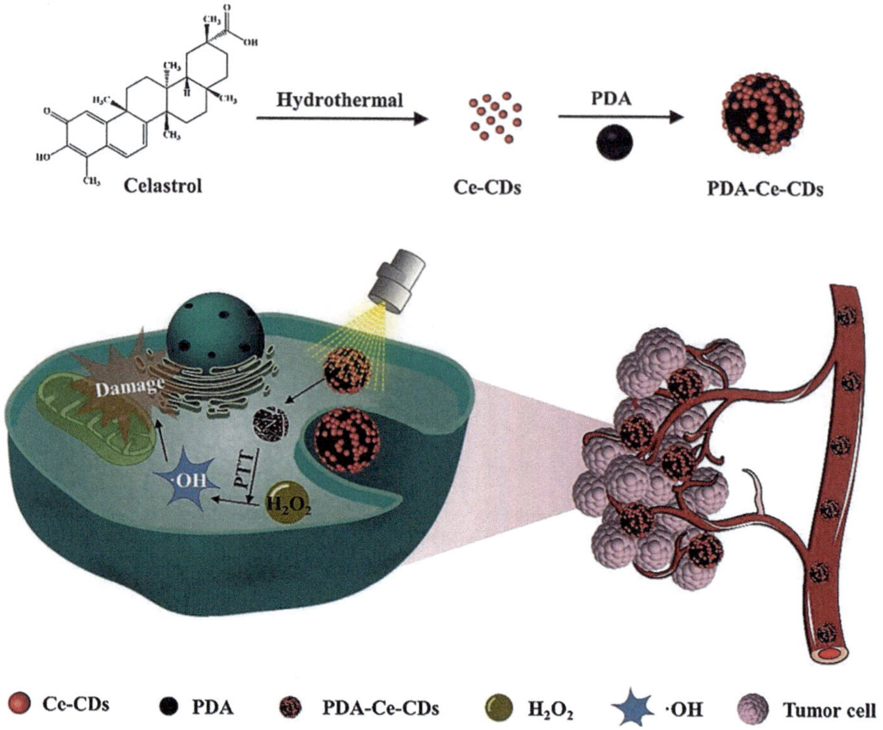

Fig. 5.18 Steps of synthesizing PDA-Ce-CDs and anti-tumor mechanism (reproduced from Sheng et al. 2023) Copyright © 2023, ACS

region (NIR-II is 1000–1700 nm) wavelengths are rarely reported for biological imaging and photothermal therapy. In 2019, the research groups of Zeng Songjun and Hao Jianhua used green plant watermelon juice as raw material to prepare NIR-II emitting carbon dots. The emission wavelength range of these carbon dots is 900–1200 nm, with a fluorescence quantum yield of 0.4% and low luminescence efficiency. Evaluate the metabolism of carbon dots by testing the content of CDs in urine. After tail vein injection, nearly 65.0% of CDs were metabolized approximately 6 h later, indicating that this carbon dot can be rapidly cleared by the kidneys. When irradiated with laser, the photothermal conversion efficiency of CDs is 30.6%, and they also have good inhibitory ability on tumor growth in mice (Fig. 5.19) (Li et al. 2019).

Carbon nanomaterials have demonstrated significant advantages and potential applications in tumor photothermal therapy. With the continuous advancement of technology and research, the application of carbon nanomaterials in PTT will become more extensive and in-depth, bringing new breakthroughs and developments to the field of tumor therapy. However, it should be noted that the application of carbon nanomaterials in PTT is still in the research stage, and their safety and effectiveness need further verification and improvement.

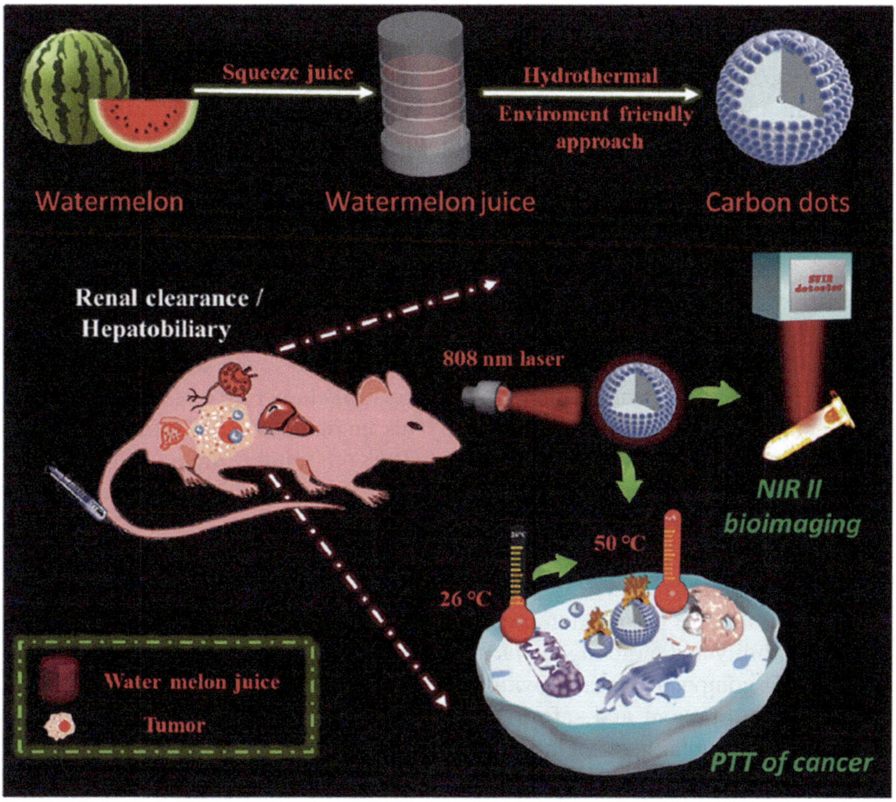

Fig. 5.19 Schematic illustration of the synthesis of NIR-II emitting CDs for rapid renal clearance NIR-II bioimaging and PTT of cancer (reproduced from Li et al. 2019) Copyright © 2019, ACS

5.4.3 Sonodynamic Therapy

Sonodynamic therapy (SDT) is a treatment method that uses ultrasound to activate sonosensitizers to produce anti-tumor effects (Geng et al. 2022). In the SDT process, carbon nanomaterials mainly play the following roles. Sonosensitizer carrier: Carbon nanomaterials can serve as carriers for sonosensitizers, and by surface modification, sonosensitizers (such as Bengal rose red, hemoglobin, and many more) can be loaded onto their surfaces to form composite nanomaterials. This composite nanomaterial can aggregate at high concentrations in tumor cells and tumor neovascularization endothelial cells, achieving targeted delivery of sonosensitizers (Wang et al. 2024a, b; Cao et al. 2023). Enhanced ultrasound imaging: Some carbon nanomaterials, such as multi-walled carbon nanotubes, have good ultrasound imaging performance and can be used as ultrasound contrast agents during SDT to improve the imaging contrast of tumor tissue and help accurately locate tumor lesions. Promoting sonochemical reactions: Under the action of ultrasound, the sonosensitizer on the surface of carbon nanomaterials can absorb energy and undergo electronic transitions, producing singlet oxygen or other reactive oxygen species (ROS). These ROS have strong oxidative power and can damage the organelles of tumor cells, leading to tumor cell death. Taking the multi walled carbon nanotubes (RB MWCNTs) modified with the sonosensitizers Bengal Rose Red as an example, the composite nanomaterials exhibited excellent performance in SDT. Research has shown that RB MWCNTs can accurately locate tumor lesions under ultrasound imaging guidance and generate a large amount of singlet oxygen under the action of ultrasound, which has a significant killing effect on tumor cells. In addition, RB MWCNTs can effectively enrich at the tumor site, further improving the therapeutic effect of SDT.

In 2024, Liu prepared RB CDs using glycine and the sound sensitizer Rose Red (RB) as raw materials, and further modified the RGD targeting $\alpha v\beta 3$ overexpressed on the surface of tumor cell membranes. This engineered RB-CDs@RGD could generate sufficient reactive oxygen species (ROS) to induce glioma cell apoptosis under ultrasound stimulation, achieving efficient sonodynamic therapy and fluorescence navigation surgery against gliomas (Ren et al. 2024).

With the continuous development of nanotechnology, the application prospects of carbon nanomaterials in SDT will be even broader. In the future, researchers will further optimize the preparation process of carbon nanomaterials to improve their biocompatibility and targeting; at the same time, more new types of sound sensitizers and their composite methods with carbon nanomaterials will also be explored to develop more efficient and safe SDT treatment systems. In addition, combining SDT with other treatment methods such as chemotherapy, radiotherapy, etc. will also become an important research direction for future cancer treatment.

5.5 Conclusions and Perspectives

After in-depth exploration of the widespread application and cutting-edge progress of carbon nanomaterials in the field of tumor diagnosis and treatment, it can be clearly stated that carbon nanomaterials have brought revolutionary changes to early diagnosis, precise treatment, and efficacy monitoring of tumors due to their unique physical and chemical properties, excellent biocompatibility, and high modifiability. The rapid development of this field not only broadens the means of cancer diagnosis and treatment, but also provides new hope for improving patient survival rates and quality of life. Firstly, the application of carbon nanomaterials in tumor imaging technology has significantly improved the sensitivity and specificity of diagnosis. As a contrast agent for magnetic resonance imaging (MRI), carbon nanomaterials can significantly enhance the contrast between tumor tissue and normal tissue by modifying loaded paramagnetic or superparamagnetic substances, thereby achieving early detection and precise localization of tumors. In addition, some carbon nanomaterials such as single-walled carbon nanotubes (SWNTs) have unique near-infrared fluorescence properties that can be effectively detected in complex biological environments, providing powerful tools for optical imaging of tumors. The combined application of these imaging technologies provides clinical doctors with multidimensional tumor information, which helps to develop more personalized treatment plans. Secondly, the role of carbon nanomaterials in tumor treatment cannot be ignored. As a drug delivery carrier, carbon nanomaterials can efficiently load therapeutic molecules such as chemotherapy drugs and photosensitizers, and achieve targeted delivery and controlled release of drugs through surface modification. This precise delivery method not only increases the concentration of drugs at the tumor site, enhances the therapeutic effect, but also significantly reduces the toxic side effects on normal tissues. In addition, carbon nanomaterials themselves also exhibit certain anti-tumor activity, such as killing tumor cells through high temperature generated by photothermal conversion effect, or inducing tumor cell apoptosis through sonodynamic therapy. The combination of these treatment methods provides a new strategy for the comprehensive treatment of tumors. Furthermore, carbon nanomaterials have shown great potential in the integration of tumor diagnosis and treatment. By integrating imaging and therapeutic functions into the same nano platform, carbon nanomaterials can accurately target tumors while monitoring their status in real time. This integrated diagnosis and treatment model not only improves the accuracy of treatment, but also helps to evaluate treatment effectiveness and adjust treatment plans in a timely manner, thereby achieving dynamic management and long-term control of tumors. However, despite significant progress in the field of cancer diagnosis and treatment using carbon nanomaterials, their widespread application still faces some challenges. For example, further in-depth research is needed on the biosafety, long-term toxicity, and pharmacokinetic properties of carbon nanomaterials. In addition, how to optimize the preparation process of carbon nanomaterials, improve their stability and biocompatibility, is also an important direction for future research.

In summary, the application of carbon nanomaterials in tumor diagnosis and treatment provides strong support for early diagnosis, precise treatment, and efficacy monitoring of tumors. With the continuous development of materials science, nanotechnology, and biomedical engineering, the application prospects of carbon nanomaterials in the field of tumor diagnosis and treatment will be even broader. In the future, through continuous research and innovation, we have reason to believe that carbon nanomaterials will bring safer, more effective, and personalized diagnosis and treatment solutions to cancer patients, making important contributions to the progress of oncology medicine.

References

Alagarsamy KN, Mathan S, Yan W et al (2021) Carbon nanomaterials for cardiovascular theranostics: promises and challenges. Bioact Mater 6(8):2261–2280

Anzar N, Hasan R, Tyagi M et al (2020) Carbon nanotube - a review on synthesis, properties and plethora of applications in the field of biomedical science. Sens Int 1:100003

Azam N, Najabat Ali M, Javaid Khan T (2021) Carbon quantum dots for biomedical applications: review and analysis. Front Mater 8:700403
Bao Y, Hua X, Li Y et al (2018) Hyperthemia-promoted cytosolic and nuclear delivery of copper/carbon quantum dot-crosslinked nanosheets: multimodal imaging-guided photothermal cancer therapy. ACS Appl Mater Interfaces 10(2):1544–1555
Baskar AV, Benzigar MR, Talapaneni SN et al (2022) Self-assembled fullerene nanostructures synthesis and applications. Adv Funct Mater 32:2106924
Beack S, Kong W, Jung H et al (2015) Photodynamic therapy of melanoma skin cancer using carbon dot - chlorin e6 - hyaluronate conjugate. Acta Biomater 26:295–305
Cao XR, Zhen MM, Li L et al (2022) Oral fullerene tablets for colorectal cancer therapy based on modulation of tumor inflammatory microenvironments. J Mater Chem B 10:9457–9465
Cao Y, Qiao B, Chen Q et al (2023) Tumor microenvironment remodeling via targeted depletion of M2-like tumor-associated macrophages for cancer immunotherapy. Acta Biomater 160:239–251
Chen BB, Liu ML, Li CM et al (2019) Fluorescent carbon dots functionalization. Adv Colloid Interf Sci 270:165–190
Chin SM, Reina G, Chau NDQ et al (2023) Functional graphene for peritumoral brain microenvironment modulation therapy in glioblastoma. Small 19(18):e2208227
Chung S, Revia RA, Zhang M (2021) Graphene quantum dots and their applications in bioimaging, biosensing, and therapy. Adv Mater 33:e1904362
Ci Q, Wang Y, Wu B et al (2023) Fe-doped carbon dots as NIR-II fluorescence probe for in vivo gastric imaging and pH detection. Adv Sci 10(7):e2206271
Cilingir E, Seven E, Zhou Y et al (2021) Metformin derived carbon dots: highly biocompatible fluorescent nanomaterials as mitochondrial targeting and blood-brain barrier penetrating biomarkers. J Colloid Interface Sci 592:485–497
Consoli G, Giuffrida ML, Stefania Z et al (2023) Green light-triggerable chemo-photothermal activity of cytarabine-loaded polymer carbon dots: mechanism and preliminary in vitro evaluation. ACS Appl Mater Interfaces 15(4):5732–5743
Desmond LJ, Phan AN, Gentile P (2021) Critical overview on the green synthesis of carbon quantum dots and their application for cancer therapy. Environ Sci Nano 8:848–862
Devi S, Kumar M, Tiwari A et al (2022) Quantum dots: an emerging approach for cancer therapy. Front Mater 8:798440
Dordevic L, Arcudi F, Cacioppo M et al (2022) A multifunctional chemical toolbox to engineer carbon dots for biomedical and energy applications. Nat Nanotechnol 17(2):112–130
Dutta SD, Hexiu J, Kim J et al (2022) Two-photon excitable membrane targeting polyphenolic carbon dots for long-term imaging and pH-responsive chemotherapeutic drug delivery for synergistic tumor therapy. Biomater Sci 10(7):1680–1696
Feng T, Ai X, An G et al (2016) Charge-convertible carbon dots for imaging-guided drug delivery with enhanced in vivo cancer therapeutic efficiency. ACS Nano 10(4):4410–4420
Feng G, Zhang G, Ding D (2020) Design of superior phototheranostic agents guided by Jablonski diagrams. Chem Soc Rev 49(22):8179–8234
Fernandes NB, Shenoy RUK, Kajampady MK et al (2022) Fullerenes for the treatment of cancer: an emerging tool. Environ Sci Pollut Res 29:58607–58627
Gao W, He J, Chen L et al (2023) Deciphering the catalytic mechanism of superoxide dismutase activity of carbon dot nanozyme. Nat Commun 14:160
Gaur M, Misra C, Yadav AB et al (2021) Biomedical applications of carbon nanomaterials: fullerenes, quantum dots, nanotubes, nanofibers, and graphene. Materials 14(20):5978
Ge JC, Lan MH, Zhou BJ et al (2014) A graphene quantum dot photodynamic therapy agent with high singlet oxygen generation. Nat Commun 5:4596
Ge J, Jia Q, Liu W et al (2015) Red-emissive carbon dots for fluorescent, photoacoustic, and thermal theranostics in living mice. Adv Mater 27(28):4169–4177
Geng B, Yang D, Pan D et al (2018) NIR-responsive carbon dots for efficient photothermal cancer therapy at low power densities. Carbon 134:153–162

Geng B, Hu J, Li Y et al (2022) Near-infrared phosphorescent carbon dots for sonodynamic precision tumor therapy. Nat Commun 13(1):5735
Getachew G, Hsiao C, Wibrianto A et al (2023) High performance carbon dots based prodrug platform: image-guided photodynamic and chemotherapy with on-demand drug release upon laser irradiation. J Colloid Interface Sci 633:396–410
Gong X, Zhang Q, Gao Y et al (2016) Phosphorus and nitrogen dual-doped hollow carbon dot as a nanocarrier for doxorubicin delivery and biological imaging. ACS Appl Mater Interfaces 8(18):11288–11297
Guo T, Tang QY, Guo YT et al (2020) Boron quantum dots for photoacoustic imaging-guided photothermal therapy. ACS Appl Mater Interfaces 13(1):306–311
Han G, Zhao J, Zhang R et al (2019) Membrane-penetrating carbon quantum dots for imaging nucleic acid structures in live organisms. Angew Chem Int Ed 58(21):7087–7091
Han Y, Liu HM, Fan M et al (2022) Near-infrared-II photothermal ultra-small carbon dots promoting anticancer efficiency by enhancing tumor penetration. J Colloid Inter Sci 15:595–604
Hani B, Cormac M, Oleksii P et al (2024) In-brain multiphoton imaging of vaterite cargoes loaded with carbon dots. Nano Lett 24(27):8232–8239
He H, Zheng X, Liu S et al (2018) Diketopyrrolopyrrole-based carbon dots for photodynamic therapy. Molecules 10(23):10991–10998
He Y, Hu C, Li ZC et al (2022) Multifunctional carbon nanomaterials for diagnostic applications in infectious diseases and tumors. Mater Today Bio 14:100231
Hua X, Bao Y, Wu F (2018) Fluorescent carbon quantum dots with intrinsic nucleolus-targeting capability for nucleolus imaging and enhanced cytosolic and nuclear drug delivery. ACS Appl Mater Interfaces 10(13):10664–10677
Huang P, Lin J, Wang X et al (2012) Light-triggered theranostics based on photosensitizer conjugated carbon dots for simultaneous enhanced fluorescence imaging and photodynamic therapy. Adv Mater 24(37):5104–5110
Huo JW, Li J, Liu Y et al (2022) Amphiphilic aminated derivatives of [60] fullerene as potent inhibitors of tumor growth and metastasis. Adv Sci 9:2201541
Iannazzo D, Celesti C, Espro C (2021) Recent advances on graphene quantum dots as multifunctional nanoplatforms for cancer treatment. Biotechnol J 16(2):e1900422
Illescas BM, Martin N (2006) [60] fullerene-based electron acceptors. C R Chim 9:1038–1050
Jaque D, Martinez Maestro L, del Rosal B et al (2014) Nanoparticles for photothermal therapies. Nanoscale 6(16):9494–9530
Jiang BP, Zhou B, Lin Z et al (2019) Recent advances in carbon nanomaterials for cancer phototherapy. Chem Eur J 25:3993
Jiang M, Mu J, Jacobson O et al (2021) Reactive oxygen species activatable heterodimeric prodrug as tumor-selective nanotheranostics. ACS Nano 14(12):16875–16886
Jiao M, Wang Y, Wang W et al (2022) Gadolinium doped red-emissive carbon dots as targeted theranostic agents for fluorescence and MR imaging guided cancer phototherapy. Chem Eng J 440(15):135965
Karagianni A, Tsierkezos NG, Prato M et al (2023) Application of carbon-based quantum dots in photodynamic therapy. Carbon 203:273–310
Kumar S, Rani R, Dilbaghi N et al (2017) Carbon nanotubes: a novel material for multifaceted applications in human healthcare. Chem Soc Rev 46:158–196
Kuo WS, Chang CY, Chuang HY et al (2023) Single-sized N-functionality graphene quantum dot in tunable dual-modality near infrared-I/II illumination detection and photodynamic therapy under multiphoton nonlinear excitation. Biosens Bioelectron 241:115648
Lan M, Zhao S, Zhang Z et al (2017) Two-photon-excited near-infrared emissive carbon dots as multifunctional agents for fluorescence imaging and photothermal therapy. Nano Res 10:3113–3123
Lee D, Beack S, Yoo J et al (2018) In vivo photoacoustic imaging of livers using biodegradable hyaluronic acid-conjugated silica nanoparticles. Adv Funct Mater 28(22):1800941

Lesani P, Singh G, Viray CM et al (2020) Two-photon dual-emissive carbon dot-based probe: deep-tissue imaging and ultrasensitive sensing of intracellular ferric ions. ACS Appl Mater Interfaces 12(16):18395–18406
Lesani P, Lu Z, Singh G et al (2021) Influence of carbon dot synthetic parameters on photophysical and biological properties. Nanoscale 13(25):11138–11149
Li Y, Zheng X, Zhang X et al (2017) Porphyrin-based carbon dots for photodynamic therapy of hepatoma. Adv Healthc Mater 6(1):1600924
Li Y, Bai G, Zeng S et al (2019) Theranostic carbon dots with innovative NIR-II emission for in vivo renal-excreted optical imaging and photothermal therapy. ACS Appl Mater Interfaces 11(5):4737–4744
Li F, Li S, Guo X et al (2020) Chiral carbon dots mimicking topoisomerase I to mediate the topological rearrangement of supercoiled DNA enantioselectively. Angew Chem Int Ed 59(27):11087–11092
Li H, Sun J, Zhu H et al (2021a) Recent advances in development of dendritic polymer-based nanomedicines for cancer diagnosis. Wiley Interdiscip Rev Nanomed Nanobiotechnol 13(2):e1670
Li B, Zhao S, Huang L et al (2021b) Recent advances and prospects of carbon dots in phototherapy. Chem Eng J 408:127245
Li Y, Biswas R, Kopcha WP et al (2023a) Structurally defined water-soluble metallofullerene derivatives towards biomedical applications. Angew Chem Int Ed 62:e202211704
Li H, Feng Y, Luo Q et al (2023b) Stimuli-activatable nanomedicine meets cancer theranostics. Theranostics 13(15):5386–5417
Li L, Zhen M, Wang H et al (2023c) Tumor microenvironment-modulated multiple nanotherapeutic system for potent cancer immunotherapy and metastasis inhibition. Nano Today 48:101702
Li L, Fu JJ, Ye JH et al (2024) Developing hypoxia-sensitive system via designing tumor-targeted fullerene-based photosensitizer for multimodal therapy of deep tumor. Adv Mater 36:2310875
Liang L, Peng X, Sun F et al (2021) A review on the cytotoxicity of graphene quantum dots: from experiment to simulation. Nanoscale Adv 3(4):904–917
Lin F, Jia C, Wu F-G et al (2022) Carbon dots for intracellular sensing. Small Struct 3:2200033
Liu S, Lu Y-J, Kappes MM, Ibers JA (1991) The structure of the C60 molecule: X-ray crystal structure determination of a twin at 110 K. Science 254:408–410
Liu H, Yang J, Li Z et al (2019) Hydrogen-bond-induced emission of carbon dots for wash-free nucleus imaging. Anal Chem 91(14):9259–9265
Liu H, Li C, Qian Y et al (2020a) Magnetic-induced graphene quantum dots for imaging-guided photothermal therapy in the second near-infrared window. Biomaterials 232:119700
Liu Q, Niu X, Zhang Y et al (2020b) Carbon dots for lysosome targeting and imaging of lysosomal pH and Cys/Hcy in living cells. Nanoscale 12(24):13010–13016
Liu Y, Zhu D, Zhu X et al (2020c) Enhancing the photodynamic therapy efficacy of black phosphorus nanosheets by covalently grafting fullerene C60. Chem Sci 11:11435–11442
Liu S, Wang J, Wang X et al (2024) Nucleus-targeted carbon dots as peroxidase nanozyme for photoacoustic imaging and phototherapy of tumor. Colloids Surf B Biointerfaces 239:113950
Lu L, Fu X, Liew Y, Zhang Y et al (2019a) Soft and MRI compatible neural electrodes from carbon nanotube fibers. Nano Lett 19(3):1577–1586
Lu G-H, Shang W-T, Deng H et al (2019b) Targeting carbon nanotubes based on IGF-1R for photothermal therapy of orthotopic pancreatic cancer guided by optical imaging. Biomaterials 195:13–22
Markovin Z, Ristic B, Arsikin K et al (2012) Graphene quantum dots as autophagy-inducing photodynamic agents. Biomaterials 33(29):7084–7092
Martelli C, King A, Simon T et al (2020) Graphene-induced transdifferentiation of cancer stem cells as a therapeutic strategy against glioblastoma. ACS Biomater Sci Eng 6(6):3258–3269
Mauro N, Cillari R, Gagliardo C et al (2023) Gadolinium-doped carbon nanodots as potential anticancer tools for multimodal image-guided photothermal therapy and tumor monitoring. ACS Appl Nano Mater 6(18):17206–17217
Murali G, Kwon B, Kang H et al (2022) Hematoporphyrin photosensitizer-linked carbon quantum dots for photodynamic therapy of cancer cells. ACS Appl Nano Mater 5(3):4376–4385

Najaflu M, Shahgolzari M, Bani F et al (2022) Green synthesis of near-infrared copper-doped carbon dots from alcea for cancer photothermal therapy. ACS Omega 7(38):34573–34582
Nasrin A, Hassan M, Gomes V et al (2020) Two-photon active nucleus-targeting carbon dots: enhanced ROS generation and photodynamic therapy for oral cancer. Nanoscale 12(40):20598–20603
Pang W, Jiang P, Ding S et al (2020) Nucleolus-targeted photodynamic anticancer therapy using renal-clearable carbon dots. Adv Healthc Mater 9(16):2000607
Qi L, Pan T, Ou L et al (2021) Biocompatible nucleus-targeted graphene quantum dots for selective killing of cancer cells via DNA damage. Commun Biol 4(1):214
Rahimi S, Chen Y, Zareian M et al (2022) Cellular and subcellular interactions of graphene-based materials with cancerous and non-cancerous cells. Adv Drug Deliv Rev 189:114467
Reina G, González-Domínguez JM, Criado A et al (2017) Promises, facts and challenges for graphene in biomedical applications. Chem Soc Rev 46:4400–4416
Ren X, Shi Y, Yang Y et al (2024) Integrin αvβ3-targeted engineered carbon dots for efficacious sonodynamic therapy and fluorescence navigation surgery against gliomas. Mater Chem Front 8:2511–2524
Riley PR, Narayan RJ (2021) Recent advances in carbon nanomaterials for biomedical applications: a review. Curr Opin Biomed Eng 17:100262
Rosenkrans ZT, Sun T, Jiang D et al (2020) Selenium-doped carbon quantum dots act as broad-spectrum antioxidants for acute kidney injury management. Adv Sci 7(12):2000420
Shen CL, Liu HR, Lou Q et al (2022) Recent progress of carbon dots in targeted bioimaging and cancer therapy. Theranostics 12(6):2860–2893
Sheng P, Bu C, Hui T et al (2023) Polydopamine-activated celastrol carbon dots for synergistic chemotherapy-photothermal therapy of tumors. Int J Pharm X 6:100218
Shi Y, Xia Y, Zhou M et al (2024) Facile synthesis of Gd/Ru-doped fluorescent carbon dots for fluorescent/MR bimodal imaging and tumor therapy. J Nanobiotechnol 22(1):88
Shu G, Jiang T, Zhang X et al (2024) Gadolinium and ytterbium-doped carbon dots for fluorescence/magnetic resonance/computed tomography multimodal imaging. Chem Eng J 494:153086
Sinha R, Jaykishan OS, Purkayastha P (2023) A curated graphene quantum dot-porphyrin-based photosensitizer for effective singlet oxygen generation through energy transfer. Langmuir 39(31):11108–11118
Su Z, Zhao Y, Huang Y et al (2022) Bi-functional nanocomposite based on phosphor and carbon nanotubes for tumor ablation in a photothermal fiber system with temperature feedback. Chem Eng J 436:134994
Sun S, Zhang L, Jiang K et al (2016) Toward high-efficient red emissive carbon dots: facile preparation, unique properties, and applications as multifunctional theranostic agents. Chem Mater 28(23):8659–8668
Szczepaniak J, Jagiello J, Wierzbicki M et al (2021) Reduced graphene oxides modulate the expression of cell receptors and voltage-dependent ion channel genes of glioblastoma multiforme. Int J Mol Sci 22(2):515
Tanzi L, Terreni M, Zhang Y (2022) Synthesis and biological application of glyco- and peptide derivatives of fullerene C60. Eur J Med Chem 230:114104
Tomczyk MM, Boncel S, Herman A et al (2020) Oxygen functional groups on MWCNT surface as critical factor boosting T2 relaxation rate of water protons: towards improved CNT-based contrast agents. Int J Nanomedicine 15:7433–7450
Tong L, Wang X, Chen Z et al (2020) One-step fabrication of functional carbon dots with 90% fluorescence quantum yield for long-term lysosome imaging. Anal Chem 92(9):6430–6436
Wang J, Xu M, Wang D et al (2019) Copper-doped carbon dots for optical bioimaging and photodynamic therapy. Inorg Chem 58(19):13394–13402
Wang C, Chen Y, Xu Z et al (2020a) Fabrication and characterization of novel cRGD modified graphene quantum dots for chemo-photothermal combination therapy. Sens Actuators B 309:127732

Wang X, Zhou W, Li X et al (2020b) Graphene oxide suppresses the growth and malignancy of glioblastoma stem cell-like spheroids via epigenetic mechanisms. J Transl Med 18(1):200
Wang B, Song H, Qu X et al (2021a) Carbon dots as a new class of nanomedicines: opportunities and challenges. Coord Chem Rev 442:214010
Wang S, Zhang L, Zhao J et al (2021b) A tumor microenvironment-induced absorption red-shifted polymer nanoparticle for simultaneously activated photoacoustic imaging and photothermal therapy. Sci Adv 7(12):3588
Wang B, Lu S et al (2022) The light of carbon dots: from mechanism to applications. Matter 5:110–149
Wang F, Fan Y, Liu Y et al (2024a) Oxygen-carrying semiconducting polymer nanoprodrugs induce sono-pyroptosis for deep-tissue tumor treatment. Exploration 4(4):20230100
Wang L, Wang W, Wang Y et al (2024b) The graphene quantum dots gated nanoplatform for photothermal-enhanced synergetic tumor therapy. Molecules 29(3):615
Wei S, Wang B, Zhang H et al (2023) Portable smartphone platform based on tunable chiral fluorescent carbon dots for visual detection of L-asp and L-Lys. Chem Eng J 466:143103
Wei F, Karges J, Gao S et al (2024) Twophoton phototriggering of ROS storm in ruthenium (II) coordinated carbon nitride for robust cancer immunotherapy. Nano Today 54:102066
Wu X, Sun S, Wang Y et al (2017) A fluorescent carbon-dots-based mitochondria-targetable nanoprobe for peroxynitrite sensing in living cells. Biosens Bioelectron 90:501–507
Wyrzykowska E, Mikolajczyk A, Lynch I et al (2022) Representing and describing nanomaterials in predictive nanoinformatics. Nat Nanotechnol 17:924–932
Xu XY, Ray R, Gu YL et al (2004) Electrophoretic analysis and purification of fluorescent single-walled carbon nanotube fragments. J Am Chem Soc 126(40):12736–12737
Xu S, He X, Huang Y et al (2020) Lysosome-targeted ratiometric fluorescent sensor for monitoring pH in living cells based on one-pot-synthesized carbon dots. Microchim Acta 187(8):478–487
Xu N, Gu Q, Du J et al (2021a) Photodynamic inheritance from methylene blue to carbon dots against reduction, aggregation, and DNA interference. Sci China Mater 64:2325–2336
Xu Y, Wang C, Ran G et al (2021b) Phosphate-assisted transformation of methylene blue to red-emissive carbon dots with enhanced singlet oxygen generation for photodynamic therapy. ACS Appl Mater Interfaces 4(5):4820–4828
Yan C, Chen C, Hou L et al (2017) Single-walled carbon nanotube-loaded doxorubicin and Gd-DTPA for targeted drug delivery and magnetic resonance imaging. J Drug Target 25:163–171
Yan J, Wu Q, Zhao Z et al (2020) Light-assisted hierarchical intratumoral penetration and programmed antitumor therapy based on tumor microenvironment (TME)-amendatory and self-adaptive polymeric nanoclusters. Biomaterials 255:120166
Yang L, Wang Z, Wang J et al (2016) Doxorubicin conjugated functionalizable carbon dots for nucleus targeted delivery and enhanced therapeutic efficacy. Nanoscale 8(12):6801–6809
Yang D, Yang G, Gai S et al (2017) Multifunctional theranostics for dual-modal photodynamic synergistic therapy via stepwise water splitting. ACS Appl Mater Interfaces 9(8):6829–6838
Yang Y, Wang B, Zhang X et al (2023) Activatable graphene quantum-dot-based nanotransformers for long-period tumor imaging and repeated photodynamic therapy. Adv Mater 35(23):e2211337
Younis MR, He G, Lin J et al (2021) Graphene-semiconductor nanocomposites for cancer phototherapy. Biomed Mater 16(2):022007
Yue L, Li H, Qi S et al (2020) Red-emissive ruthenium-containing carbon dots for bioimaging and photodynamic cancer therapy. ACS Appl Nano Mater 3(1):869–876
Yue J, Miao P, Li L et al (2022) Injectable carbon dots-based hydrogel for combined photothermal therapy and photodynamic therapy of cancer. ACS Appl Mater Interfaces 14(44):49582–49591
Zare H, Ahmadi S, Ghasemi A et al (2021) Carbon nanotubes: smart drug/gene delivery carriers. Int J Nanomedicine 16:1681–1706

Zhang Y, Jia Q, Nan F et al (2023) Carbon dots nanophotosensitizers with tunable reactive oxygen species generation for mitochondrion-targeted type I/II photodynamic therapy. Biomaterials 293:121953

Zhang W, Fan M, Yang R et al (2024) New features of edge-selectively hydroxylated graphene nanosheets as NIR-II photothermal agent and sonothermal agent for tumor therapy. J Mater Chem B 12(32):7892–7904

Zhao S, Yang K, Jiang L et al (2021) Polythiophene-based carbon dots for imaging-guided photodynamic therapy. ACS Appl Nano Mater 4(10):10528–10533

Zhao WB, Chen DD, Liu KK et al (2023) Near-infrared I/II emission and absorption carbon dots via constructing localized excited/charge transfer state for multiphoton imaging and photothermal therapy. Chem Eng J 452:139231

Zheng D, Li B, Li C et al (2016) Carbon-dot-decorated carbon nitride nanoparticles for enhanced photodynamic therapy against hypoxic tumor via water splitting. ACS Nano 10(9):8715–8722

Zhou ZJ, Yang LJ, Gao JH et al (2019) Structure-relaxivity relationships of magnetic nanoparticles for magnetic resonance imaging. Adv Mater 31(8):e1804567

Zhu X, Zhang T, Jiang D et al (2018) Stabilizing black phosphorus nanosheets via edge-selective bonding of sacrificial C60 molecules. Nat Commun 9:4177

Chapter 6
Application of Functionalized Nanomaterials Based on Microfluidic Chips in Tumor Diagnosis and Treatment

Min Gao

Abstract The function of biofunctionalized microscale/nanoparticles depends on the synergistic effect of the composite material composition and fine structure. However, how to achieve the controllable preparation of multiple material compositions and fine structures simultaneously and accurately in the micro–nanoscale space is still an important challenge facing functionalized microscale/nanomaterials. Microfluidics chips are made based on microfabrication technology. They can integrate multiple components and functional units on a chip, including sampling, synthesis, testing, and data acquisition modules. They can automatically analyze complex biological samples and build targeted drug treatment models. Compared with conventional preparation methods, it also has significant advantages such as low cost, portability, stable microenvironment, controllable reaction process, and high throughput. At present, microfluidic chips are used in many fields such as early cancer detection, drug and gene delivery, synthesis of micro–nanoparticles and drug carriers, cell analysis, tissue engineering, and many more. This chapter reviews the new research progress in the construction of functionalized micro–nanomaterials based on microfluidic chips and their application in tumor marker detection and tumor diagnosis and treatment, focusing on the application progress in two aspects: (1) microfluidic technology for the capture, detection, and analysis of three tumor markers in cancer liquid biopsy applications; (2) construction of functionalized micro–nanomaterials based on microfluidic chips and their application as targeted drug carriers in tumor diagnosis and treatment. It provides research references and support for researchers in related fields in the structural expansion and innovation of micro–nanofunctionalized materials, clinical application of liquid biopsy technology, integrated tumor diagnosis and treatment, and large-scale preparation of targeted drugs.

M. Gao (✉)
Shandong Provincial Key Laboratory of Tumor Imaging Equipment Development and Tumor Diagnosis and Treatment Integration Technology, School of Chemistry and Chemical Engineering, Linyi University, Linyi, Shandong, People's Republic of China
e-mail: gaomin@lyu.edu.cn

X. Song et al. (eds.), *Functional Inorganic Materials for Tumor Diagnostics and Therapy*, https://doi.org/10.1007/978-981-95-0528-9_6

Keywords Micro–nanomaterials · Drug carriers · Tumor markers · Liquid biopsy · Theranostics

6.1 Microfluidic Technology

6.1.1 Microfluidics Technology Overview

Microfluidics is a technology that studies the control, behavior, and manipulation of fluids at the micrometer scale and submicrometer spatial scales. It is also called Lab on a chip. The microfluidic chip, is a type of miniaturized devices mostly produced by the microfabrication technique, has developed rapidly in the last few decades, since its advent in the 1990s (Tavakoli et al. 2019). The microfluidic pipeline structure is at the micrometer to nanometer scale. The fluid at this scale exhibits different characteristics from those at the macroscale (for example, laminar flow, microdroplets). By utilizing the characteristics of fluid at the microscale, microfluidics can perform some micro-operations that are difficult to achieve using conventional methods. Compared with traditional analytical techniques, this technique has the several advantages for cellular assays: (1) Less sample and reagent consumption. The chip has microscale channels, which requires only a small amount of samples and regent consumption, and this aspect is especially important in bioanalytical applications since biological samples are often limited in quantity; (2) Miniaturization and integration. It can not only miniaturize analytical instruments but also realize multifunctional analysis on a single chip. Microfluidic technique can integrate multiple functions such as cell sampling, fluid control, single-cell capture, cell lysis, mixing, and detection on a single device; (3) High throughput. Microfluidic chips can process a large number of samples in parallel with a fast analysis speed. The area of microfluidics has four branches: molecular analysis, microelectronics, biodefence, and molecular biology (Whitesides 2006). Microfluidics systems are used in many areas such as clinical diagnostics (Schulte et al. 2002), anticancer drug studies (Valente et al. 2017), drug discoveries (Pihl et al. 2005), protein crystallization (Li and Ismagilov 2010), point of care diagnostics (Myers and Lee 2008), single-cell analyzes (Murphy et al. 2018), and cell-to-cell interaction (Rothbauer et al. 2018).

The material of the chip determines the performance of the chip. To make a microfluidic chip, the chip material must follow the following principles: good biocompatibility, good electrical insulation and heat dissipation, good optical properties, good surface modifiability, simple manufacturing process, and low manufacturing cost. At present, the commonly used materials for making microfluidic chips are: single crystal silicon wafers, quartz, glass, organic polymers (PMMA, PDMS, PC, COP), paper, and many more (Fig. 6.1) (Ren et al. 2014). Among them, silicon is the earliest material used to make microfluidic chips. It has strong chemical inertness, is not easy to react with other substances, has good heat resistance,

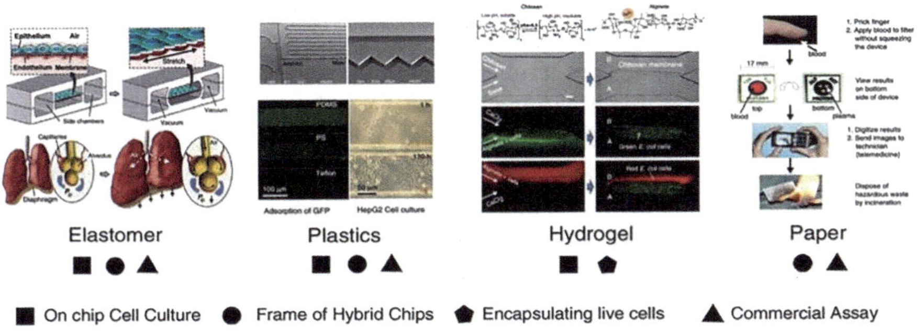

Fig. 6.1 New materials for microfluidics in biology (Ren et al. 2014) (Reproduced from Ren et al. 2014)

and is not easy to deform or deteriorate when processed at high temperatures. However, the price of silicon is high and it is fragile, which increases the processing cost of the chip. In addition, silicon has poor optical properties and is easy to conduct electricity, so it is subject to many restrictions in actual use. In contrast, glass is low in cost and easy to process, has excellent optical properties, is easy to modify the surface, and has strong chemical inertness. It is an ideal chip material. Given that quartz is relatively expensive, it is generally used only when there are special detection needs. Organic polymers are easy to process and mold, and the cost is relatively low. They are suitable for mass production. Polydimethylsiloxane (PDMS) is currently a representative organic polymer material. As one of the most commonly used chip processing materials, PDMS has the advantages of being cheap and durable, easy to process, light-transmitting and breathable, and has good biocompatibility and can be modified in a variety of surfaces.

The preparation methods of early microfluidic chips were mainly based on micro–nanoprocessing technology, and most of them used common technologies in the field of semiconductor processing, such as lithography, etching, chemical corrosion, film removal, and metal sputtering (Toepke and Kenis 2005; Hannes et al. 2008) to process functional structures such as microfluidic channels and microvalves and micropumps on silicon wafers or glass substrates (Tse et al. 2003; Saitoh et al. 2002) With the development of microfluidic technology, various new materials and new processing technologies have emerged, among which a variety of polymer materials have gradually replaced silicon and glass as the main substrate materials for microfluidic chips. At present, the most commonly used preparation methods for microfluidic chips include soft etching (Carlborg et al. 2011), laser processing (Cai et al. 2017), micro–nanoimprinting (Wang et al. 2016), injection molding (Iwai et al. 2014), and 3D printing.

With the continuous expansion of the application of microfluidic chip technology in cell biology, microfluidic chips have the advantage of integrating sample pretreatment and analysis, and have shown great application potential in research fields such as DNA sequencing (Hashimoto et al. 2007; Liu et al. 2007), protein detection (Kim et al. 2006; Emrich et al. 2007), cell manipulation and intracellular component analysis (Bhagat et al. 2010; VanDijken et al. 2007). It is worth mentioning that research on organ chips based on microfluidics is also gradually being carried out. In previous studies, experimental animals have always been an important model for preclinical research on tumors, but animal models are expensive, have long experimental cycles, and have species differences with humans, which greatly limits the development of basic tumor research and the development of antitumor drugs. With the development of materials science and microfabrication technology, microfluidic technology is being used more and more widely in bionic models. In contrast, tumor bionic models based on microfluidic chips can overcome the defects of single-cell types and simple functions in the past, and gradually develop into multi-organ and multifunctional system models. Microfluidic technology makes organoids closer to patients in structure and function, which not only provides the possibility of studying individual differences in disease manifestations, but also provides more reliable in vitro models for application fields such as disease mechanism research and drug evaluation through high-throughput drug screening and dynamic detection methods.

6.1.2 Droplet-Based Microfluidics

Droplet-based microfluidics, a key subset of microfluidic technology, enables the generation and manipulation of highly uniform droplets, ranging from picoliters to nanoliters, within microchannels under precise control. These droplets offer advantages such as excellent monodispersity, minimal cross-contamination, and high repeatability. What distinguishes droplet microfluidics is its ability to produce droplets at an extremely high rate, enabling thousands of assays to be performed in parallel. Additionally, droplet microfluidics provides significantly higher analytical resolution compared to conventional methods. The droplet sizes can be adjusted, allowing them to be reduced to levels comparable to individual cells. This small-scale confinement allows for the trapping of single cells within droplets, facilitating high-throughput single-cell analysis. Droplet generator chips are robust, capable of operating for extended periods with minimal supervision, and their discrete nature eliminates the risk of cross-contamination. The applications of droplet microfluidics are diverse, ranging from microparticle synthesis to molecular biology and microbiology studies.

6.1.2.1 Types of Droplet Microfluidics

To date, the types of microdroplets reported in the literature primarily include gas–liquid phase droplets and liquid–liquid phase droplets. Gas–liquid phase droplets are easily volatilized and cause cross-contamination in microchannels, which limits their application. Liquid–liquid phase microdroplets have the advantages of small volume, no diffusion between droplet samples, avoidance of cross-contamination between samples, stable reaction conditions, and rapid mixing under proper manipulation. Therefore, it is the focus of the development of droplet microfluidics. It is an ideal microreactor and has been used in fields such as chemistry and life sciences to study many reactions and processes under microscale conditions.

The key to forming droplets in a microfluidic chip lies in applying a force sufficient to disrupt the interfacial tension between the continuous and dispersed phases, causing instability. When the force exerted at a specific point on the dispersed phase exceeds the interfacial tension, a small amount of liquid breaks through the interface and enters the continuous phase, forming droplets. The capillary number Ca, an important kinetic constant, is usually introduced to describe this physical process. $Ca = \mu U/\gamma$, that is, the ratio of system viscosity to interfacial tension, where μ represents the viscosity of the continuous phase, U represents the fluid velocity of the continuous phase, and γ represents the interfacial tension between the two phases. At low capillary numbers, interfacial tension dominates, and droplets tend to form spherical shapes during transport, minimizing surface area. In contrast, at higher capillary numbers, viscosity becomes more influential, and droplets are more likely to deform and stretch into asymmetric shapes during transport. Additionally, the hydrophilicity or hydrophobicity of the microfluidic channel walls, along with the presence of surfactants in both phases, plays a crucial role in determining whether O/W (oil-in-water), W/O (water-in-oil), W/O/W (water/oil/water), or O/W/O (oil/water/oil) droplets are formed.

There are many methods and techniques for generating droplets, which can be divided into passive and active methods depending on whether external energy is applied during the droplet generation process. Passive droplet generation refers to the use of different microchannel structures to allow immiscible dispersed phase and continuous phase fluids to meet at the channel junction. By adjusting the channel structure, the flow rate of the two phases and the flow rate ratio, droplets of controllable size can be generated downstream of the channel junction (Sartipzadeh et al. 2020). Depending on the channel geometry, passive droplet generation (Table 6.1) can be divided into T-junction (Zhang et al. 2019), flow-focusing (Feng et al. 2010), and coaxial flow (Zhou et al. 2018). Active droplet generation refers to

Table 6.1 Characteristics and microchannel geometry of the passive method for droplets formation

Passive method	Channel characteristics	Channel geometry
T-junction	The device is simple to fabricate, but the droplets generated have low monodispersity and narrow size range	
Flow-focusing	The droplets' sizes are well controllable and can be down to a few hundred nanometers	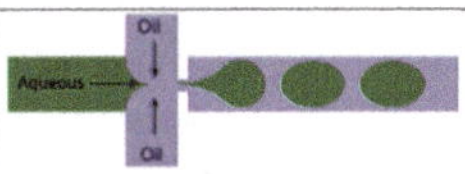
Co-flow	The co-flow geometry avoids the droplets contacting the channel wall and prevents the effect of shear force on the droplets	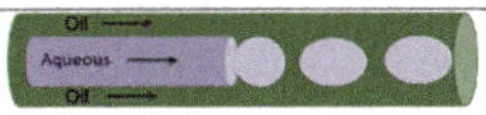

Table 6.2 Principles and characteristics of the active method for droplets formation

Active method	Principle	Characteristic
Electric force	Droplets are generated by the interface displacement of two phases under the action of electric force	Easier control; low efficiency; harmful to biological samples
Magnetic force	Droplets are generated by the continuous phase shearing the dispersed phase under the action of magnetic force	Fast response speed; good biocompatibility; complex operating process; high cost
Centrifugal force	Droplets are generated under the combined action of centrifugal force and surface tension	High efficiency; capable of generating complex droplets; low monodispersity

the control of droplet generation by locally applying external forces such as electric field (Teo et al. 2020), magnetic field (Liu et al. 2011) and centrifugal force (Morimoto et al. 2017) during the droplet generation process (Table 6.2).

Passive Droplet Generation

The generation of microdroplets requires changing the state of the continuously flowing liquid to divide it into discrete individual tiny droplets. The key to effective droplet generation and manipulation lies in controlling the surface energy within the microfluidic environment. Passive droplet generation relies solely on the hydrodynamic pressure of the fluid during the droplet formation process, without the need for external energy input. By carefully designing the microchannel structure of the microfluidic chip, droplets can be generated by injecting immiscible two-phase or multi-phase fluids into the system. Common methods for droplet generation include the T-junction method, the flow-focusing method, and the capillary flow confocal method.

T-junction

The T-channel method uses the geometric characteristics of the intersection of two microchannels to make the front of the dispersed phase fluid turn at the intersection and become unstable due to the momentum change caused by the shear force of the continuous phase, thus generating droplets.

In 2001, Thorsen et al. (2001) first proposed a T-channel microfluidic chip with a constricted neck for microdroplet generation, as shown in Fig. 6.2. The oil and water phases are introduced through the corresponding inlets of the chip and flow toward the intersection of the T-junction. At this intersection, an oil–water interface is formed. When the interfacial tension between the oil and water phases is insufficient to resist the shear force exerted by the oil phase, the water phase is sheared off, forming droplets. Thorsen et al. proposed that the shear force between the oil/water interface is approximately equal to the Laplace force to predict the droplet size. For

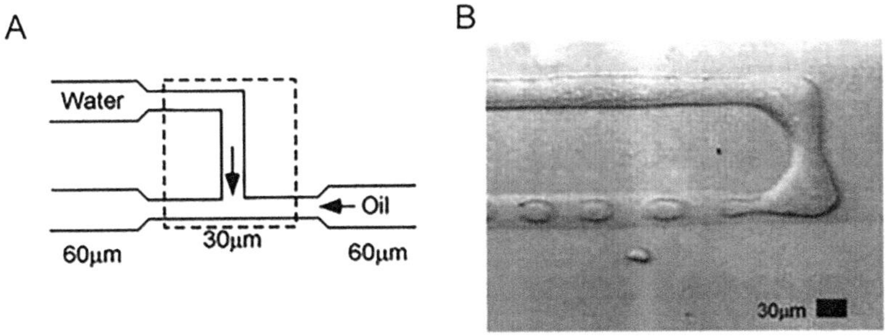

Fig. 6.2 T-type structure chip. (**a**) Structure diagram of T-type microfluidic chip. (**b**) Oil water interface under microscope (Thorsen et al. 2001) (reproduced from Thorsen et al. 2001)

monodisperse droplets generated under oil/water pressures in the range of 55.16–154.44 kPa, the predicted droplet size is within two times the actual droplet size. Nisisako et al. (2002) used water as the discrete phase and oil as the continuous phase, and changed the flow rate of the continuous phase (0.01–0.15 ms^{-1}) to generate micro-droplets with a diameter of 100–380 μm in a T-shaped channel. The larger the flow rate ratio of the continuous phase to the discrete phase, the faster the frequency of micro-droplet generation.

At present, researchers have developed dual T-channel or multi-T-channel configurations based on a single T-channel configuration, and even integrated micropump/microvalve devices on the T-channel. For example, Zeng et al. (2009) integrated a pneumatic pump microvalve on the T-channel and accurately controlled the size of the droplet by controlling the opening/closing time of the microvalve (Fig. 6.3). The development of this type of technology is conducive to the miniaturization and integration of T-channel microdroplet generation systems, and also helps to generate droplets on demand.

In summary, the T-type microchannel has a simple structure, but the necking structure in the chip is difficult to process. In addition, due to the influence of shear force and surface tension, the droplet generation stability in the T-type microchannel is poor, and the droplet size control range is narrow.

Flow-focusing

Unlike the T-junction method, which squeezes the discrete phase fluid from one side, the flow-focusing method applies pressure to the discrete phase from both sides. This results in instability at the oil/water interface in the downstream neck channel, leading to the formation of droplets.

The droplet generation process of flow focusing can be divided into three stages (Garstecki et al. 2005): (1) The dispersed phase and the continuous phase form an interface at the intersection, flowing toward the "neck" and downstream; (2) The dispersed phase forms a "contraction neck" under the coordinated constraints of the specific geometric structure at the neck position and the pressure of the continuous

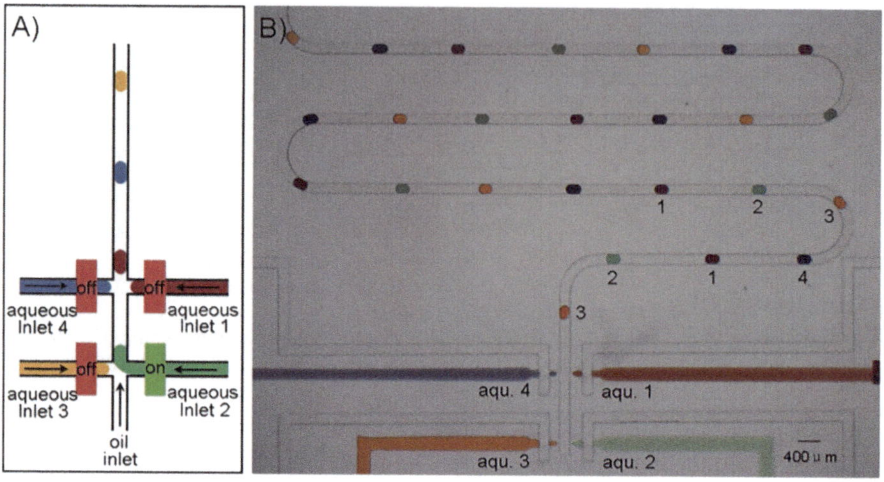

Fig. 6.3 Microvalve-based on–off control enables sequential generation of different reagent droplets (Zeng et al. 2009) (Reproduced from Zeng et al. (2009). Reproduced from Zeng et al. (2009) with permission from the Royal Society of Chemistry

phase; (3) The increasing extrusion pressure causes the liquid at the front end of the contraction neck to completely become unstable and break, thus forming a single droplet. Compared with the T-junction channel, the flow focusing channel improves the controllability of droplet size. By regulating the two-phase flow rate, droplets ranging from femtoliters to nanoliters can be formed on the same chip, and droplets can be generated at a high throughput in a short time. The monodispersity, size, and generation frequency of droplets prepared by the flow focusing channel method are also controlled by factors such as the capillary number of the continuous phase and the flow rate of the dispersed phase, the properties of the liquid, and the shape of the device. For example, Yobas et al. (2006) found that the droplet size decreases with the increase of the continuous phase flow rate, while the droplet generation frequency increases accordingly. In addition, surface tension is also a key factor (Fig. 6.4a). Peng's research group (Peng et al. 2011) explained that changing the concentration of surfactant can change the interfacial tension, which in turn can change the droplet size, generation rate, and monodispersity (Fig. 6.4b, c).

Compared to the T-junction method, the flow focusing technique produces more stable droplets with a wider range of size control, and it is capable of generating droplets much smaller than the channel size. However, it requires a highly symmetrical chip design, a small neck size, and precise fabrication techniques (Nie et al. 2008).

Coaxial Flow

In the coaxial flow focusing method, a capillary with a pointed tip is positioned along the central axis of the air channel, with the dispersed and continuous phases flowing parallel to each other. As the dispersed phase enters the continuous phase

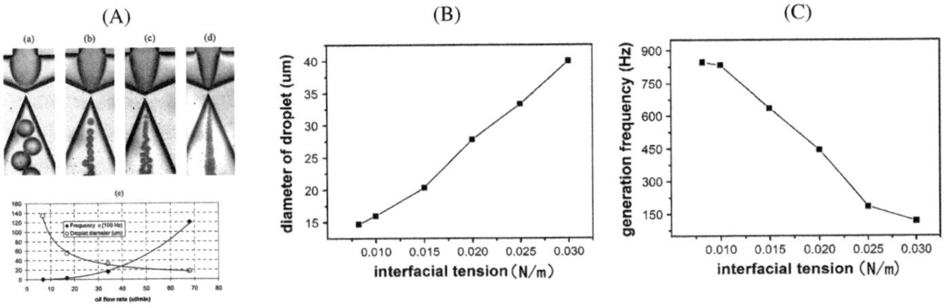

Fig. 6.4 Relevant parameters influencing droplet formation in flow-focused chips. (**a**) Droplet size and frequency of formation as a function of flow velocity (Yobas et al. 2006) (reproduced from Yobas et al. 2006). Reproduced from Yobas et al. (2006) with permission from the Royal Society of Chemistry. (**b**) Droplet size and interfacial tension (Peng et al. 2011). (**c**) Relationship between generation frequency and interfacial tension (Peng et al. 2011) (reproduced from Peng et al. 2011)

channel, it is squeezed and forms droplets due to the shear force exerted by the continuous phase fluid (Utada et al. 2005).

In theory, the main principle of droplet generation by coaxial flow is caused by the Kelvin-Helmholtz instability of the two-phase interface: when the dispersed phase and the continuous phase move forward at different speeds inside and outside the capillary, the two-phase interface will produce certain fluctuations due to the Kelvin-Helmholtz instability. Within a certain speed range, the interfacial tension between the two phases can keep the interface stable; the instability of waves with smaller wavelengths increases, and finally droplets are formed beyond this speed range. Finally, droplets are formed. The droplets generated based on microfluidic droplet technology are transported through microchannels and enter the downstream liquid storage tank in sequence. At this time, a large number of droplets are wrapped by the oil phase, but they still remain independent and separated from each other.

In summary, the coaxial flow method offers significant advantages over the previous two methods in terms of stable droplet generation and a wider range of droplet size control. However, its main drawback is the difficulty in injecting the two-phase flow.

Active Droplet Generation

Active control involves modifying the surface energy of the microfluidic channel by applying external energy during the droplet generation process. Several active droplet generation methods exist, including magnetic control (Lee et al. 2014), mechanical control (Churski et al. 2010), thermal control (Nguyen et al. 2007; Baroud et al. 2007), and electrical control (Gu et al. 2010; Link et al. 2006). The following is a brief description of these types of active droplet generation methods.

Magnetic Control

The application of magnetic force in microdroplet generation and control primarily relies on the dynamic response of special fluids, known as magnetic fluids, to magnetic fields (Lee et al. 2014). Magnetic fluids, such as ferrofluids, contain suspended magnetic particles and exhibit superparamagnetism, meaning they can be magnetized in the presence of a magnetic field but lose their magnetization once the external field is removed. Ferrofluids can be either water-based or oil-based and are used as both discrete and continuous phases in droplet generation. In microchannels, the formation of microdroplets through magnetic control is typically achieved using T-type or flow-focusing microchip structures in passive control systems.

Figure 6.5 illustrates a schematic of a water-based magnetic fluid droplet generation device featuring a T-shaped structure, as proposed by Tan et al. (2010). In this design, permanent magnets can be positioned both upstream and downstream of the T-shaped structure. The droplet diameter is influenced by the magnetic field gradient, the magnetization intensity of the magnetic fluid, and the relative positioning of the magnet. In the absence of a permanent magnet, the droplet diameter is 230 μm. When a permanent magnet is placed upstream of the T-shaped structure, the droplet diameter increases to 280 μm as the magnetic flux density rises. Conversely, when the magnet is placed downstream, the droplet diameter decreases to 190 μm as the magnetic flux density increases. It is important to note that magnetic control of droplet generation is specific to magnetic fluids and cannot be applied to most non-magnetic fluids, as it requires the unique properties of magnetic fluids.

Mechanical Control

The process of mechanically controlled microdroplet generation involves the physical deformation of the fluid interface. The power sources that induce the deformation of the fluid interface include hydraulic, pneumatic, piezoelectric, and so on

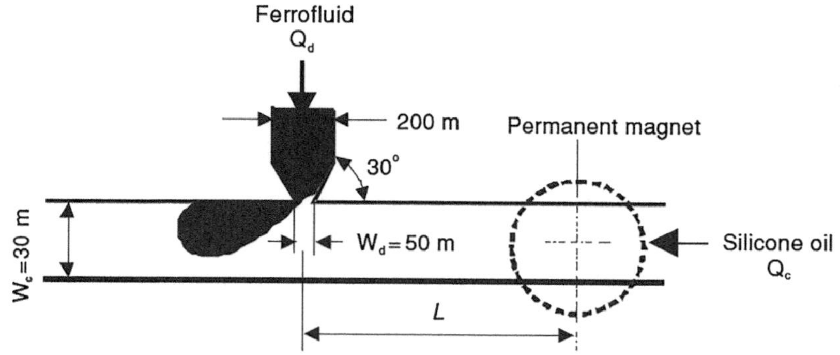

Fig. 6.5 Formation of magnetically controlled droplets in T-shaped structure (Tan et al. 2010) (reproduced from Kangning et al. 2014; Tan et al. 2010) with permission from the Royal Society of Chemistry

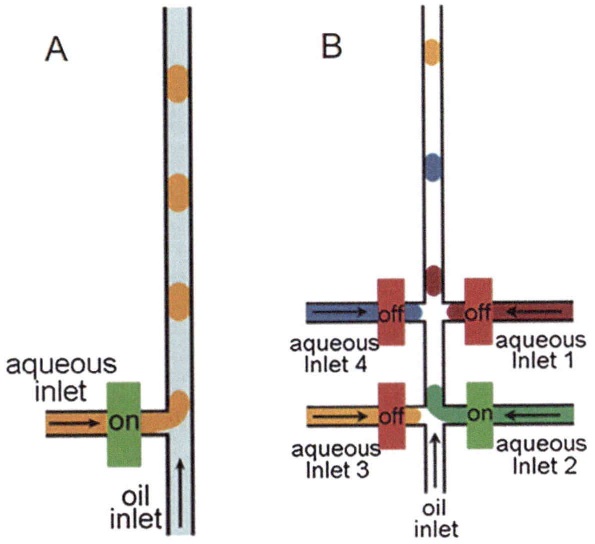

Fig. 6.6 Schematic diagram of micro-valve device (Zeng et al. 2009) (Reproduced from Zeng et al. 2009)

(Churski et al. 2010). For instance, the mechanical components in the droplet generation process can be controlled using hydraulic and pneumatic systems. In 2009, Zeng et al. (2009) proposed a design of an integrated pneumatic PDMS microvalve based on a T-shaped structure, as shown in Fig. 6.6, which is used to control the aqueous phase fluid to achieve droplet generation. The minimum volume of the droplet that can be generated is 1.3 nL. In this device, single droplets can be generated on demand by opening or closing the microvalves in sequence.

Thermal Control

Energy sources for thermally controlled droplet generation and control include resistive heating at the nodes and local heating by a focused laser beam, which essentially exploits the temperature dependence of the fluid. Specifically, as the temperature increases, the viscosity and interfacial tension of most fluids decrease, which is directly reflected in changes to the capillary number (Ca). Nguyen et al. (2007) first introduced to apply resistive heating to control droplet size of generation, with their structural design shown in Fig. 6.7a. By integrating a microheater and temperature sensor into the microfluidic device, they were able to precisely control the application of heat, thus regulating both droplet generation and size. By using relatively low heating temperatures (25–70 °C), the droplet diameter could be increased to more than twice its original size. Figure 6.7b presents the device used by Baroud et al. (2007), which utilizes a focused laser beam to locally heat the microchannel for droplet generation. Compared to Nguyen's method, laser heating offers greater flexibility, as the position of the focused laser point can be easily adjusted to achieve precise local heating at the droplet generation site.

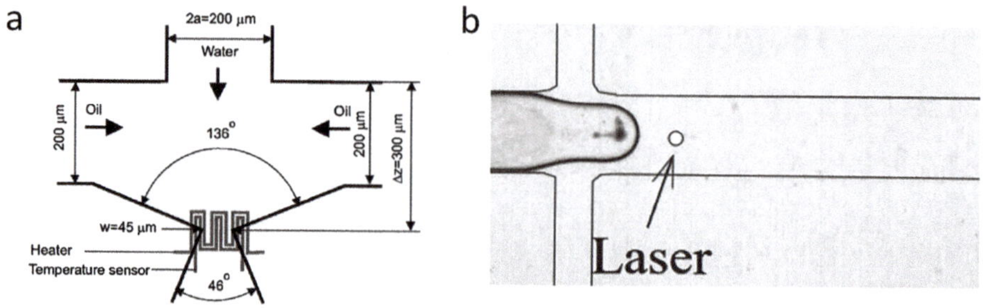

Fig. 6.7 (**a**) Droplet formation controlled by resistance heating (Nguyen et al. 2007). (**b**) Schematic diagram of droplet generation controlled by laser local heating (Baroud et al. 2007) (reproduced from Baroud et al. 2007)

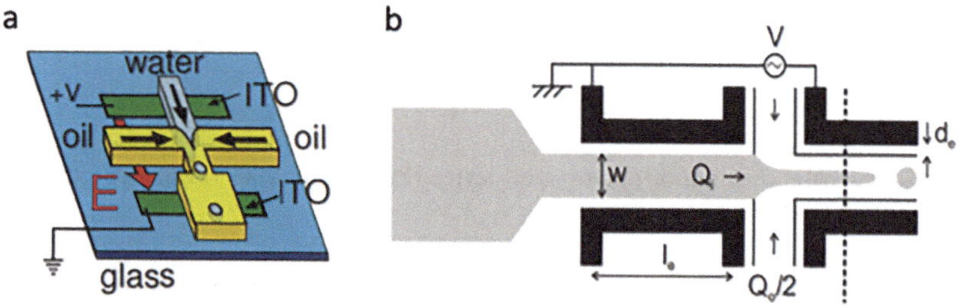

Fig. 6.8 (**a**) Generation of charged droplets (Link et al. 2006) (reproduced from Link et al. 2006). (**b**) Flow focusing interface under AC electric field (Castro-Hernandez et al. 2015) (reproduced from Castro-Hernandez et al. 2015)

Electrical Control

The electrical control method for microdroplet generation involves applying a voltage to the fluid within the microchannel to initiate and control droplet formation (Gu et al. 2010). The voltage source can be either AC or DC. In DC control, the voltage remains constant throughout the droplet generation process. In contrast, AC control involves voltage fluctuations, with the frequency of these fluctuations differing from the droplet generation frequency. For high-frequency AC control, the control signal frequency is significantly higher than the droplet generation frequency. Link et al. (2006) proposed a droplet generation chip design as shown in Fig. 6.8, in which DC voltage is used to control droplet generation. The bottom of the chip contains two indium tin oxide (ITO) electrodes for driving voltage access. The electrodes are in direct contact with the fluid, and the size of the droplets can be controlled by adjusting the voltage. Similarly, the device can also control the generation of droplets with AC voltage. The biggest disadvantage of this device design is that the electrodes are easily contaminated when in contact with the fluid, and the generated droplets are charged droplets, which is not conducive to application in biochemical samples. At present, in conventional electrical control methods, the electrodes are still in contact with the liquid, so dielectric materials must be introduced to separate the fluid to prevent electrode contamination or droplet charge.

Castro-Hernandez et al. (2015) proposed a device design that is different from the traditional electrical control method. As shown in Fig. 6.8, all electrodes are not in contact with the liquid, and droplets are generated by electrical control. The device uses alternating current with a frequency of up to 50 kHz to control droplet generation. The higher the voltage, the longer the jet length and the smaller the volume of the generated droplets. In addition, direct current can also be used in the electrowetting on dielectric (EWOD) effect, and alternating current has the advantage of reducing contact angle hysteresis. The experimental results show that the size of the droplet is related to the voltage difference between the oil–water interface and the downstream electrode.

6.1.3 Summary

Microfluidic chips have unique advantages and good application prospects in the research and clinical fields of tumor marker detection and tumor diagnosis and treatment and are becoming a new platform and new technology in the next generation of tumor diagnosis and treatment. Looking at the progress and trends of related research work at home and abroad, it is expected that microfluidic chips will have the following developments and breakthroughs in the field of tumor marker detection and tumor diagnosis and treatment in the future:

1. Develop cytological diagnosis technology based on microfluidic chips to achieve efficient sorting and identification of diseased cells in whole blood.
2. Develop nucleic acid molecular diagnosis technology based on microfluidic chips to complete nucleic acid extraction, separation, amplification, sequencing, and detection functions on microfluidic chips.
3. Develop integrated microfluidic chip whole blood analysis technology, combine sample pretreatment, sample separation and subsequent detection technology, conduct simultaneous analysis of multi-component biomarkers, and develop "one drop of blood" tumor marker diagnosis technology.
4. Develop the preparation of functionalized targeted micro–nano-drug delivery systems based on microfluidic chips to provide new materials and technologies for the realization of integrated tumor diagnosis and treatment.

6.2 Construction of Functionalized Micro–Nanointerfaces Based on Microchips and their Applications in Liquid Biopsy of Tumor Markers

Liquid biopsy refers to minimally invasive tests used to detect and analyze tumor-derived biomarkers in patients' blood and other body fluids (Alix-Panabières and Pantel 2021). These tests can provide diagnostic, prognostic, and predictive information about the disease. Liquid biopsy-based diagnostic methods have gained significant attention in recent years due to their non-invasive nature and the potential to detect cancer-specific biomarkers (Ma et al. 2023). More importantly, this technology can effectively overcome tumor heterogeneity and solve the pain points of precision medicine. Therefore, liquid biopsy is not only a breakthrough technology for the early diagnosis of solid tumors but also an important means to achieve personalized precision medicine for tumor patients. It is expected to become an ideal technology for early diagnosis, real-time monitoring, and efficacy evaluation of cancer.

Liquid biopsy leverages a variety of biomarkers from different sources. Tumor markers are primarily categorized into three types: circulating tumor cells (CTCs), extracellular vesicles (EVs) from tumor cells, and circulating tumor DNA (ctDNA) (Lone et al. 2022). These biomarkers provide real-time insights into the disease's status. Consequently, liquid biopsy can be used to detect early signs of cancer, monitor disease progression, and assess the effectiveness of treatment (Fig. 6.9). In recent years, numerous innovative technologies have been developed to enhance the detection of trace tumor markers in the early stages of cancer, significantly improving both sensitivity and specificity. Among these, microchip-based detection of trace tumor markers in liquid biopsy has emerged as one of the most efficient methods. This section focuses on blood specimens, highlighting the latest advancements in the detection and analysis of two key tumor markers (CTCs and EVs) in whole blood samples using microfluidic chips.

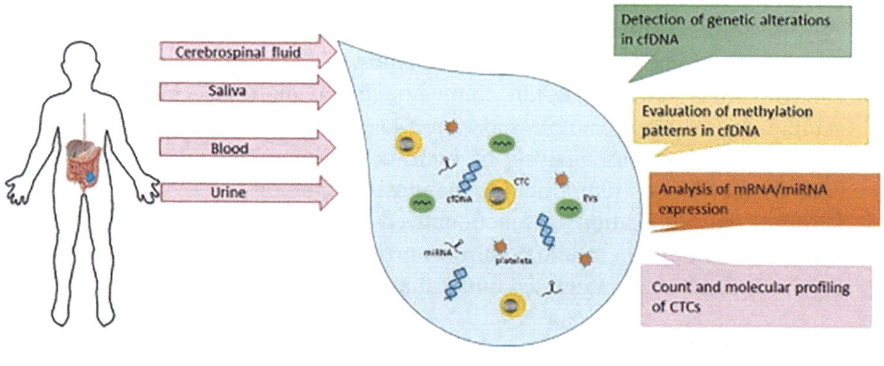

Fig. 6.9 Applications of liquid biopsy in CRC patients. Tumor biomarkers can be isolated from different body fluids including blood, saliva, urine, and cerebrospinal fluid (CSF). Blood is a source of cell-free (cf) DNA, miRNA, CTCs, platelets, and extracellular vescicles (EVs). For cfDNA isolation, plasma is preferred to serum because the clotting process leads to lysis of white blood cells and consequent release of genomic DNA with further dilution of tumor-derived DNA. Nucleic acids and proteins are also contained in CTCs and EVs that protect from degradation. Urine, saliva, and CSF are mainly sources of cfDNA, miRNA, and EVs. The current applications of liquid biopsy are listed (Normanno et al. 2018) (reproduced from Normanno et al. 2018)

6.2.1 *Detection of Circulating Tumor Cells Based on Microchips*

CTCs carry valuable information related to tumor onset, progression, metastasis, and drug resistance, making them crucial for monitoring the effectiveness of cancer treatments. Although CTCs sorting has great guiding value for clinical cancer treatment, there are still many challenges and difficulties in capturing and analyzing CTCs from whole blood samples (Alix-Panabières and Pantel 2014). This is mainly due to three typical characteristics of CTCs: (1) Scarcity: CTCs are present in extremely low numbers. While the number of blood cells in each milliliter of peripheral blood can exceed 10^9, CTCs typically number only a few to a few hundred (Arya et al. 2013). This requires that the detection method must be efficient and accurate in detecting very few CTC from a large number of blood cells; (2) Heterogeneity: CTCs exhibit significant antigen expression variability. Cancer cells of the same type may express different antigens, and even tumor cells within the same patient can have distinct antigen profiles; (Bishop 1987) (3) Size similarity to white blood cells: CTCs are roughly the same size as white blood cells, which makes their separation from the rest of the blood cells more difficult.

The most commonly used CTCs enrichment and analysis technologies currently include membrane filtration, immunomagnetic bead positive/negative enrichment, RT-PCR, immunofluorescence staining, microfluidics, and many more. Among them, the CTCs separation and analysis method based on microfluidic chips has significant advantages, such as the small amount of sample required and the absence of the need for pre-treatment of blood samples; the flowing fluid environment is conducive to the removal of other non-specifically adsorbed cells, thereby greatly improving the separation purity.

The principles of CTCs sorting and enrichment by microfluidic chips can be divided into two categories: (a) Sorting by immunoaffinity property (Fig. 6.10). Immunoaffinity-based enrichment technologies involve positive selection by specifically targeting epithelial biomarkers present on CTCs. This method relies on the binding of antibodies or ligands to these biomarkers to isolate CTCs from other blood constituents. (b) Sorting by biophysical property. Biophysical methods are label-free and exploit differences in cell size, deformability, density, and electric charge between CTCs and other blood cells (Yang et al. 2019). Each method has its own characteristics, advantages, and disadvantages. In addition, the researchers also integrated different functional modules into the microfluidic chip, which not only enables efficient capture and separation of CTCS, but also enables non-destructive release and recovery of separated CTCS for downstream analysis such as cell analysis and DNA (or mRNA) sequencing. These analyzes can even be performed at the single-cell level. The downstream analysis of CTCs generally adopts two strategies: one is to simply transfer the target CTCs from on-chip to off-chip for downstream processing and analysis of traditional steps, which is also the most commonly used analysis strategy at present. The other is the downstream analysis of the captured CTCs directly on the chip, which is crucial for the promotion and application of

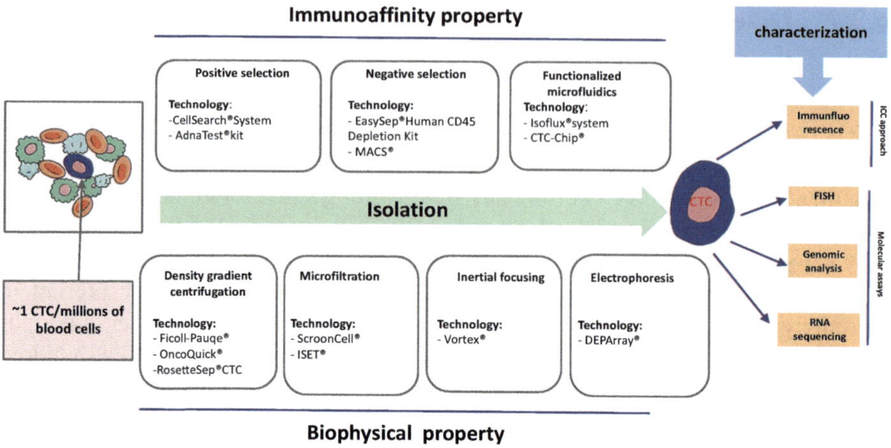

Fig. 6.10 Overview of technologies for CTCs capture, enrichment, and characterization. Immunoaffinity-based enrichment technologies capture CTC by positive or negative selection, typically using antibodies bound to the device surface or to magnetic beads. Positive selection is based on the specific targeting of CTCs epithelial biomarkers, whereas negative selection depletes hematopoietic cells by targeting cell surface antigens not expressed in CTCs. Functionalized microfluidics platforms can combine the advantages of microfluidic and the characters of positive capture and negative enrichment. Biophysical methods are label-free technologies relying on cell size, shape, density, and electric charge differences between CTC and other blood constituents. Density gradient centrifugation relies in the separation of different cell populations based on their relative densities. Microfiltration consists on size-based cell separation using pores or three-dimensional geometries. Inertial focusing relies on the passive separation of cells by size, through the application of inertial forces that affect positioning within the flow channel in microfluidics devices. Electrophoresis separates cells based on their electrical signatures, using an electric field. The methods of CTCs characterization include immunocytochemistry (ICC)-based approaches and molecular assays. Of which, ICC-based approaches are consist of immunofluorescence and immunohistochemistry technology, and molecular assays are consist of fluorescent in situ hybridization (FISH), real-time polymerase chain reaction (RT-PCR), genomic analysis, and RNA sequencing (Yang et al. 2019) (reproduced from Yang et al. 2019)

liquid biopsy technology in the clinic. Based on different CTCs sorting principles, this article summarizes the research status and latest progress of CTCs separation and enrichment using microfluidic chips in recent years.

6.2.1.1 Isolation of CTCs Based on Immunoaffinity Property

Affinity reaction is one of the most classical and commonly used methods for cell sorting on microfluidic chips. The specific step is to modify specific antibodies or aptamers that can bind to antigens on the surface of target cells on microchannels or microstructures inside the chip, such as epithelial cell adhesion molecules. When the sample flows through the microchannel, the target cell surface antigen binds to the specific antibody or aptamer on the microchannel or microstructure, and then the cells are fixed in the chip, other cells flow out of the chip with the buffer. Target cells are then collected for downstream analysis using appropriate methods, such as buffer replacement and elution.

Antibody Recognition

Among the technical methods for separating CTCs, antibody-based affinity separation is one of the most widely used techniques. Epithelial cell adhesion molecule (EpCAM) plays a crucial role in epithelial cell carcinogenesis and serves as a key biomarker for distinguishing CTCs from other blood cells. In microfluidic chip-based separation, the EpCAM antibody is integrated into the chip's internal structure. When the sample flows through the chip, the EpCAM antibody specifically recognizes and binds to the target CTCs through non-covalent interactions, effectively capturing the CTCs within the chip. Meanwhile, other blood cells are carried away by the buffer flow (Ahmed et al. 2017).

However, since the fluid in the microchannel usually flows in a laminar flow with a low Reynolds number, the mixing of substances can only rely on lateral diffusion, and only CTCs on the contact channel surface can be captured. The combination of slow diffusion rates and limited contact area results in a low net capture rate of CTCs. To address these challenges, microstructures are often added to the chip, or the channel design is specially modified to enhance the capture efficiency.

Sun et al. (2019a) developed an antibody-based microfluidic device with separate capture and flow zones (ZonesChip) for highly efficient capture of CTCs from human blood (Fig. 6.11). In this device, an external force, such as dielectrophoresis (DEP), is applied to move cells from the flow zone into the capture zone. The capture zone is characterized by a lower flow rate, which not only facilitates the binding of CTCs to antibodies immobilized on the microcolumns but also minimizes the risk of captured cells being dislodged by flow-induced forces. At high flow rates ($\geq$0.58 mm/s), the separation of the capture and flow zones in the ZonesChip significantly enhances capture efficiency, increasing it from approximately 0% (in conventional devices lacking separation) to nearly 100%.

However, improving the chip structure often presents challenges related to processing technology. The precise etching of microstructures and micro-grooves requires expensive equipment and specialized expertise. To overcome these challenges while maintaining high CTC capture efficiency, alternative approaches have been proposed, such as replacing microstructures with polymer molecular chains or natural biofilms. Yu et al. (2019) employed electrospinning technology to create poly(lactic-co-glycolic acid) (PLGA) nanofibrous arrays, which were arranged either randomly or in an aligned orientation on glass slips (Fig. 6.12). PLGA is known for its excellent biocompatibility, and its terminal hydroxyl groups can be easily modified to link with biotin via a condensation reaction. The biotinylated PLGA can then be functionalized with EpCAM antibodies through the interaction between biotin and avidin. The randomly arranged PLGA nanofibers form a complex network that not only prevents nonspecific adsorption by reducing direct

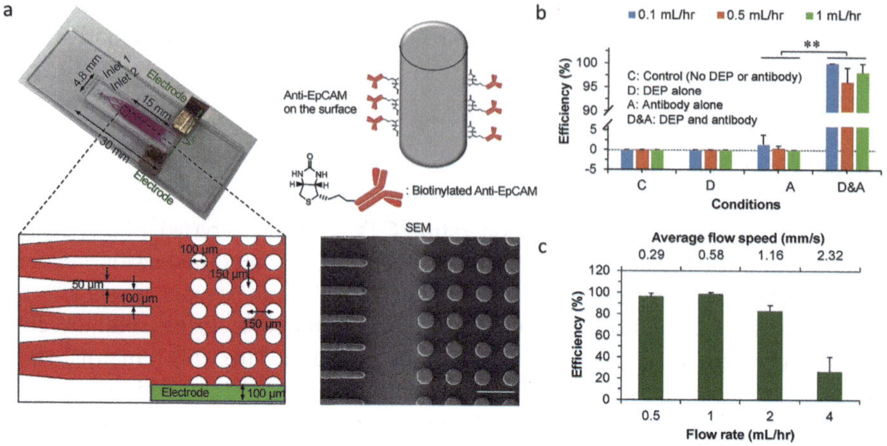

Fig. 6.11 (**a**) A real image of the microfluidic device on a glass slide together with a scheme showing the chemistry of anti-EpCAM modification on the micropost in the device, and a sketch and scanning electron microscopy (SEM) image showing the placement and design of the electrodes (green) and microposts (100 rows and 32 columns in the horizontal and vertical directions, respectively) in the main channel of the device. Scale bar: 200 μm. (**b**) The enhancement of capture efficiency by separating the capture and flow zones. (**c**) The capture efficiency at various flow rates (or various average flow speed in the main channel) (Sun et al. 2019a) (reproduced from Sun et al. 2019a). Reprinted with permission from Sun et al. (2019a). Copyright (2020) American Chemical Society

contact with the channel surface, but also significantly increases the available reaction area. This structural modification enhances the net capture rate of CTCs. Using spike recovery assays, it was shown that the capture rate of CTCs from human whole blood samples using this chip can reach up to 90%.

Sun et al. (2019b) proposed a modular CTC capture chamber for efficient cancer cell separation. The core components of the CTC capture module consist of polydimethylsiloxane (PDMS) films with cell-replication topologies and an anti-EpCAM antibody coating. The CTC chamber, composed of three capture modules, was used to isolate cancer cells from spiked blood samples representing late-stage (3 mL blood, 10 cancer cells/mL) and early- to mid-stage (9 mL blood, 1 cancer cell/mL) cancer. The results demonstrated high capture rates (EpCAM positive: ~80%, EpCAM negative: ~65%) and high purity (EpCAM positive: ~90%, EpCAM negative: ~80%) within 1 h (Fig. 6.13). These findings suggest that this simple, modular CTC chamber could pave the way for the development of next-generation CTC detection devices capable of diagnosing cancer at various stages.

Wu et al. (2020) utilized EpCAM aptamer-functionalized leukocyte nanovesicles to modify a microfluidic chip, creating a fluidic multivalent nanointerface for high-performance isolation of CTCs (Fig. 6.14). Compared to conventional aptamer-functionalized chips, this approach increased capture efficiency by sevenfold and significantly improved capture purity. The natural biomembrane of the leukocyte

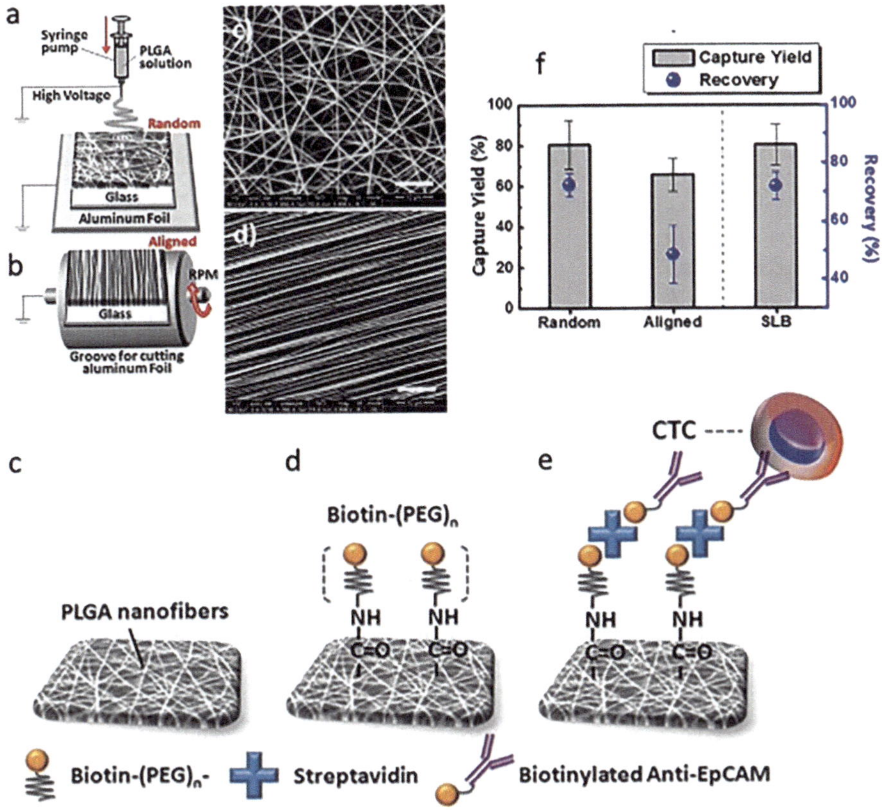

Fig. 6.12 (**a**, **b**) Schematic representation of the setups for electrospinning (**a**) random PLGA nanofiber arrays on a stationary glass cover slip and (**b**) aligned PLGA nanofiber arrays collected on a glass cover slip when using a spinning drum. (**c–e**) Schematic representations of the bioconjugation of biotinylated anti-EpCAM to PLGA nanofiber arrays for CTC isolation: (**c**) morphology of the nanofibers from SEM imaging; (**d**) EDC/NHS-mediated bonding of the biotin–(PEG)$_n$ to the surface of the PLGA nanofibers; (**e**) SA bonding to form biotinylated anti-EpCAM antibodies for capturing CTCs. (**f**) Capture and recovery yields of the microfluidic device with antibody-conjugated PLGA nanofiber arrays or a supported lipid bilayer (SLB) for EpCAM-positive HCT116 cells (Yu et al. 2019) (reproduced from Yu et al. 2019)

nanovesicles helps reduce nonspecific adsorption of blood cells, enhancing the specificity of CTC capture.

Cheng et al. (2021) integrated a 3D conductive scaffold coated with gold nanoparticles into a microfluidic chip, enabling highly efficient capture and electrochemical release of CTCs with high purity. The Au-S bond facilitates both antibody functionalization and electrochemical release. The macroporous structure of the scaffold enhances fluid mixing, while the high-density nano-gold particle interface improves cell interaction. This system not only allows for the reversible capture and release of CTCs, but also depletes leukocytes through nonspecific adsorption, thereby increasing the purity of the recovered CTCs (Fig. 6.15).

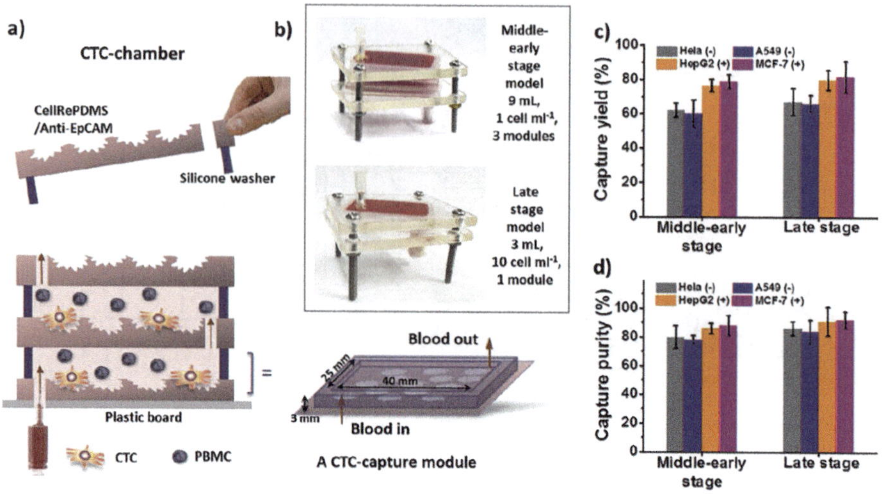

Fig. 6.13 Manufacture and applications of the CTC chamber for capturing CTCs from blood spiked with cancer cells. (**a**) Schematic representation of the CTC chamber assembled with CTC capture modules. A CTC capture module consisting of a CellRePDMS/Anti-EpCAM and a silicone washer is shown on the right. (**b**) Digital photographs of CTC chambers with variable number of modules. The one consisting of three modules in the top was used to capture cancer cells from 9 mL of blood containing 1 cancer cell per mL and the other one consisting of 1 module in the bottom was used to capture cancer cells from 3 mL of blood containing ten cancer cells per mL. Capture yield (**c**) and capture purity (**d**) of cancer cells obtained with the above two CTC chambers, which simulated the capture of CTCs from the blood of a middle-early-stage and late-stage cancer patient, respectively (Sun et al. 2019b) (reproduced from Sun et al. 2019b). Reprinted with permission from Sun et al. (2019b). Copyright (2020) American Chemical Society

Aptamer Recognition

Although antibodies can specifically recognize target CTCs, their cost is relatively high. As a more affordable alternative, nucleic acid aptamers can also be used for the separation of CTCs. Aptamers are short single-stranded DNA or RNA molecules selected through in vitro screening. They can form non-covalent bonds with target ligands, enabling specific binding (Wang et al. 2019). Unlike antibodies, which primarily bind to antigen clusters on the surface of target cells, aptamers interact with their targets through helical folding and binding to complementary sequences on the target ligand.

Chaoyong Yang et al. developed a microfluidic chip (ApTDN-Chip), which uses a sub-10-nm three-dimensional DNA structures (TDNs) as a framework and modifies an aptamer on the top (Zhang et al. 2020). The vertical nanoframe ensures that the aptamer is in a highly ordered upright orientation, which is more conducive to the contact between the aptamer on the microfluidic chip and the CTC in the sample and improves the capture efficiency of CTC (Fig. 6.16). Additionally, the tetrahedral nanostructure reduces local overcrowding effects. Compared to a monovalent aptamer-modified chip, the ApTDN-Chip demonstrates a nearly 60% increase in

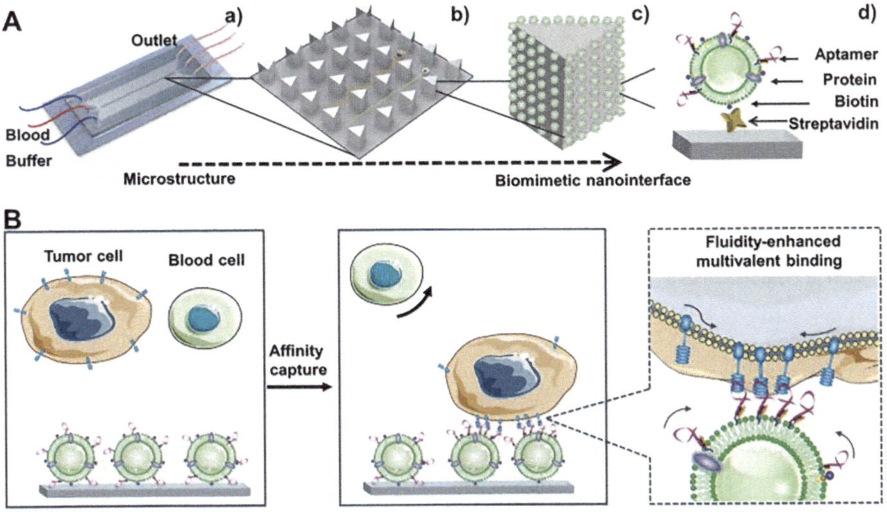

Fig. 6.14 (**a**) Schematic diagram for the fabrication of FLASHChip: (a) the appearance of FLASH-Chip; (b) the movement paths of tumor cell and blood cell along DLD-patterned microarray in FLASH-Chip; (c) nanovesicle-cloaked micropillar interface; (d) schematic diagram for assembling nanovesicles on-chip substrate through biotin-streptavidin interaction. (**b**) Working principle and schematic diagram of the fluidity-enhanced multivalent binding on biomimetic membrane nanointerface (Wu et al. 2020) (reproduced from Yang et al. 2020). Reprinted with permission from Yang et al. (2020). Copyright (2020) American Chemical Society

capture efficiency, thanks to the rigid framework of the 3D DNA structure. Furthermore, the ApTDN-Chip facilitates easier access for DNase I to the aptamer, achieving a release efficiency of nearly 85% with a cell survival rate of 91%, making it suitable for downstream molecular analysis (Fig. 6.17).

However, relying solely on the specific recognition of Ep-CAM antibodies to capture CTCs has a major drawback, as highly metastatic CTCs may undergo epithelial-mesenchymal transition, causing a significant downregulation of Ep-CAM expression, which ultimately leads to the neglect of these CTCs, resulting in false-negative results. To address this issue, Wu et al. (2021) modified two aptamers, EpCAM and N-cadherin, on the poly(lactic-co-glycolic acid) (PLGA) nanofiber-based microfluidic devices at the same time for highly efficient capture and specific release of epithelial and mesenchymal CTCs of ovarian cancer. The PLGA nanofiber-based microfluidic device demonstrated excellent capture efficiency (91% for A2780 cells, 89% for OVCAR-3 cells), release efficiency (95% for A2780 cells, 88% for OVCAR-3 cells), and sensitivity for rare cells (92% for A2780 cells, 88% for OVCAR-3 cells). This dual-aptamer-modified PLGA nanofiber microfluidic device shows significant potential in improving overall capture rates and reducing false-negative results.

Affinity sorting methods offer high specificity and can effectively separate different types of cells with similar shapes and sizes. However, a trade-off exists

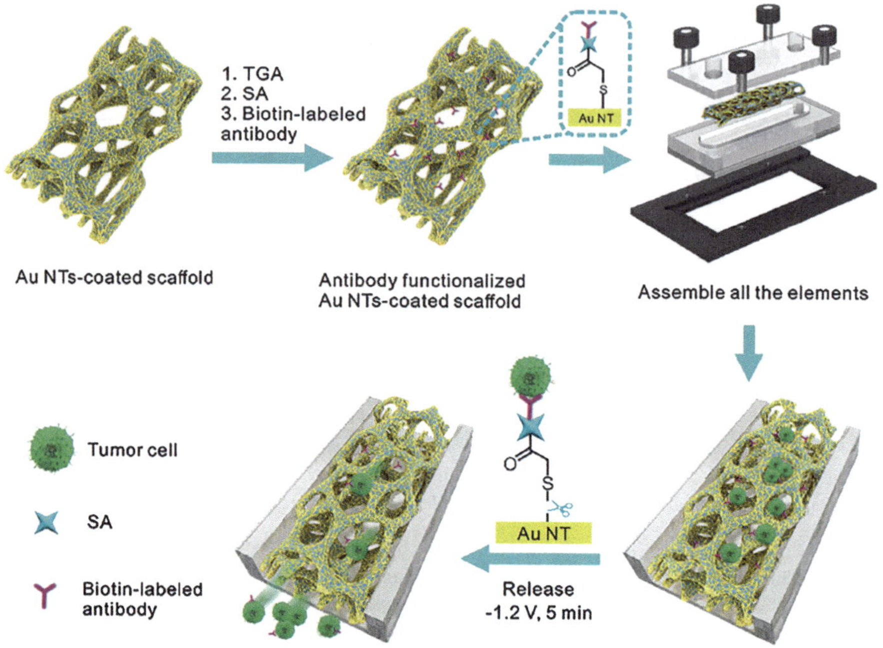

Fig. 6.15 Capture and release of CTCs by the 3D conductive scaffold microchip (Cheng et al. 2021) (reproduced from Cheng et al. 2021). Reprinted with permission from Cheng et al. (2021). Copyright (2021) American Chemical Society

between flux and efficiency. As the flow rate increases, the time available for CTCs and antibodies to interact decreases, leading to lower sorting efficiency. To address this, enhancing the contact area and increasing the probability of interaction between cells and the chip's internal structure, as well as incorporating multiple antibodies to capture CTCs expressing different proteins, are key strategies in the development of microfluidic chips that use affinity-based separation to isolate CTCs.

6.2.1.2 Isolation of CTCs Based on Biophysical Property

CTC sorting based on physical properties leverages differences in size, density, deformability, and adhesion between CTCs and normal blood cells to isolate CTCs from other cells in the blood. Compared to immunoaffinity methods, physical separation techniques are often simpler, as they do not require chemical modifications or biomarkers, resulting in less impact on cell viability. In microfluidic systems, physical separation typically involves the application of external physical fields or the use of size-based separation principles, such as pore sieving. This is achieved by incorporating various microstructural units into the chip. Common microstructures used for physical sorting include micropores, microfilters, and microcolumns.

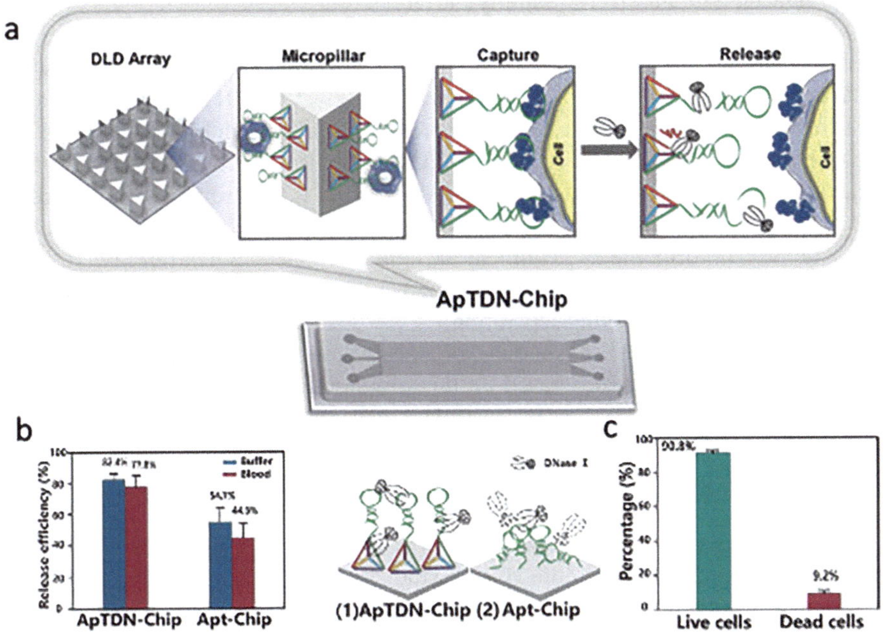

Fig. 6.16 (**a**) Working principle of the ApTDN-chip. Engineered aptamer SYL3C-TDN was used as a nano-scaffold to anchor the microfluidic interface to form an intrinsically highly ordered and upright orientation nano-topography. Under the synergy of size-selective enhanced and trans-scale interaction, CTCs are more accessible to SYL3C-TDN as compared to blood cells. After capture, the enriched CTCs can be efficiently released by deoxyribonuclease I (DNase I), which is compatible with downstream molecular analysis. (**b**) Release efficiency of captured model CTCs using the ApTDNChip and Apt-Chip. (**c**) Cell viability of recovered SW480 cells (Zhang et al. 2020) (reproduced from Zhang et al. 2020)

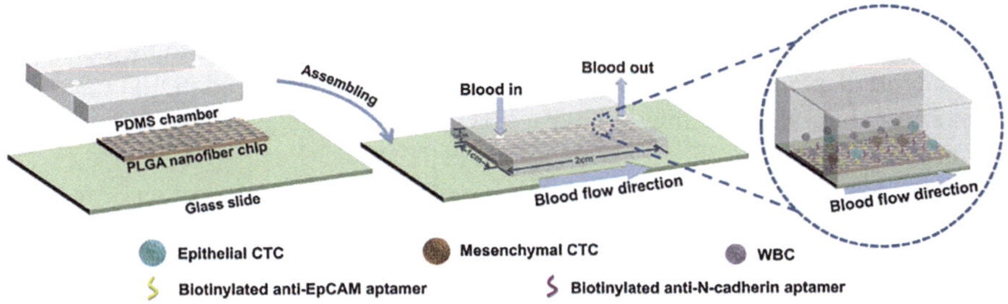

Fig. 6.17 Schematic illustration of the dual aptamer-modified PLGA nanofiber-based microfluidic device (2 cm × 1 cm × 1 mm) for the different phenotypic CTC capture (Wu et al. 2021). Reproduced from Wu et al. (2021) with permission from the Royal Society of Chemistry

Based on Cell Size and Deformability Differences

Most epithelial-derived CTCs have a diameter ranging from 14 to 26 μm, while leukocytes typically range from 8 to 20 μm (Gorges et al. 2012). By incorporating microstructural elements such as micropores, microfilters, and microcolumns that are smaller than the diameter of CTCs into the chip, the CTCs are trapped in these structures due to their larger size, while blood cells flow through with the buffer. Larger white blood cells may also be captured by the microstructures. Because CTCs are less deformable than white blood cells, increasing the buffer flow rate will wash away the more deformable white blood cells, while the CTCs remain trapped in the chip, thereby achieving separation.

ClearCell® CX System of Abnova Company (Tan et al. 2009) is a representative platform for CTC separation based on this principle, and it also enables dynamic monitoring of the CTC capture process. The main structure of the chip consists of cylindrical microcolumns, with each capture unit comprising three cylinders arranged in a "claw" configuration. After capture, CTCs can be stained, identified, and quantified. They can also be collected and cultured for subsequent experiments, such as studying the molecular biological characteristics of CTCs, establishing animal models, and conducting clinical case studies.

Chen et al. (2021) constructed a pore-sieve chip system based on the biomimetic splenic sinus microstructure. The system uses ultra-thin silicon nitride materials to create slit filters for CTC separation, taking advantage of the excellent properties of silicon nitride, such as high structural strength, transparency, and good biocompatibility. The slit-structured filter pores offer lower flow resistance compared to traditional circular structures. By optimizing the flow rate and slit width, efficient separation can be achieved while preserving the high activity of captured CTCs. However, because white blood cells and CTCs are similar in size, this method has relatively low screening accuracy and may yield false positives. Additionally, placing a filter pore in the cell flow path can lead to clogging, which increases negative pressure and reduces separation efficiency. To address these issues while maintaining the advantages of the method, researchers have optimized CTC screening based on cell size (Fig. 6.18a–d). Liu et al. (2021) integrated filtration with the deterministic lateral displacement (DLD) principle and designed a cascade DLD microcolumn array chip, which achieves a 96% CTC recovery rate at a high throughput of 1 mL/min and removes 99% of white blood cells.

There are several advantages to capturing CTCs by sorting based on differences in cell size and deformability: (1) Simple operation—The sample only needs to be diluted with buffer, without the need for staining or marking. The diluted whole blood is directly passed into the chip during detection, and the cells are collected at the exit. (2) High capture efficiency; (3) High-throughput sorting; (4) Lower cost compared to CellSearch; (5) no reliance on surface markers—The captured CTCs can be identified using multiple antibodies. However, this method has some limitations. Filtration sorting is based solely on cell size and deformability. Because CTCs and white blood cells often overlap in size, some CTCs may pass through the filter or microcolumn gaps. Additionally, due to epithelial-to-mesenchymal transition

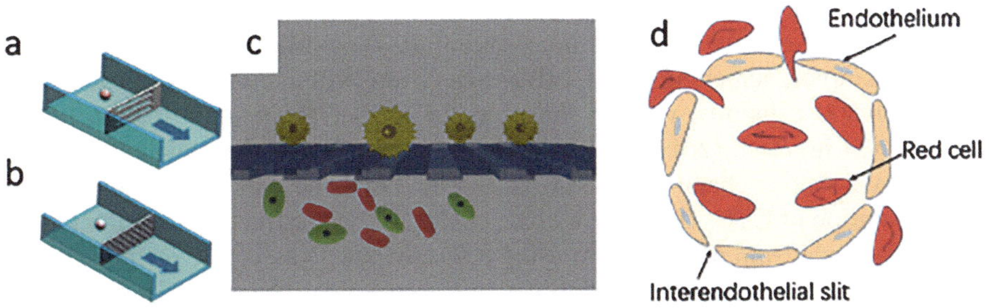

Fig. 6.18 (**a–d**) Schematic diagram of the microfluidic device of bionic spleen IES for CTCs sorting. (**a**) Simulation model of the IES structure. (**b**) Simulation model of circular pores. (**c**) Schematic of the CTCs physical filtration method using the bionic IES microfluidic device. (**d**) Schematic of the structure of the spleen sinus (Chen et al. 2021) (reproduced from Chen et al. 2021). (**e–g**) Filter deterministic lateral displacement (filter-DLD) structure. (**e**) Schematic design of the CFD chip. The device is composed of a blood depletion module, followed by a cell-size-separation module. The blood depletion module consists of three sets of symmetric filter-DLD arrays. The cell-size-separation module consists of a filter-DLD array with gradually increasing inclination angle. (**f**) SEM image of the filter-DLD structure. Scale bar, 20 µm. (**g**) Image shows the movement of blood cells in the filter-DLD array. Red arrows indicate the trajectories of zigzag mode, green arrows indicate the trajectories of bumping mode (Liu et al. 2021) (reproduced from Liu et al. 2021)

(EMT), these CTCs may resemble interstitial cells, which are more likely to metastasize and exhibit higher malignancy (Kim et al. 2012). Moreover, under the action of large mechanical forces, CTCs are easy to break when the buffer flows through the microcolumn or filter screen (Li et al. 2013). These factors can affect both separation purity and cell viability. To address these issues, the design of the internal capture unit should avoid angular microcolumns, such as triangles, rectangles, or squares, and instead use rounded or oval microcolumns to minimize mechanical damage to the cells.

Based on Differences in Cellular Mechanical Properties

Microfluidic chips used for screening CTCs based on cell size typically incorporate various microstructures that act as physical barriers. However, fabricating these precision structures can be challenging. With advances in microscale fluid mechanics and a deeper understanding of particle motion in microfluidic environments, a new method for separating and enriching CTCs has emerged. This method eliminates the need for microstructures and can be achieved simply by regulating the fluid dynamics.

Kulasinghe et al. (2019) designed a simple square-channel chip based on the inertial effect at the microscale. In this chip, the combined effects of Dean vortices and inertial lift generate an equilibrium point near the midpoint of the long side of the channel. As a result, CTCs are enriched at the center of the channel, while other components are pushed toward the periphery. This chip was able to effectively capture CTCs and circulating tumor microemboli (CTMs) from head and neck cancer (HNC) samples, achieving a high cell recovery rate within clinically relevant numbers (10–500 cells/mL) without the need for further purification (Fig. 6.19a–c). Similarly, Lim et al. (2019) leveraged the inertial effect by introducing tangential flow in a T-shaped channel, which isolates blood cells near the tube wall, enabling the separation of CTCs (Fig. 6.19a). Zhu et al. (2021) constructed a spiral microfluidic chip made from polymer film using a jigsaw puzzle assembly method. By utilizing inertial forces and Dean vortices in a trapezoidal channel, this system was able to separate CTCs with a minimum size difference of 2 μm at a high throughput of 3000 μL/min, achieving a recovery rate of 90–94% (Fig. 6.19b). Lu et al. (2020) designed a microfluidic chip integrating a triangular prism array-based capture device with a streamline-based focus-separation mechanism. The inertial effect is first used to separate blood cells, followed by capture of CTCs using the triangular microcolumn array. This system achieved a high capture rate of 94.8% with a throughput of 40 mL/h (Fig. 6.19c–e).

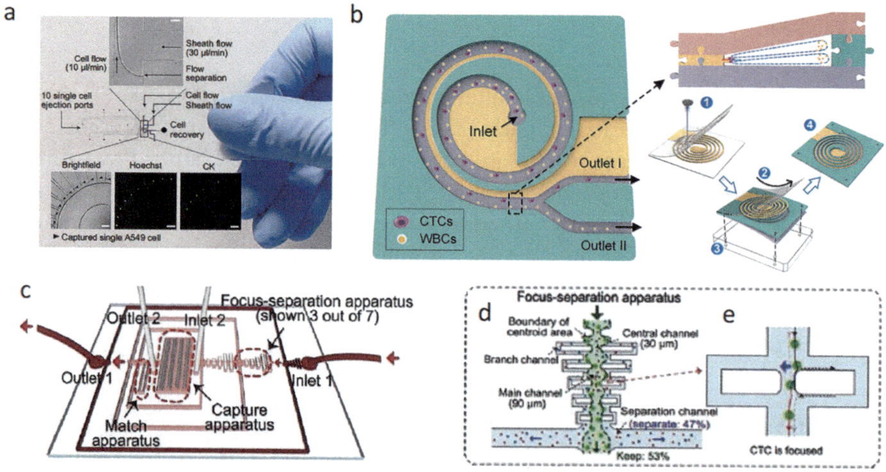

Fig. 6.19 (**a**) Microfluidic enrichment for single-cell analysis. Inertial focusing, label-free capture of single cancer cells using a microfluidic device. (top) Hydrodynamic focusing of cell flow (A549 lung adenocarcinoma) by sheath flow (glycerol). (bottom) Bright-field and immunofluorescent images of isolated single A549 cells (scale bar, 100 μm) (Lim et al. 2019) (reproduced from Lim et al. 2019). (**b**) Conceptual design of the polymer film inertial microfluidic jigsaw (PIMJ) sorter fabricated by using the jigsaw puzzle method. The microfluidic jigsaw scheme for flexible and rapid fabrication of trapezoidal spiral channels (Zhu et al. 2021). Reproduced from Zhu et al. (2021) with permission from the Royal Society of Chemistry. (**c**) A schematic illustration of the chip. (**d**) An enlarged view of the focus-separation apparatus. (**e**) An enlarged view indicates a CTC whose centroid is near the boundary that could be pushed from its original trajectory to a new trajectory by the sidewall of the central channel (Lu et al. 2020) (reproduced from Lu et al. 2020)

6.2.2 Detection of Extracellular Vesicles Based on Microchips

EVs are hemispherical vesicles surrounded by a lipid bilayer. They are released by cells into the extracellular space and can be found in various body fluids, including blood, saliva, amniotic fluid, breast milk, and urine. EVs typically contain cytosolic components from the secreting cells, including high levels of cytoskeletal proteins and heat shock proteins (Yáñez-Mó et al. 2015). Their surfaces are covered with integrins, glycoproteins, and transmembrane proteins, which are involved in vesicle transport (Raposo and Stoorvogel 2013) According to different formation mechanisms and physiological characteristics, EVs can be divided into three categories: exosomes, microvesicles, and apoptotic bodies. Exosome formation begins with cell endocytosis, and forms multivesicular inclusion bodies by "budding" inward, and releases small vesicles after fusing with the cell membrane; while microvesicles are formed by the mother body "budding" outward and shedding; apoptotic bodies are microvesicles released by programmed necrosis or late apoptosis of cells. More and more studies have shown that EVs have important biological functions. They not only participate in the transfer of substances and information between cells but also play an important role in physiological and pathological processes such as cell migration, angiogenesis, immune response, and tumor infiltration and metastasis (Théry et al. 2002; Colombo et al. 2014). And the amount, types, and compositional changes of proteins, lipids, and ribonucleic acids are closely related to diseases.

Despite the wide variety of body fluids, especially blood, containing complex compositions, achieving efficient and rapid separation of EVs remains a significant challenge (Shao et al. 2018). Methods for separating EVs include ultracentrifugation, gel exclusion chromatography, filtration, precipitation, immunomagnetic beads, and microfluidic chips. Among them, traditional methods such as ultracentrifugation, gel exclusion chromatography, and filtration that separate based on physical properties such as size, density, and surface charge have problems such as low throughput, cumbersomeness, lack of standardized methods, low recovery rate, and low purity (Tauro et al. 2012). In addition, furthermore, commercial kits that rely on precipitation or immunomagnetic bead-based methods (Wunsch et al. 2016) are prone to contamination with nonexosome materials (Gholizadeh et al. 2017). These methods also involve significant manual handling, suffer from low throughput, and can lead to EV deformation or contamination. As a result, they are not well-suited for large-scale scientific research or clinical applications.

Efficient and highly selective methods for EV separation are urgently needed. To address this challenge, advanced microfluidic chip technology—characterized by miniaturization, integration, and automation—offers a promising solution. The integration of microfluidic chips in EV separation has become a major focus of research. This section reviews and summarizes the latest advancements in microfluidic-based EV separation methods. These methods can be broadly categorized into passive and active approaches. Passive methods include surface binding capture techniques (e.g., antibody-based capture), sieving technologies using porous membranes, and capture methods based on sieving structures. Active methods include optical tweezers (leveraging optical polarizability), ultrasonic resonance, dielectrophoresis, electrophoresis, flow sorting, and magnetic sorting. From the perspective of separation mechanisms, microfluidic-based EV separation techniques can be further classified into two categories: physical and biological (see Table 6.3).

6.2.2.1 Physical Separation Technology

The mechanism of physical separation is mainly based on the differences in the physical properties of EVs (such as size, deformability, and fluid dynamics). EVs are separated by setting up different microstructure units including micropores, microfilters (Liang et al. 2017; Woo et al. 2017) and microcolumns in the chip, and using inertial force and lateral displacement methods (Liga et al. 2015), sono-nanofilter systems (Wu et al. 2017), electrokinetic concentration technology (Cheung et al. 2018), and concentration polarization electrophoresis (Marczak et al. 2018). Physical technology provides a method to separate exosomes without labeling or derivatization, relying only on the chip's own structure or other physical method.

Mechanical Filtering with Micro- and Nanostructures

Traditionally, filtration relies on a porous mechanical barrier to separate larger particles from smaller ones in a fluid. With micro- and nanoscale devices, particles the size of vesicles can be efficiently separated. However, clogging remains a significant challenge, particularly with biological samples that are viscous and contain particles spanning several orders of magnitude in size. This issue can be mitigated

Table 6.3 Classification by separation mechanism

Separation mechanism	Method	Advantages	Disadvantages
Based on physical properties	Micropore or micro-filtration	Easy to realize	Easy saturation, low recovery and difficulty to dispose
	Microcolumn	Easy to realize	Easy saturation
	Lateral displacement	High throughput	Poor specificity
	Acoustics	High throughput	High technical requirement and poor selectivity
Based on biological and chemical properties	Immunomagnetic beads	Good selectivity	Expensive, only extracellular vesicles with specific components on the surface can be separated
	Aptamer	Good selectivity	Highly selective aptamers are required
	Molecular imprinting	Low price	Target extracellular vesicles are required as the templates

by designs that allow larger particles to pass through or by employing sequential filters. Yeh et al. (2020) developed a microdevice consisting of three-dimensional carbon nanotube arrays for the rapid capture of EVs based on their size. This label-free capture method preserves the structural integrity and biological properties of the EVs, enabling the study of their internalization by neurons and other cells. This platform facilitates size-based analysis of EVs and provides insights into extracellular communication. As such, it holds potential for early diagnosis of neurological diseases and the development of novel therapeutic strategies (Fig. 6.20).

Physical Field Sorting

Acoustophoresis is a technique used to manipulate particles in microfluidic systems through primary radiation forces and scattered acoustic interactions. It is a gentle method for separating, enriching, and washing particles and vesicles without the need for labeling, making it particularly suitable for biological samples. The acoustic wave that generates the acoustic force is a mechanical wave, which can be classified into surface acoustic waves (SAW) and bulk acoustic waves (BAW). SAW devices typically operate at higher frequencies than BAW devices, resulting in stronger radiation forces that are effective in isolating EVs from continuous flows.

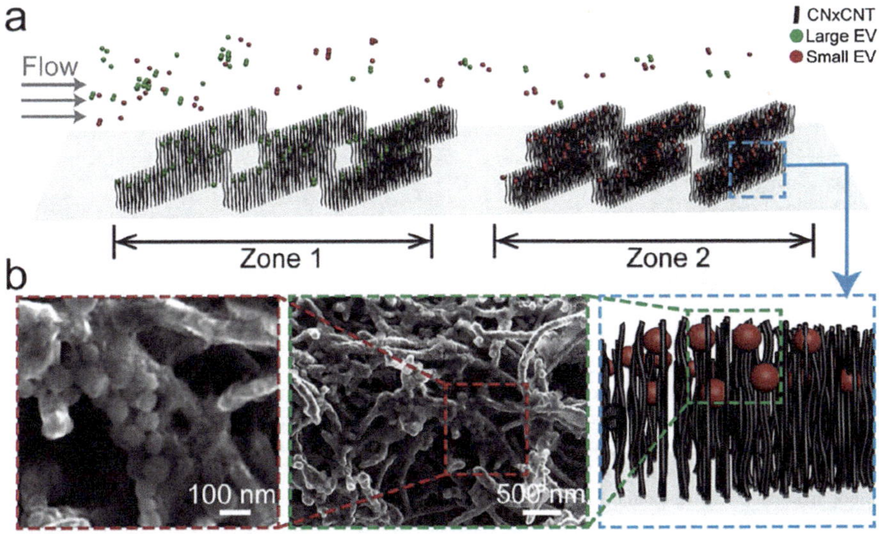

Fig. 6.20 Aligned Carbon nanotube arrays for size-based exosome capture. (**a**) Illustration of the size-based capture by different ITD zones of the CNxCNT arrays. (**b**) Scanning electron microscopy images of the aligned CNxCNT arrays after capture (Yeh et al. 2020) (reproduced from Yeh et al. 2020). Reprinted with permission from Yeh et al. (2020). Copyright (2020) American Chemical Society

However, BAW devices have a higher acoustic energy density and generally offer higher throughput.

Acoustic trapping is a technique used to retain particles in an acoustic field. A novel acoustic trapping channel, utilizing a larger capillary, has been developed to generate multiple pressure nodes (Fig. 6.21a, b). This design enables the enrichment and washing of EVs at a flow rate of 500 μL/min. Several milliliters of urine were processed, yielding sufficient protein for subsequent EV proteome profiling using mass spectrometry (Broman et al. 2021).

Acoustic trapping can also be performed using surface acoustic wave (SAW) actuation. In this method, a packed bed of seed particles, confined by micropillar posts, can be actuated by two opposing interdigital transducers (IDTs) at the resonance frequency of the seed particles (Habibi and Neild 2019). The scattered sound interactions between the seed particles and nanoparticles cause the nanoparticles to aggregate within the packed bed, effectively retaining them against the flow (Fig. 6.21c, d).

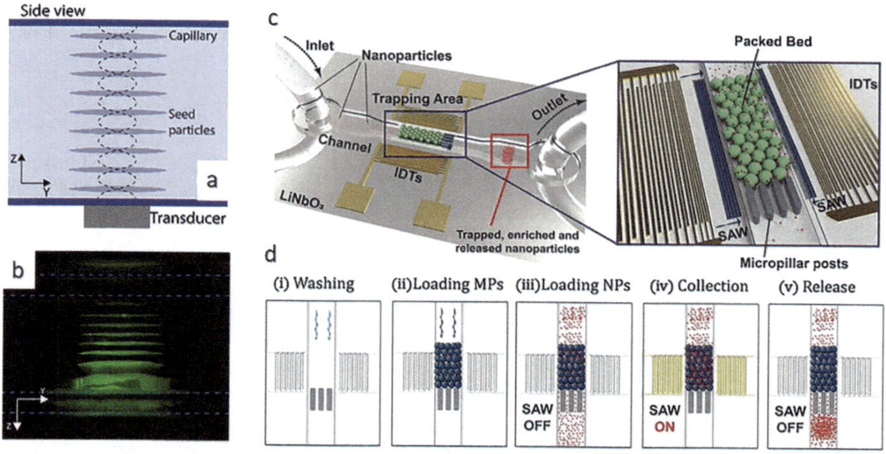

Fig. 6.21 (**a**) Model of the standing acoustic wave inside the channel viewed from the side. Particle clusters are shown, collected in the nine nodes created by the standing wave (Broman et al. 2021). (**b**) Seed particle cluster enriched with 500 nm fluorescent polystyrene particles, viewed from the side. Blue dashed lines indicate the capillary wall (Broman et al. 2021) (reproduced from Broman et al. 2021). Reprinted with permission from Broman et al. (2021). Copyright (2021) American Chemical Society. (**c**) Schematic of the microfluidic device for trapping and enriching nanoparticles using resonance of a passively-trapped packed bed of 10-μm polystyrene beads excited by surface acoustic wave (SAW). The enlarged view of the trapping area shows two opposing IDTs used to generate the standing SAW and the micropillars that trapped the microbeads (Habibi and Neild 2019). (**d**) Schematic illustration of the system that shows the sequence of the loading of microparticles (MPs) and then nanoparticles (NPs). Upon activating the SAW (here IDTs are highlighted in yellow), nanoparticles can be collected on demand and the high concentration sample can be released by switching off the SAW (Habibi and Neild 2019) (Reproduced from Habibi et al. 2019)

In addition to acoustic and magnetic forces, electric fields can also be used to manipulate and separate EVs. A common technique is dielectrophoresis (DEP), in which dielectric particles are polarized by an electric field gradient and migrate to either the electric field minimum or maximum, depending on the dielectric properties of the particles and the surrounding medium (Liu et al. 2019b). Chen et al. (2022) demonstrated the separation of exosomes from breast cancer cells and bovine/human breast milk based on differences in membrane capacitance. Exosomes from bovine and human milk exhibited similar DEP properties, whereas exosomes from cancer cells showed distinct differences. These results suggest that DEP can be effectively used to isolate exosomes released from cancer cells for liquid biopsy applications.

Insulator-based dielectrophoresis (iDEP) utilizes a series of insulating pillars to create high electric field gradients that can trap particles. Figure 6.22 illustrates how the spacing between these pillars can be optimized to separate purified exosomes into two distinct populations with diameters of 113 ± 10 nm and 73 ± 9 nm, respectively. One advantage of this system is that it is driven by electroosmotic flow, eliminating the need for a syringe pump. However, this device is not suitable for samples containing cell debris, as the small gaps between the pillars may become clogged (Ayala et al. 2019).

6.2.2.2 Biological and Chemical Separation Technologies

Current research on the separation of exosomes based on their biochemical characteristics primarily focuses on differences in surface protein expression. Exosomes are commonly isolated from samples using immunocapture techniques. Immunocapture is a method that involves adsorbing antibodies specific to exosome

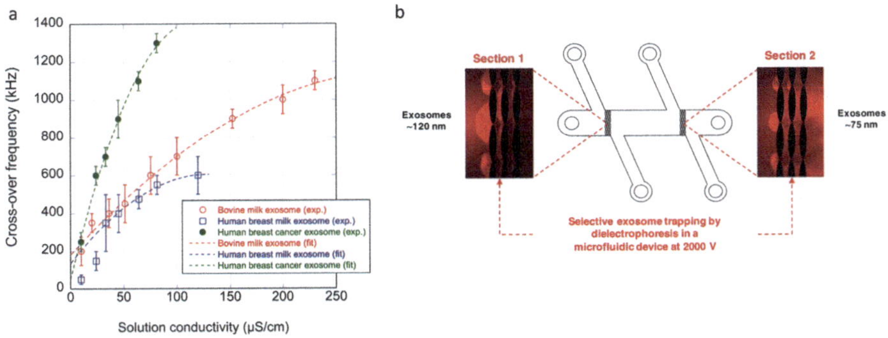

Fig. 6.22 (**a**) Size-dependent isolation of exosomes using dielectrophoresis with arrays of insulating posts (Ayala et al. 2019) (reproduced from Ayala et al. 2019). Reprinted with permission from Ayala et al. (2019). Copyright (2019) American Chemical Society

markers onto the surface of a solid-phase carrier, enabling the selective capture of exosomes. This technique offers several advantages, including speed, sensitivity, simplicity, and ease of standardization of the carrier, making it widely used in exosome research. Methods for exosome separation using biological and chemical technologies are summarized in Table 6.4.

One of the main challenges in extracellular vesicle (EV) protein analysis is the small size and low concentration of proteins within the EV membrane. Protein detection and analysis must be both highly sensitive to capture faint signals and specific enough to distinguish these signals from background noise (Ramirez et al. 2018). To overcome these challenges, researchers have employed various strategies, including signal amplification to increase detection sensitivity and additional washing steps to reduce background signal while preserving the integrity of the EVs (Hilton and White 2021).

Liu et al. (2019a, b) developed a microfluidic co-flow system capable of sorting EVs into three distinct size ranges (Fig. 6.23). In this system, DNA aptamers specific to HER2 and EpCAM are first conjugated to the EV surface, followed by size-selective separation of EV subpopulations using λ-DNA-mediated viscoelastic microfluidics. This approach enables two-dimensional analysis of individual EVs, allowing for characterization by both size and marker expression. The platform provides valuable insights into the heterogeneity of single EVs and has been used to differentiate breast cell lines and breast cancer patients based on EV signatures. This system offers a powerful tool for evaluating EV heterogeneity at the single-vesicle level, with promising applications in clinical diagnostics.

In a different approach, Beekman et al. (2019) present a multi-modal analysis platform designed for the specific capture of tumor-derived EVs on antibody-functionalized stainless steel substrates, followed by in situ multi-modal characterization using scanning electron microscopy (SEM), Raman spectroscopy, and atomic force microscopy (AFM) imaging (Fig. 6.24a). The process begins with the covalent attachment of anti-EpCAM (epithelial cell adhesion molecule) antibodies to the stainless steel substrates. EV samples derived from the LNCaP prostate cancer cell line were then introduced into a microfluidic device incorporating these antibody-functionalized substrates for capture (Fig. 6.24b–e). For sample preparation, three different polydimethylsiloxane (PDMS) devices were employed: a 6-mm-diameter reservoir, a xurography channel, and a navigation marker device (Fig. 6.24f). Following successful capture of EVs from the LNCaP cell line, the isolated vesicles were analyzed using a combination of Raman spectroscopy, SEM, and AFM imaging.

Yang et al. (2020) developed an integrated microfluidic device designed for the isolation and in situ detection of lung cancer-specific exosomes collected from patient urine. The system features a nanoporous gold (Au) nanocluster membrane (with 100 nm pores) functionalized with a capture antibody, supported by a polymethyl methacrylate (PMMA) base (Fig. 6.25). To detect and quantify the lung cancer-specific exosomes, a second antibody-conjugated Au nanorod probe was used in conjunction with dark-field microscopy. This method is not only fast and

Table 6.4 Methods based on EV analysis by biological and chemical separation technologies

Name	Target	Type	Sample	Biological application	Year	References
λ-DNA- and aptamer-mediated sorting and analysis of extracellular vesicles	Membrane protein/ size sorting	Dual	Conditioned media and blood plasma	Heterogeneity assessment and cancer typification	2019	Liu et al. (2019a)
Immuno-capture of extracellular vesicles for individual multi-modal characterization using AFM, SEM and Raman spectroscopy	Membrane protein	Single	Conditioned media	Identification, size and density surface analysis of specific EVs (EpCAM.)	2019	Beekman et al. (2019)
An integrative microfluidic device for isolation and ultrasensitive detection of lung cancer-specific exosomes from patient urine	Membrane protein	Single	Blood plasma	Isolation and biomarker identification for lung cancer	2020	Yang et al. (2020)
Isolation and digital counting of extracellular vesicles from blood via membrane-integrated microfluidics	Membrane protein	Single	Blood	Isolation and quantification of specific EVs	2022	Lu et al. (2022)
Integrated microfluidic system for isolating exosome and analyzing protein marker PD-L1	Membrane protein	Single	Conditioned media and blood plasma	Isolation and quantification of PDL1 EVs	2022	Kim et al. (2020)
EV-ident: Identifying tumor-specific extracellular vesicles by size fractionation and single-vesicle analysis	Membrane protein	Double	Conditioned media and blood plasma	Size fractionation and quantification of EVs (HER2 and PSMA)	2020	Zhang et al. (2019)
Multiplexed immunophenotyping of circulating exosomes on nanoengineered ExoProfile chip toward early diagnosis of cancer	Membrane protein	Multiplex	Blood plasma	Identification of 7 biomarkers for diagnosis and staging of cancer	2019	Chen et al. (2022)

(continued)

Table 6.4 (continued)

Name	Target	Type	Sample	Biological application	Year	References
Single extracellular vesicle protein analysis using immuno-droplet digital polymerase chain reaction amplification	Membrane protein	Multiplex	Conditioned media and blood plasma	Identification of 3 biomarkers in immune-therapy monitoring	2020	Ko et al. (2020)
Ultrasensitive single extracellular vesicle detection using high-throughput droplet digital enzyme-linked immunosorbent assay	Membrane protein	Single	Conditioned media and human plasma	Identification of tetraspanins in EVs	2022	Ko et al. (2021)
Sequencing-based protein analysis of single extracellular vesicles	Membrane protein	Multiplex	Conditioned media and plasma	Identification of EV-associated signatures as disease biomarker	2021	Yang et al. (2022)

ultrasensitive but also adaptable for isolating and detecting exosomes from other cancer types, making it a versatile tool for cancer diagnostics.

Lu et al. (2022) developed an integrated microfluidic exosome isolation and detection system (EXID system) for isolating and characterizing PD-L1 on individual EVs from both cell lines and lung cancer patients, with the goal of monitoring immunotherapy responses. The microfluidic device features a serpentine-shaped channel, an incubator chamber for exosome isolation and labeling, and micropillar arrays for analyzing PD-L1 protein markers (Fig. 6.26). EV samples, along with CD9 antibody-functionalized beads, were introduced into the chip, where they flowed through the serpentine channel to the incubator chamber. In the chamber, microbeads and CD9-positive EVs were captured by a magnetic field. The EVs were then washed, labeled with a PD-L1-specific immunofluorescent marker, and the magnetic beads were released. Finally, the beads, now decorated with PD-L1-positive EVs, were captured in the micropillar array and scanned using fluorescence microscopy to identify the PD-L1-positive EVs.

In 2020, Kim et al. (2020) developed a microfluidic chip (EV-Ident) to isolate EVs into three size fractions and characterize individual EVs (Fig. 6.27). The device employs multiple anodized aluminum membrane filters with pore sizes of 200, 100, and 20 nm in a disc chamber to separate large EVs (EV200 nm; 200–1000 nm), medium EVs (EV100 nm; 100–200 nm), and small EVs (EV20 nm; 20–100 nm). The presence of human epidermal growth factor receptor 2 (HER2) and prostate-specific membrane antigen (PSMA) was then analyzed in EVs derived from breast and prostate cancer cells. HER2-positive breast cancer cells exhibited the highest

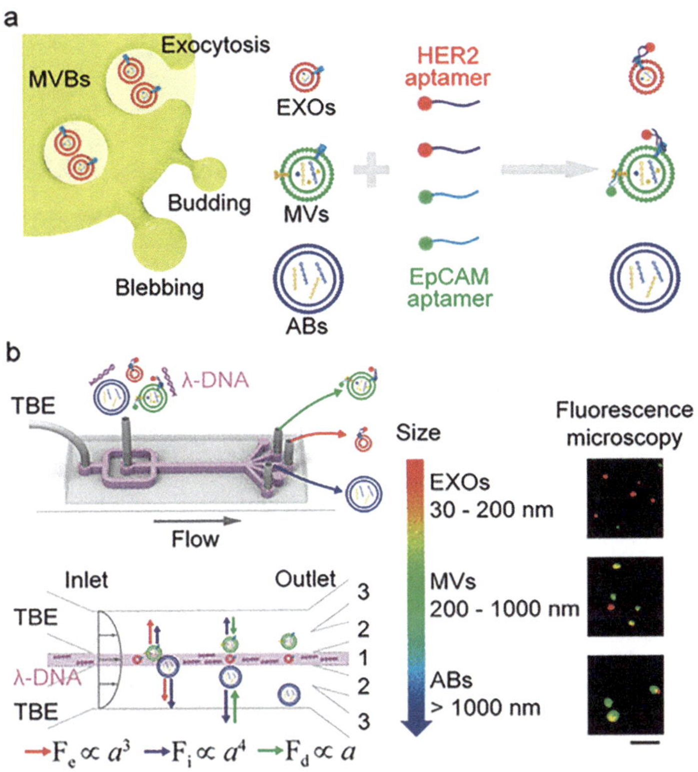

Fig. 6.23 Schematic of λ-DNA-mediated sorting of EV subpopulations and aptamer based analysis of individual EVs. (**a**) Labeling of cell-originating EVs including exosomes (EXOs, red), microvesicles (MVs, green), and apoptotic bodies (ABs, blue) with fluorescent HER2 and EpCAM aptamers. (**b**) Size-selective separation of EV subpopulations by λ-DNA mediated viscoelastic microfluidics. Fluorescence microscopy images showed HER2 (red) and EpCAM (green) expression of isolated individual EVs (Liu et al. 2019a) (reproduced from Sun et al. 2019b). Reprinted with permission from Sun et al. (2019b). Copyright (2019) American Chemical Society

HER2 expression in the $EV_{20\,nm}$ fraction, while PSMA expression was most prominent in the $EV_{200\,nm}$ fraction from PSMA-expressing prostate cancer cells.

Zhang et al. (2019) constructed 3D porous serpentine nanostructures through patterned colloidal self-assembly to provide extensive reaction sites, thereby enhancing the biosensing efficiency for exosomes (Fig. 6.28). This multichannel microdevice allowed for the simultaneous detection of eight markers from a single addition of exosome samples. The chip was validated using purified exosomes from SKOV3 cells, achieving a limit of detection of 21 exosomes per μL. Circulating exosomes from the plasma of 15 ovarian cancer patients and 5 benign controls were analyzed for the presence of biomarkers including EGFR, HER2, CA125, FRα, CD24, EpCAM, and CD9 + CD63. Significant heterogeneity in the exosomal expression of these biomarkers was observed among patients. The study

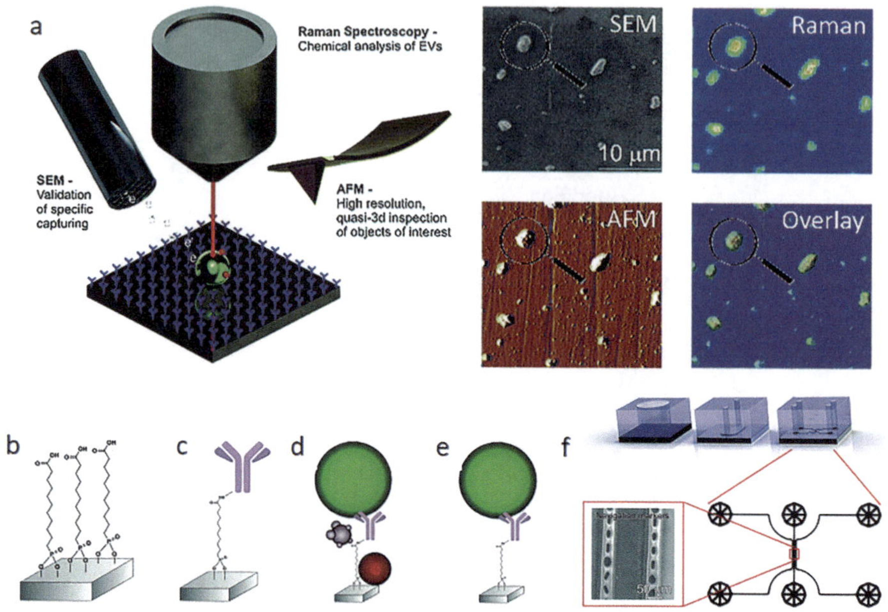

Fig. 6.24 (**a**) Multi-modal analysis of tdEVs on antibody-functionalized stainless steel substrates. After their selective capture on a stainless steel substrate functionalized with antibodies of interest (here anti-EpCAM antibodies targeting tdEVs), EVs are successively imaged using Raman spectroscopy, SEM and AFM, and information collected from these different imaging modalities correlated to get a comprehensive picture on the captured objects. (**b–e**) Different steps of surface modification and capture of the tdEVs on stainless steel substrates. (**b**) An oxygen plasma-treated stainless steel substrate is functionalized with a carboxydecyl phosphonic acid (CDPA) monolayer. (**c**) Anti-EpCAM antibodies are conjugated to the CDPA monolayer using NHS/carbodiimide chemistry. (**d**) tdEVs (in green here) are specifically immuno-captured on the antibody-functionalized surface. (**e**) The substrate is washed to remove non-specifically bound materials (in red and purple here), before retained EVs are fixed and dehydrated. (**f**) Various PDMS devices used in this work. Top, and from left to right: 6-mm-diameter reservoir used for the immobilization of antibodies under static conditions on a CDPA monolayer; xurography microchannel ($6 \times 3 \times 0.2$ mm^3) used for the capture of EVs and subsequent washing under mild flow conditions (400 µl min^{-1}); microfluidic device used for the fabrication of navigation markers next to a 50 µm × 1 mm sample region, subsequently considered for analysis using SEM, Raman and AFM. Bottom, right: Design of the microfluidic device used to fabricate the navigation markers, consisting of three microfluidic channels, two microchannels comprising pillars with various geometries, flanking one sample microchannel, all channels being 30 µm height × 50 µm width × 1 mm length, and (left) actual SEM image showing the navigation markers fabricated from cyanoacrylate glue injected in the side-channels and used to retrace the captured objects in the different imaging instruments (Beekman et al. 2019) (reproduced from Beekman et al. 2019)

demonstrated that multiplex analysis of combined biomarker panels significantly improves diagnostic performance compared to single biomarkers, enabling the classification of early-stage and late-stage tumors.

An interesting approach is presented in the study by Chen et al. (2022), which describes a membrane-based EV isolation and counting (mEVic) microfluidic

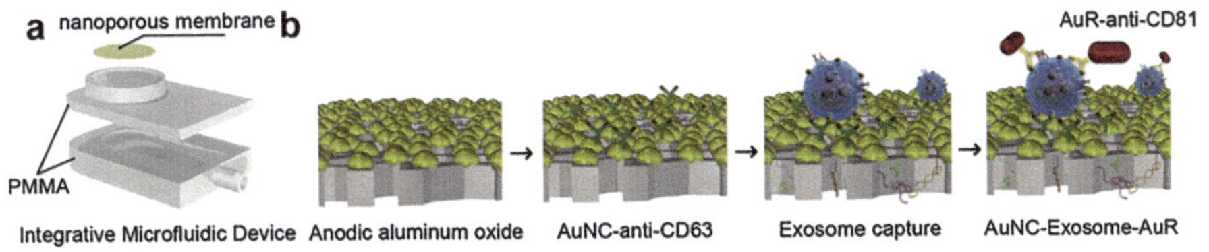

Fig. 6.25 Integrative microfluidic device for exosome isolation and detection. (**a**) Design of the integrative microfluidic device. (**b**) Schematic illustration of in-situ detection of exosome (Yang et al. 2020) (reproduced from Yang et al. 2020)

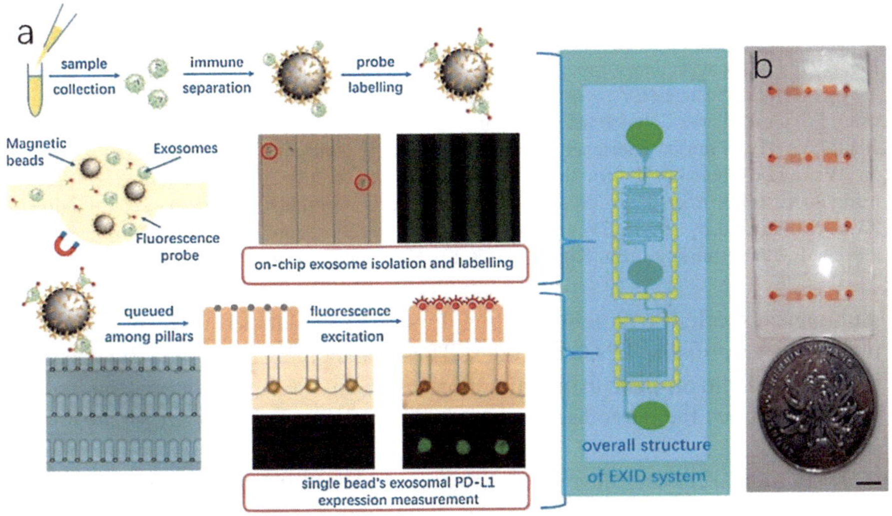

Fig. 6.26 Structure and operation principles of the EXID system. (**a**) Schematic diagram of on-chip exosome processing, during which exosome isolation, labeling and single-bead-level exosomal PD-L1 measurements were conducted in separate regions. (**b**) Photo of the prototype microfluidic system (Lu et al. 2022) (reproduced from Lu et al. 2022)

platform. This platform integrates two membrane filters for EV isolation from blood, followed by the quantification of immuno-stained particles on a single chip (Fig. 6.29a). A polycarbonate (PC) membrane was used for EV isolation via stirrer-enhanced filtration, achieving a 99.4% isolation rate for small EVs (sEVs), while an aluminum oxide (AO) membrane preserved the sEVs with a recovery rate of 81.1%. The platform also includes on-chip CD63 immunostaining for visualizing fluorescence-labeled CD63+ EVs (Fig. 6.29b). This novel mEVic platform holds promise as a powerful tool for isolating and quantifying specific EVs and could be integrated into fingertip-based disease diagnostic applications.

Another approach for EV protein analysis is the encapsulation of single EVs using droplet generator microfluidic chips. In this strategy, Ko et al. (2020) describe an immuno-droplet digital polymerase chain reaction (iddPCR) method for

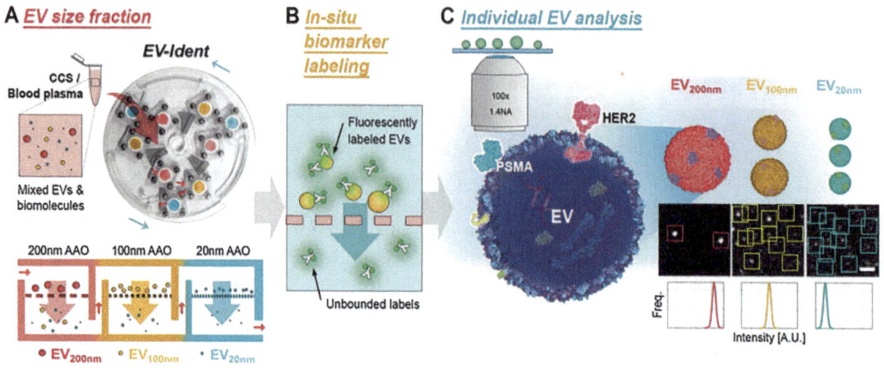

Fig. 6.27 EV-Ident prepares three size fractions of nanoparticles for analysis of size-specific single extracellular vesicle (EV). (**a**) Cell culture supernatant (CCS) or blood plasma are processed using EV-Ident. Three EV size fractions: large EVs (EV200 nm), medium EVs (EV100 nm), and small EVs (EV20 nm) are enriched through three serially configured membranes. (**b**) Biomarkers on EVs can be labeled in situ (on the device) for each size fraction. (**c**) EVs can then be visualized using fluorescence microscopy. Scale bar: 2 μm. The expression levels of biomarkers in individual EVs are analyzed by single EV analysis (Kim et al. 2020) (reproduced from Kim et al. 2020) reprinted with permission from Kim et al. (2020). Copyright (2019) American Chemical Society

multiplexed single EV protein profiling. First, EVs are labeled with antibodies conjugated to amplifiable DNA barcodes. These labeled EVs are then stochastically incorporated into droplets using microfluidics. In situ PCR with fluorescent reporter probes amplifies the barcode signals, which are then read out via droplet imaging (Fig. 6.30).

Ko et al. (2021) also developed an antibody-based immunosequencing method for multiplexed protein analysis from individual nanometer-sized EVs. Droplet microfluidics were employed to compartmentalize and barcode individual EVs. The barcodes, along with antibody-conjugated DNA, were then sequenced to determine the protein composition of each EV. This highly sensitive technology enables the detection of specific proteins at the single EV level (Fig. 6.31).

Yang et al. (2022) developed the droplet digital enzyme-linked immunosorbent assay (ddELISA) to quantify specific subpopulations of EVs with high throughput. This innovative platform allows for the quantification of individual EV subpopulations at rates approaching 20 million droplets per minute, a throughput that is 100 times greater than previous methods (Fig. 6.32). The platform is composed of three main sections: The first section contains multiple droplet generators; the second section features a long spiral channel that collects and stabilizes the droplets; and the third section arrays the droplets for detection. This system was evaluated for potential use with clinical samples by quantifying human neuron-derived EVs spiked into a complex media matrix, with a background EV concentration comparable to that of 2 μL of human blood. The assay achieved a limit of detection (LOD) of 11 EVs/μL.

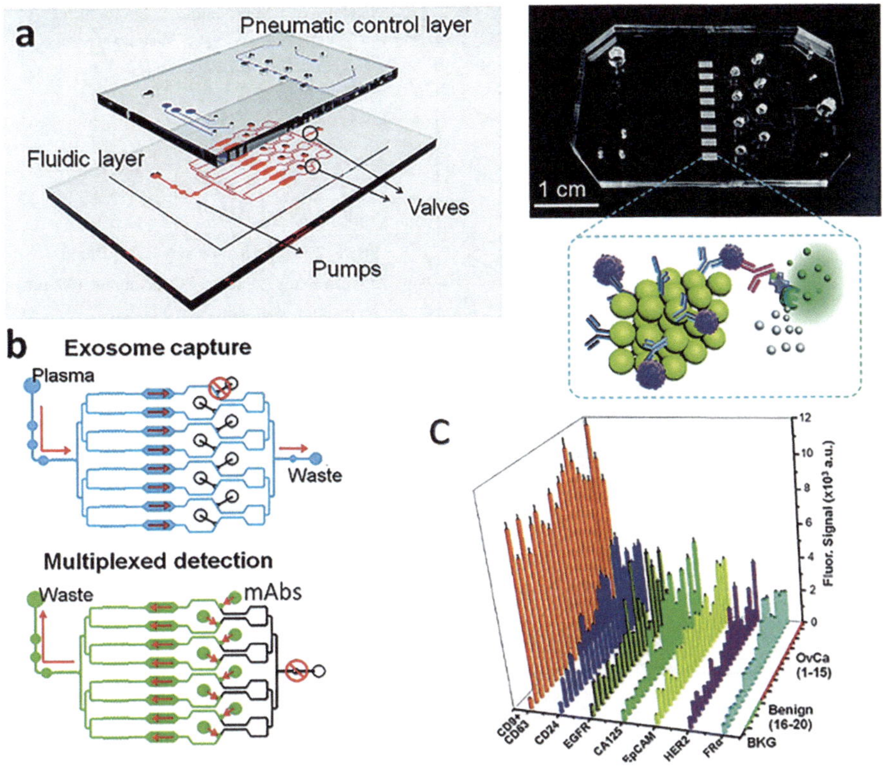

Fig. 6.28 ExoProfile chip. (**a**) Schematic illustration and photo of the ExoProfile chip, consists of pneumatic and fluidic layer and 3D serpentine nanostructures, for in situ immunophenotyping of exosomes. (**b**) The working principle and flow manipulation of ExoProfile chip in the exosome immunocapture mode and the multiplexed immunodetection mode. Arrows indicated the flow direction. Blue channel: sample injection; green channel: injection of monoclonal detection antibodies (mAbs); black valve/channel: closed valve/channel. (**c**) Quantification of six tumor markers (EGFR, HER2, CA125, FRα, CD24, EpCAM) and exosomal markers (CD9 + CD63) from 15 patients and 5 benign controls plasma samples. Error bars: S.D. (Zhang et al. 2019) (reproduced from Zhang et al. 2019)

6.2.3 Summary

Although great progress has been made in the research of exosome separation in recent years, each method has its advantages and disadvantages. The centrifugation and filtration methods are still the most commonly used and effective methods. A large number of EVs can be obtained at one time, but it is not suitable for trace and precious samples. The purity, recovery rate, and biological activity of EVs extracted by centrifugation are affected by factors such as centrifugal force and centrifugation time. In addition, complex preparation work is required before separation, which

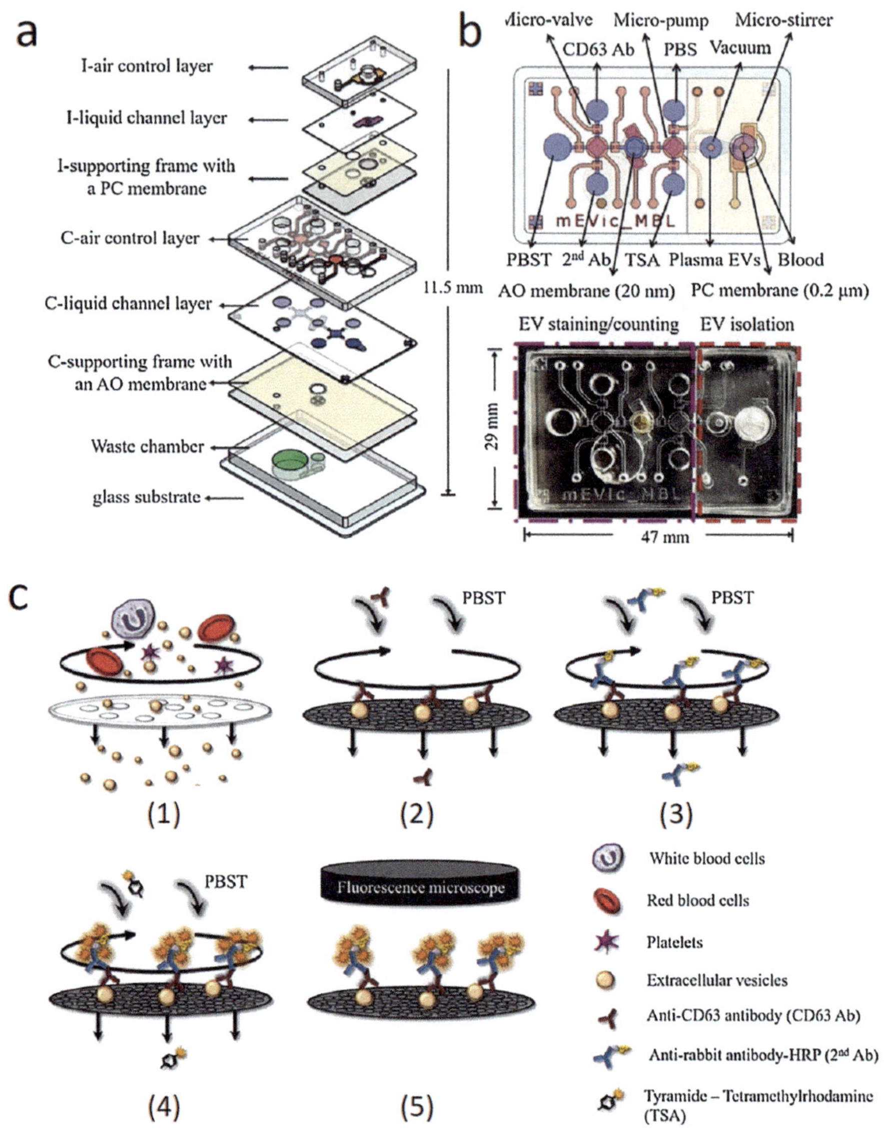

Fig. 6.29 (**a**) An exploded view of the membrane-integrated EV isolation/counting (mEVic) chip, which was composed of two individual air control layers, liquid channel layers, supporting frames with a polycarbonate (PC) membrane (pore size = 0.2 μm) for EV isolation (I-), an aluminum oxide (AO) membrane (pore size = 20 nm) for EV staining and counting (C-), and a waste chamber on a glass substrate. (**b**) A top view of the mEVic chip illustrating the micro-stirrers, micropumps, microvalves, and corresponding chambers for blood, filtered plasma-derived EVs (Plasma EVs), and the following reagents: PBS, PBS with 0.05% Tween-20 (PBST), anti-CD63 antibody, anti-rabbit antibody-HRP, and tyramide-TAMRA. (**c**) Membrane-based extracellular vesicle (EV) isolation, staining, digital counting, and quantification. (1) Plasma EV isolation from blood cells and platelets via stirring-enhanced filtration through a 0.2-μm PC membrane. (2–4) EV staining by anti-CD63 antibody, anti-rabbit antibody-HRP, and tyramidetetramethylrhodamine on a 20-nm blackened aluminum oxide (AO) membrane. (5) Observation of fluorescent CD63 + EVs directly on an AO membrane with a fluorescence microscope (Chen et al. 2022) (reproduced from Chen et al. 2022)

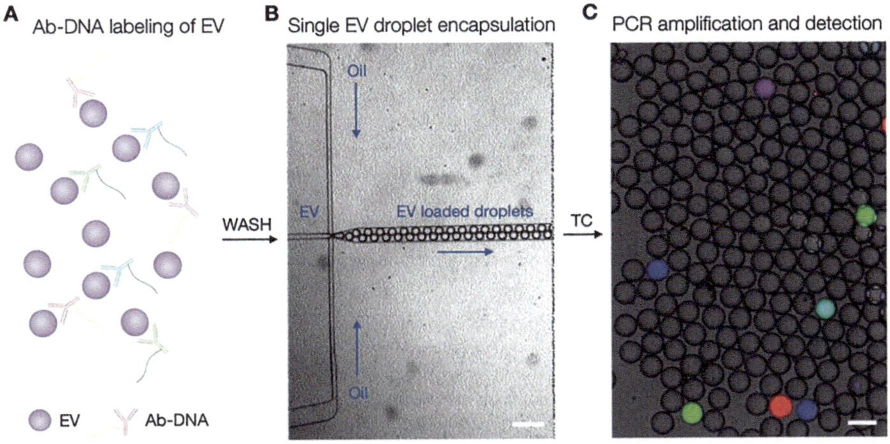

Fig. 6.30 Schematic of droplet-based single EV detection. Single EV detection is achieved via three steps. (**a**) First, EV are labeled with Ab–DNA conjugates for multiplexing and DNA-based detection of protein expressions. Remaining free Ab–DNA conjugates are then removed using size exclusion chromatography (SEC) before proceeding to the next step. (**b**) Second, single EV are encapsulated in droplets by creating EV and PCR master mix containing water droplets in oil (scale bar = 300 μm). (**c**) After thermal cycling (TC), EV that are labeled with targets of interest amplify and fluoresce for detection using imaging (Ko et al. 2020) (reproduced from Ko et al. 2020)

takes a long time (2–6 h) (Van et al. 2014; Linares et al. 2015; Yang et al. 2017). The multi-step filtration method still requires centrifugation in the early preparation, the steps are cumbersome, and a large amount of sample (30–100 mL) is required, and the yield is very low. The separation effect of the immunomagnetic bead separation method is heavily dependent on specific antibodies, which is costly and can only separate exosomes that are affinity with antibodies, resulting in the loss of certain types of exosomes.

In recent years, microfluidics has been widely used in the separation of EVs. The separation method based on microfluidics has the advantages of faster separation rate, higher separation yield and purity, and less sample demand. Compared with the above-mentioned traditional methods, it is more suitable for trace sample analysis and is expected to be widely used in future personal medical care and precision treatment. However, the current microfluidics-based extracellular vesicle separation method is still in the experimental research stage and needs further development and research. With the development of exosome separation methods and technologies, the role and mechanism of exosomes in important life activities such as cell communication and disease occurrence will be further revealed, promoting the development of disease detection and related treatment methods.

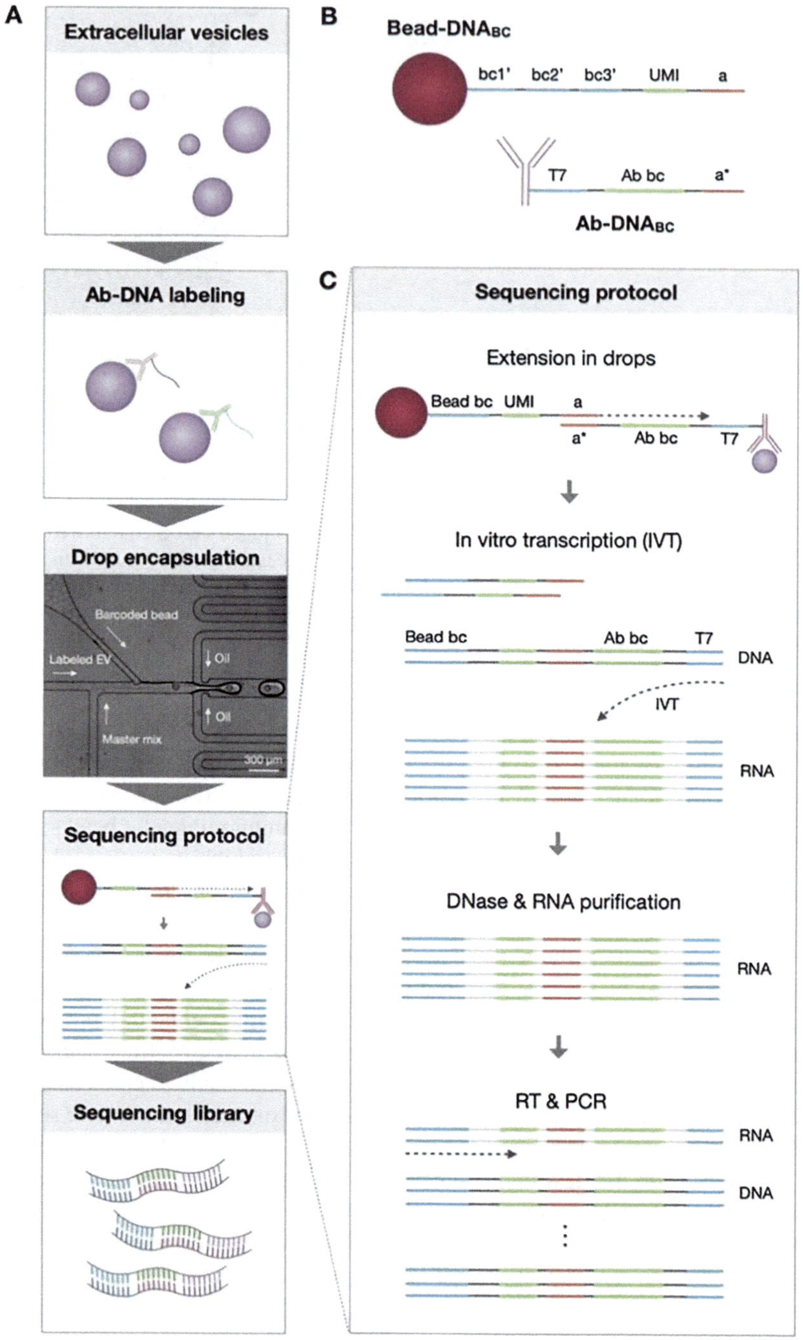
A
Extracellular vesicles
Ab-DNA labeling
Drop encapsulation
Barcoded bead
Labeled EV
Oil
Oil
Master mix
Sequencing protocol
Sequencing library
B
Bead-DNABC
bc1' bc2' bc3' UMI a
T7 Ab bc a*
Ab-DNABC
C
Sequencing protocol
Extension in drops
Bead bc UMI a
a* Ab bc T7
In vitro transcription (IVT)
Bead bc Ab bc T7
DNA
IVT
RNA
DNase & RNA purification
RNA
RT & PCR
RNA
DNA

6.3 Construction of Functionalized Micro–Nanodrug Carriers Based on Microchips and their Applications in Tumor Diagnosis and Treatment

Drug carriers represent a promising approach to enhance drug delivery and efficacy (Han et al. 2019; Lim et al. 2015; Wang et al. 2009). These carriers are characterized by their strong targeting ability, precise drug release mechanisms, and improved pharmacokinetics. Microfluidic platforms offer significant advantages in this regard, as their geometry and flow rates are easily adjustable. Drug carriers prepared using microfluidic chips can not only facilitate high-throughput, low-cost production but also enable the creation of carriers with diverse components, complex structures, and high monodispersity. This capability significantly enhances the effectiveness, stability, and controllability of the drugs they deliver. As a result, microfluidic chips provide a powerful technical platform for the development and screening of new drug carriers, particularly those with controlled-release functionalities (Neves et al. 2013). This section offers an overview of newly developed functionalized micro- and nano-sized drug carriers fabricated using droplet microfluidic chips. These carriers fall into three main categories: microcapsules, nanoemulsions, and nanoparticles. The focus will be on the on-chip synthesis methods for each type of drug carrier, as well as their applications in drug-controlled release, tumor diagnosis, and treatment.

6.3.1 Microcapsules

Microcapsules are micro-sized containers or capsules, typically with diameters ranging from 2 to 1000 μm, and are enclosed by either natural or synthetic polymer shells. Through various methods and devices, drug molecules, protein molecules, or

Fig. 6.31 Schematic of seiSEQ. (**a**) The pipeline includes EV labeling with Ab-DNABC constructs, drop encapsulation with barcoded beads, and single EV sequencing protocol. The drop encapsulation step includes a microscopic image of a droplet generator with four input channels for oil, barcoded beads, labeled EV, and master mix, and one output channel to collect individual droplets (scale bar = 300 μm). (**b**) DNA sequence composition on barcoded beads (bc1′, bc2′, bc3′ = three subbarcoded regions created using a split-pool approach during bead synthesis; UMI = unique molecular identifier; a = hybridizing sequence to Ab-DNABC) and antibodies (T7 = T7 promoter sequence; Ab bc = antibody barcode; a* = complementary strand to 'a' on the bead-DNABC). (**c**) Schematic on the sequencing protocol. Ab-DNABC and Bead-DNABC are hybridized at the a/a* sequence region. After hybridization, extension is performed within droplets. The extended product consists of bead barcode (Bead bc), antibody barcode (Ab bc), UMI, and T7 promoter sequence. The T7 promoter sequence is used to efficiently amplify RNA. Then, DNase is treated to remove any remaining DNA and RNA is purified and converted to cDNA using RT. The cDNA undergoes PCR for amplification and post-PCR purification for sequencing (Ko et al. 2021) (reproduced from Ko et al. 2021) reprinted with permission from Ko et al. (2021). Copyright (2021) American Chemical Society

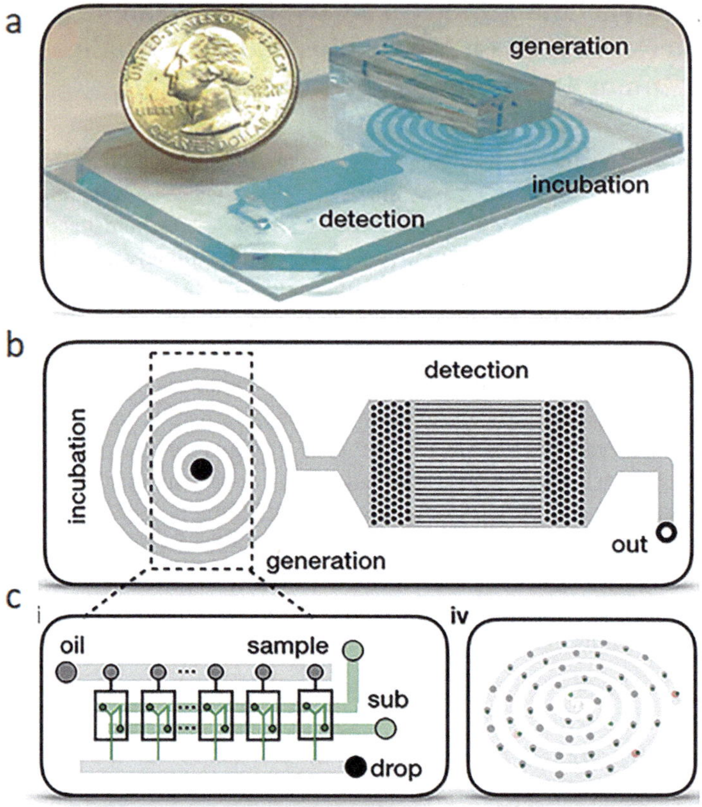

Fig. 6.32 (**a**) Miniaturized microfluidic platform for DEVA that integrates droplet generation, incubation, and detection. (**b**) All droplets flow through a delay line chamber for incubation. Afterward, all droplets flow through a detection section consisting of 90 parallelized microfluidic channels. (**c**) A parallelized flow focusing droplet generator encapsulates the beads into droplets (Yang et al. 2022) (reproduced from Yang et al. 2022). Reprinted with permission from Yang et al. (2022). Copyright (2022) American Chemical Society

chemical reagents can be encapsulated within these polymer shells, forming independent microcapsules. These capsules can take on different shapes, including spherical, granular, or rod-like forms (Azadi et al. 2016). As drug carriers, microcapsules generally have pore sizes on the order of microns. They offer excellent biocompatibility and biodegradability, as well as the ability to enable targeted or controlled drug release. Due to these advantages, microcapsules hold significant potential for a wide range of drug delivery applications (Wei et al. 2014).

Traditional methods for preparing microcapsules typically follow a two-step process: first, the carrier is prepared, and then the drug is loaded. However, this approach has several drawbacks, including a complex preparation process, long production times, and low encapsulation efficiency. Additionally, traditional methods often rely on mechanical forces, such as stirring or shaking, to form a dispersed emulsion.

This can lead to several issues, such as significant variations in microcapsule structure, poor monodispersity, and uneven drug encapsulation. As a result, these methods may not meet the clinical requirements for controlled or slow drug release. These limitations highlight the need for the development of new drug carriers and more efficient preparation strategies.

In contrast, the double emulsion preparation method based on droplet microfluidics offers an effective and innovative strategy for constructing new microcapsule drug carriers. A double emulsion system, such as oil/water/oil (O/W/O) or water/oil/water (W/O/W), involves dispersed droplets that can themselves contain one or more types of smaller, monodisperse droplets, forming a multiple emulsion system. Droplet microfluidic chips enable precise control over the size distribution of microcapsules, allowing for the high-throughput production of drug carriers with complex structures and high monodispersity. Microfluidic technology provides a flexible and reliable tool for preparing novel microcapsule drug carriers, (Bawazer et al. 2016) and it shows significant potential for targeted delivery and controlled drug release (Iqbal et al. 2015; Ramazani et al. 2016). This section will focus on recent advances in using microfluidics to prepare microcapsules as drug carriers for controlled drug delivery.

Navi et al. (2020) presented an all-aqueous droplet microfluidics platform for generating magnetic polyelectrolyte microcapsules (PEMCs). These PEMCs feature biocompatible microcompartments and have potential applications in various industries, including cosmetics, food, and pharmaceuticals. In this platform, water-in-water droplets are generated using an aqueous two-phase system (ATPS) composed of polyethylene glycol (PEG) and dextran (Dex), and are magnetically functionalized using a ferrofluid. Strong polyelectrolytes (PEs) with opposite charges are employed in each ATPS phase. The magnetic droplets, which contain ferrofluid and polycations (positively charged polyelectrolytes, PE^+), act as templates for the complexation of polyelectrolytes at the interface. Using a magnetic field, the magnetic droplet templates are transferred to a PEG phase containing polyanions (negatively charged polyelectrolytes, PE^-), where the two PEs meet at the interface to form a shell. These magnetic PEMCs can passively release large pseudo-drugs, and their release can be triggered by external stimuli such as osmotic shock or pH changes (Fig. 6.33). This biocompatible all-water platform, based on magnetic PEMCs, holds significant promise for the fabrication of functionalized drug carriers for targeted drug delivery.

Chang-Hyung Choi et al. present a hydrogel microcapsule with an intermediate thin oil layer designed for smart release of a wide range of cargoes, triggered by various stimuli (Jeong et al. 2021). A microfluidic technique is used to produce triple emulsion droplets, where a thin oil layer separates the innermost aqueous phase from the hydrogel prepolymer phase. This prepolymer phase undergoes photopolymerization to form a hydrogel shell. The intermediate oil layer within the hydrogel microcapsule serves as an effective diffusion barrier, enabling the encapsulation of various small cargoes within a porous hydrogel shell until a stimulus is applied to destabilize the oil layer (Fig. 6.34). The system allows for the release of encapsulated cargo in response to diverse stimuli, including chemical dissolution, mechanical stress, and

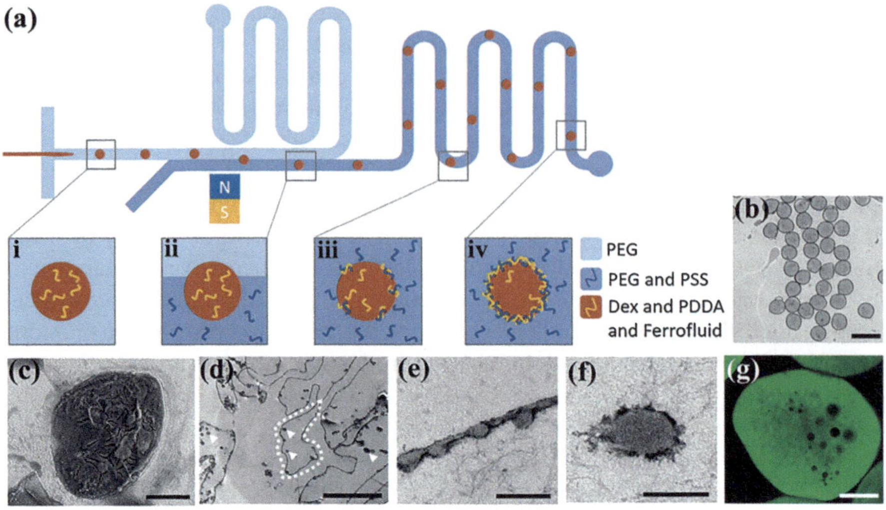

Fig. 6.33 (**a**) Schematic diagram showing the microchannel design and the steps of microcapsule generation. (i) Magnetic Dex droplet containing ferrofluid and PDDA is generated at the first junction. (ii) In the magnetic field of a permanent magnet, the magnetic Dex droplet is transferred to a PEG phase containing PSS. (iii) Once the Dex droplet is in the PEG–PSS phase, the two PEs will start partitioning to their desired phase. The PEs will meet at the interface and start forming the shell. (iv) By the time, the droplets reach the outlet, the PEMCs are fully formed. (**b**) Bright-field image of the PEMCs collected in a Petri dish. Scale bar shows 150 μm. (**c**) SEM image of an air-dried PEMC. Scale bar shows 50 μm. (**d**) TEM image of a thin cross-section of PEMCs. One full PEMC at the center (indicated by the white dotted loop) and parts of six other PEMCs are visible in the image. The PEMCs deform in TEM sample preparation steps. The complexations of the polyelectrolytes inside the microcapsules are indicated by white arrows. (**e**) TEM image of the thickness of one PEMC shell. Scale bar shows 200 nm. (**f**) TEM image of one of the polyelectrolyte complexations inside the microcapsules [shown with white arrows in (**d**)]. The scale bar indicates 200 nm. (**g**) Confocal microscopy image of a PEMC showing the smaller inner PE complexations inside an inner PEG droplet (Navi et al. 2020) (reproduced from Navi et al. 2020) Reproduced from Navi et al. (2020) with permission from the Royal Society of Chemistry

osmotic pressure. Additionally, the osmotic pressure and hydrogel shell thickness can be independently adjusted to control both the timing and release behavior of the cargo. This flexibility allows for either simultaneous or selective release of multiple cargos encapsulated within the hydrogel microcapsules.

Tumor immunotherapy is a treatment approach that targets and eliminates malignant tumor cells by reactivating and sustaining the tumor-immune cycle, thereby enhancing and restoring the body's natural antitumor immune response. While tumor immunotherapy has shown significant success in treating hematologic cancers, progress in solid tumor treatment has been slower. This delay is largely due to the complex tumor microenvironment and the tumors' ability to evade immune detection. To address these challenges, recent research has focused on combining biomaterials with tumor immunotherapy to modulate the tumor microenvironment or developing tumor vaccines to improve treatment efficacy (Wu et al. 2019; Wang and Mooney 2018).

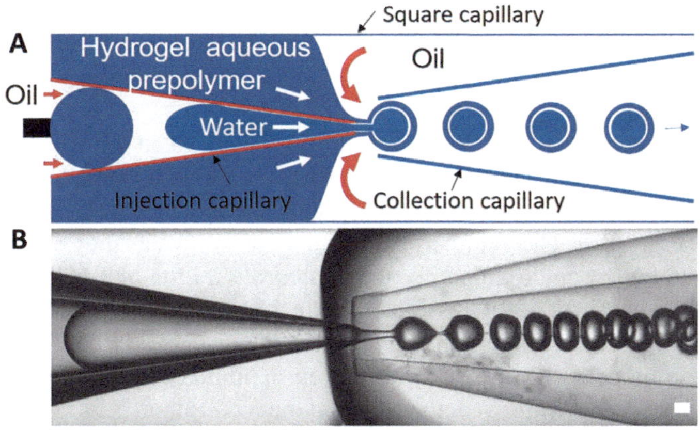

Fig. 6.34 Microfluidic production of hydrogel microcapsules with a thin oil layer to encapsulate various hydrophilic cargoes. (**a**) A schematic showing a glass capillary microfluidic device for generation of water-in-oil-in-water-in-oil (W/O/W/O) triple emulsion droplets. (**b**) Bright-field micrograph showing continuous production of emulsion droplets (Jeong et al. 2021) (reproduced from Jeong et al. 2021)

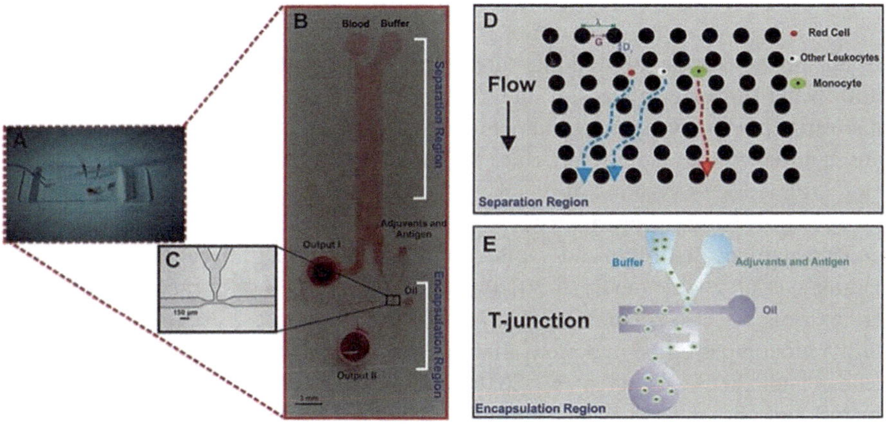

Fig. 6.35 The chip design and working mechanism. (**a**) A photograph of the device with inlets and outlets connected with tubing (chip length: 40 mm; width: 10 mm). (**b**) A top-down view of the chip (channels shown with red ink). (**c**) A microscopy image of the T-junction. (**d**) The monocyte isolation procedure (pillars height: 30 μm; diameter Dp: 30 μm; horizontal distance between two pillars G: 20 μm; vertical distance between pillars Dy: 10 μm, constituting a certain offset channel with an inclination angle θ of 5.7°). (**e**) An illustration of the cell encapsulation process (Tian et al. 2021). Reproduced from Tian et al. (2021) permission from the Royal Society of Chemistry

Tian et al. (2021) designed a "separation-encapsulation integrated" microfluidic chip that allows for the simultaneous separation of blood monocytes and the formation of cell gel microspheres, enabling the rapid preparation of a monocyte-based tumor vaccine (Fig. 6.35). In this system, alginate gel microspheres enhance the antigen uptake capacity of monocytes and stimulate the production of antigen-specific $CD8^+$ T cells in the spleen, resulting in a robust cytotoxic T lymphocyte (CTL) response. Furthermore, encapsulating immune adjuvants with specific functions within the microspheres for combined treatment can further amplify antitumor effects.

Later, this research group (Tian et al. 2022) encapsulated the adjuvants TLR1/2 and IL-21 along with $CD8^+$ T cells derived from the peripheral blood of tumor-bearing hosts, into gel microspheres to prepare adoptive T cell therapy (ATCT) injections (Fig. 6.36). The gel microspheres and the adjuvants they encapsulate create an optimal microenvironment for the rapid proliferation of T cells, significantly increasing the proportion of $CD8^+$ T cells. Notably, the proportion of poorly differentiated T cells with strong antitumor capabilities was greatly enhanced. In melanoma mouse models, this injection effectively inhibited tumor growth, stimulated CTL responses, and improved the survival rate of tumor-bearing mice. Tumor immunotherapy using gel microspheres holds promise for advancing the treatment of solid tumors and expanding the application of cell gel microspheres in personalized therapies.

6.3.2 Nanoemulsions

Nanoemulsions are classified into three types based on their composition and structure: water-in-oil, oil-in-water, and bicontinuous. To maintain the stability of the nanostructure in all types of nanoemulsions, various surfactants or co-surfactants are required to stabilize the interface. Compared to other dosage forms, nanoemulsion drug delivery systems offer several advantages, including a high surface area, unique transparent appearance, adjustable rheology, excellent processability, enhanced bioavailability, and rapid, efficient drug penetration (Tan et al. 2015; Durukan et al. 2016; Sun et al. 2016). Nanoemulsions not only serve as drug delivery carriers but also enhance several key properties of drugs, such as stability, solubility, biocompatibility, and bioavailability (Lu and Park 2013; Shakeel et al. 2012; Choi et al. 2013; Zheng et al. 2016). Given these benefits, nanoemulsions are increasingly being used as an important platform for developing drug formulations and offer promising solutions for tumor diagnosis and treatment.

Currently, the main techniques for preparing nanoemulsions include ultrasound, high-pressure valve homogenization, and microfluidics. The principle behind microfluidization technology involves using high-pressure liquid flow (water and oil phases) to generate high shear forces, resulting in fine emulsions in the submicron range (Lu and Park 2013). Compared to other nanoemulsification methods, microfluidics offers the advantage of achieving the smallest particle sizes and the highest preparation efficiency under similar conditions. This section will focus on the latest advancements in controlled drug delivery using nanoemulsions prepared via microfluidic chips as drug carriers.

Ahmad et al. (2024) presented a microfluidic-based nanoemulsion of *Ocimum basilicum* extract, demonstrating enhanced antimicrobial and anticancer properties

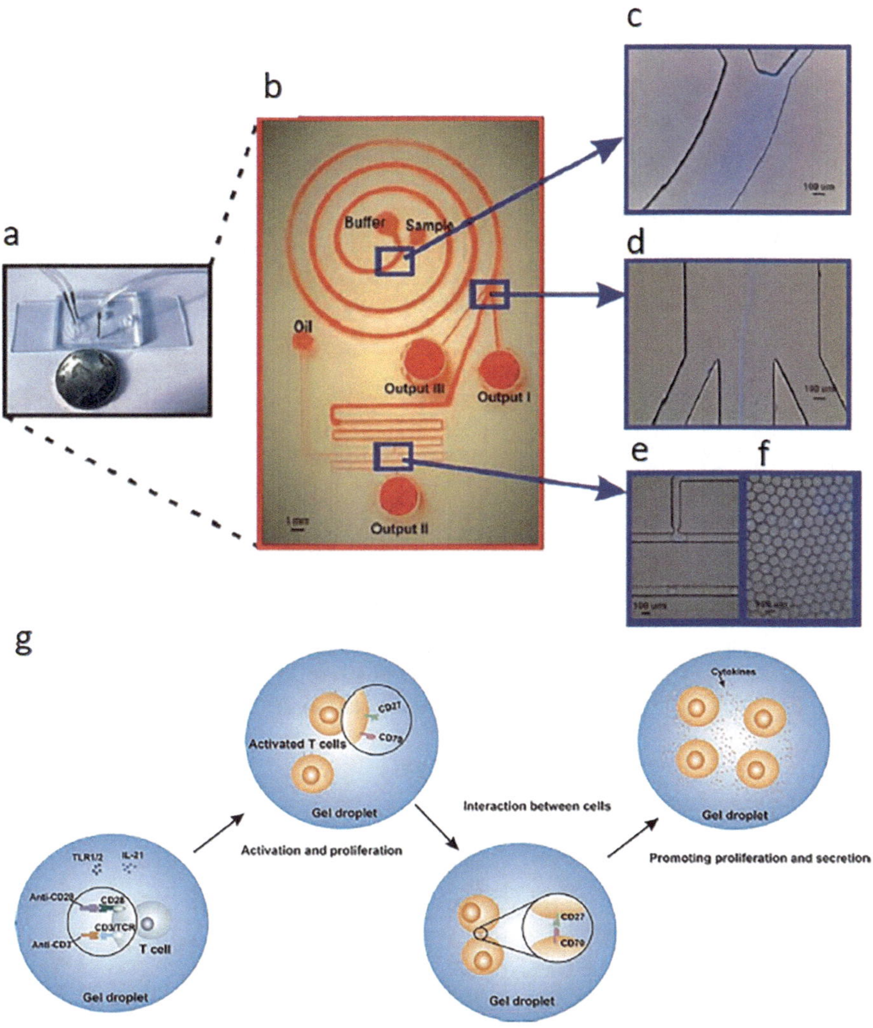

Fig. 6.36 (**a**) Photograph of device with inlets and outlets connected with tubing. (**b**) Top view of the chip (channel with red ink). (**c**) Inlet part of the chip separation region. (**d**) Outlet part of the chip separation region (**e**) Cell encapsulation using T-junction. (**f**) Gel droplet encapsulation of cells. T cells were stained with Hoechst 33342. (**g**) Principle of T cell proliferation in gel droplets (Tian et al. 2022) (reproduced from Tina et al. 2022)

(Fig. 6.37). The nanoemulsion featured a small, uniform particle size distribution, negative zeta potential, and high encapsulation efficiency for the basil extract. In vitro release studies showed that the nanoemulsion formulation exhibited sustained release of the extract, compared to the free extract solution. Additionally, the developed nanoemulsion displayed improved anticancer activity against K562 cells, while showing low toxicity toward normal HEK293 cells. These results highlight the promising potential of the nanoemulsion formulation for biomedical applications.

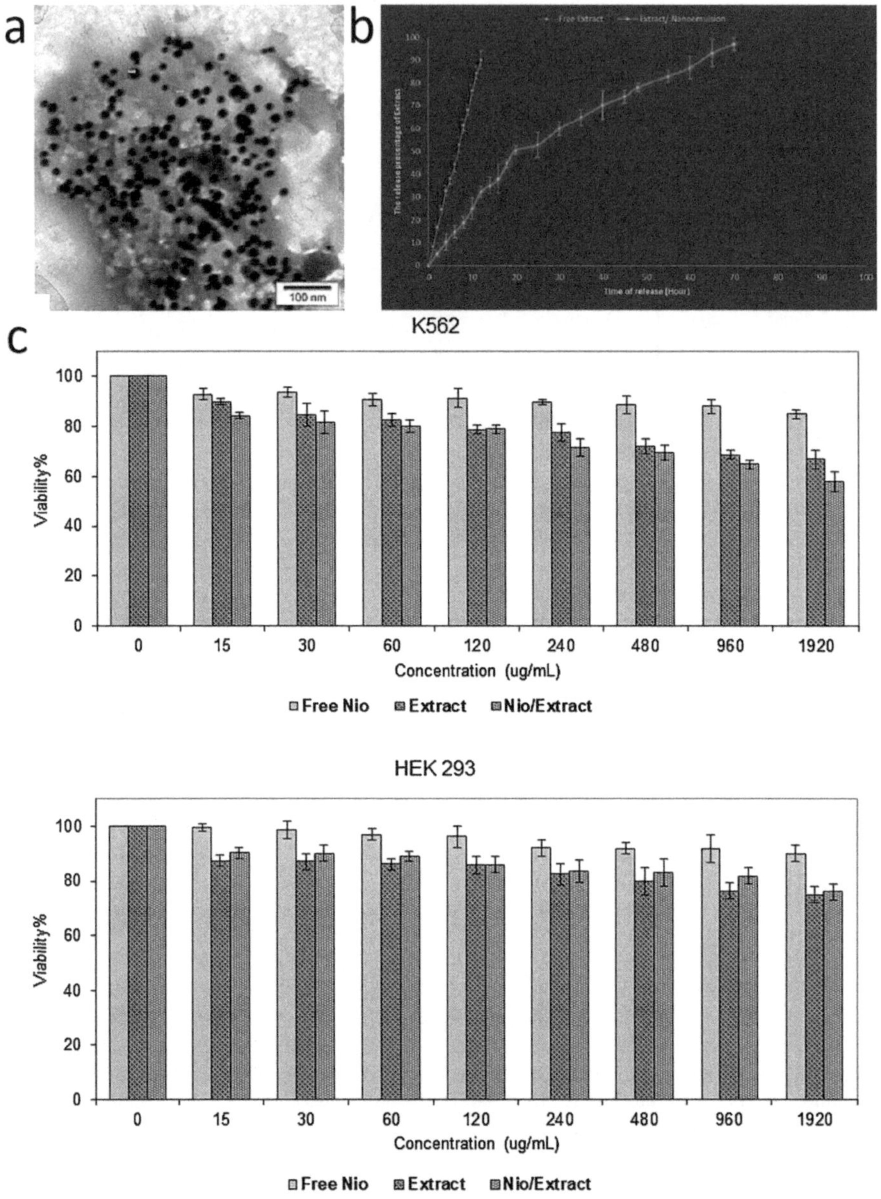

Fig. 6.37 (**a**) The effect of free nanoemulsion, extract and Nio/extract samples on the viability of K562 and HEK 293 cells. Data are represented as mean ± SD and $n = 3$. (**b**) Electron microscopy of nanoemulsion formulation prepared by microfluidic approach. (**c**) Release profile of free extract and extract loaded nanoemulsion through dialysis membrane up to 70 h (Ahmad et al. 2024) (reproduced from Ahmad et al. 2024)

6.3.3 Nanoparticles

Effective cancer treatment often requires the combination of multiple therapeutic agents, such as small molecules, protein drugs, and nucleic acid-based drugs, to overcome tumor cell resistance through synergistic effects. Functionalized nanoparticles, as an emerging drug delivery system, not only serve as carriers to enhance the encapsulation and delivery efficiency of these drugs, but also leverage the unique optical, thermal, magnetic, and electrical properties of nanoparticles for anticancer therapy. Consequently, the development of multifunctional nanoparticles for combined cancer therapy has become a key focus in nanotechnology. In addition to being stable and biodegradable, nanodrug carriers for combination therapy must be compatible with both hydrophobic and hydrophilic drugs and capable of releasing them at therapeutic levels. Thus, developing a simple, efficient, and stable method for preparing drug-loaded nanoparticles that enables complex assembly and controlled release of multiple drugs remains a significant technical challenge in this field.

Currently, conventional methods for preparing monodisperse nanoparticles with specific sizes and morphologies, such as overall mixing and nanoprecipitation, face significant technical challenges, including slow processing speeds, low efficiency, and poor controllability. In contrast, microfluidics offers precise control over fluid flow and mixing within microchannels, greatly enhancing the uniformity, diversity, and structure of nanoparticles. This approach also improves the encapsulation efficiency of anticancer drugs and enables high-throughput production. Researchers have developed a variety of novel nano-delivery systems based on microfluidics, including liposomes, polymer nanoparticles, and composite nanoparticles. This section will classify and discuss these systems in detail.

The use of microfluidics to create NPs is a relatively new area of research, and as such, biological studies are still in the early stages. As given in Table 6.5, in vitro studies are more prevalent than in vivo studies in the literature. Notably, all the in vivo studies conducted so far have been performed in Balb/c mice.

Tahir et al. (2020) prepared lipid-polymer hybrid nanoparticles (LPHNPs) using PLGA, phosphatidylcholine, and DSPE-PEG 2000 via both the bulk nanoprecipitation method and microfluidic (MF) co-flow nanoprecipitation. Sorafenib (SFN), an anticancer drug that inhibits multiple receptor tyrosine kinases (RTKs), particularly vascular endothelial growth factor receptors (VEGF-R2 and VEGF-R3) and platelet-derived growth factor receptor-β (PDGFR-β), was used as a model drug. Morphological analysis of the LPHNPs revealed that the bulk nanoprecipitation method produced spherical nanoparticles with clear boundaries, while the MF method resulted in core–shell morphology. The MF-prepared formulations (MF1-MF3) exhibited higher encapsulation efficiencies (%EE) of 95.0%, 93.8%, and 88.7%, along with a more controlled release of SFN compared to the LPHNPs prepared by the bulk nanoprecipitation method. This difference is attributed to the smaller particle size and lower polydispersity index (PDI) observed in the MF-produced NPs, which results from the reduced mixing time within the confined

Table 6.5 Examples of MF-fabricated CSNPs for cancer diagnostics and/or therapy in preclinical trials

NP	Cancer	Active agent	In vitro	In vivo	Application	References
LNP	Liver	SFN and siRNA	HepG-2	BCm	Chemotherapy	Younis et al. (2021)
LPNP	Glioblastoma	IRI and Gd-DTPA	U87 MG	–	Theranostics	Roffo et al. (2022)
CH-magnetic NP	Ovaries	Cisplatin	A2780	–	Chemotherapy	Siavashy et al. (2021)
PEI-b-PCL@ lipids	Prostate	siRNA	PC-3	BCm	Chemotherapy	Wei et al. (2020)
FeTPt@ CCM	Breast	FeTPt	4T1	BCm	Ferroptosis	Zhang et al. (2023)
Cu(DDC)2@ BSA	Breast	Cu(DDC)2	4T1	BCm	Chemotherapy	Chang et al. (2020)
LPNP	Colon	SFN	HT29-MTX	–	Chemotherapy	Tahir et al. (2020)
LPNP	Breast	SFN	(MDA-MB-231	–	Chemotherapy	Tahir et al. (2020)
LPNP	Prostate	SFN	PC3-MM2	–	Chemotherapy	Tahir et al. (2020)

BCm balb-C mice, *BSA* bovine serum albumin, *CCM* cancer cell membrane, *CH* chitosan, *DDC* diethyldithiocarbamate, *IRI* irinotecan, *LNP* lipid NP, *LPNP* lipid-polymer NP, *MF* MF method, *PCL* polycaprolactone, *pDNA* plasmid DNA, *SFN* sorafenib, *SHM* staggered herringbone micromixing

space of the MF channels. Additionally, colloidal stability, in vitro cytotoxicity, and cell growth inhibition studies confirmed the stability, biocompatibility, and safety of the LPHNPs in biological media. They also demonstrated stronger antiproliferative effects than free SFN in breast and prostate cancer cells. In conclusion, LPHNPs represent a promising platform for cancer chemotherapy and provide valuable insights into the fabrication and optimization of hybrid nanoparticles (Fig. 6.38).

Siavashy et al. (2021) prepared magnetic core/chitosan shell nanoparticles (NPs) containing cisplatin using two different methods: a conventional batch flow method and a microfluidic approach. Tripolyphosphate (TPP) was used as a cross-linker due to its negative charge, which leads to ionic gelation and facilitates self-assembly with the positively charged amino groups on chitosan. Additionally, TPP plays a crucial role in preventing the formation of strong bonds between cisplatin and chitosan molecules, which could hinder drug release. An integrated microfluidic device consisting of three stages was developed to achieve precise and highly controllable mixing (Fig. 6.39). The nanoparticles synthesized in the microfluidic device exhibited a monodispersity of 104 ± 14.59 nm ($n = 3$) and optimal morphological characteristics, whereas the nanoparticles produced by the conventional batch method

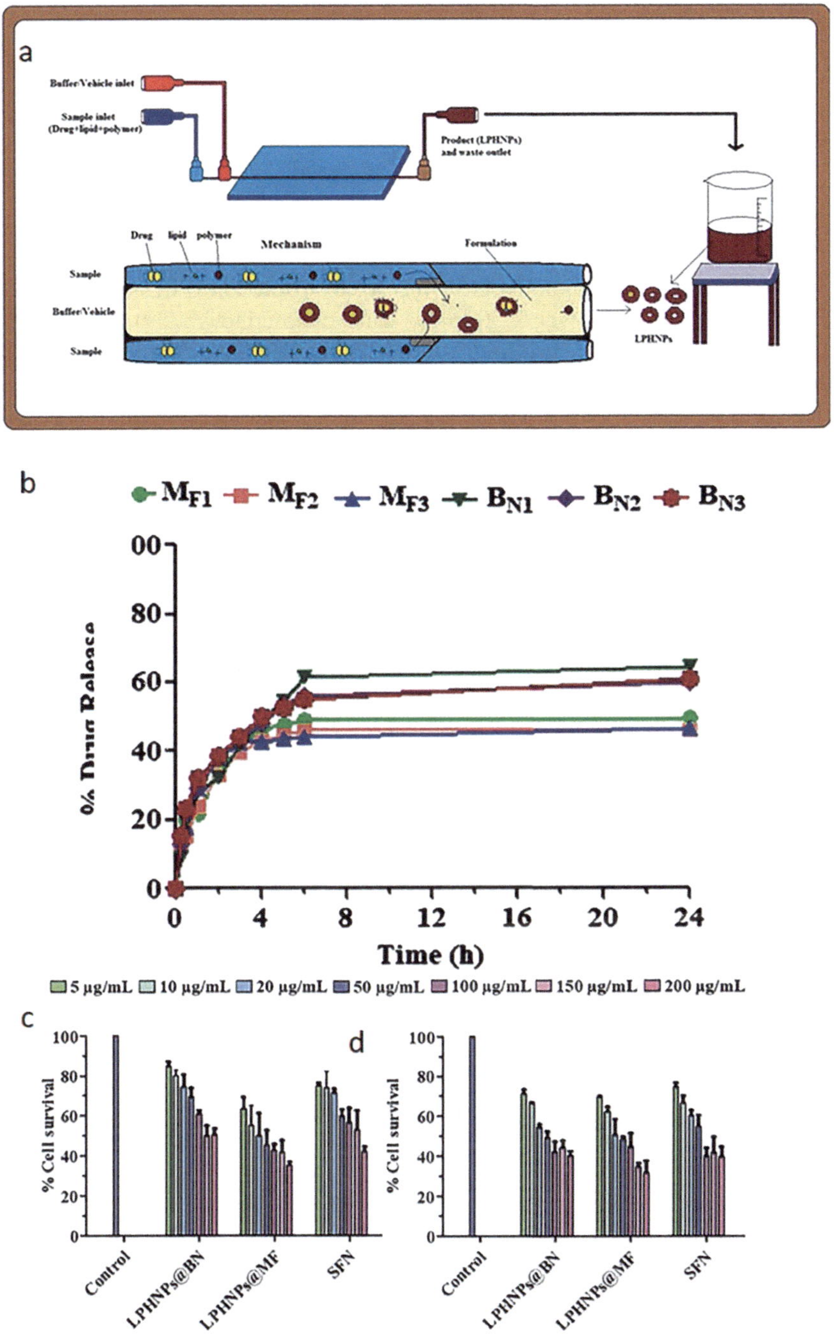

Fig. 6.38 (**a**) Schematic diagram demonstrating the flow of fluids and preparation of LPHNPs by MF method. (**b**) The release profile of all the LPHNPs prepared by MF and bulk nanoprecipitation method. (**c**) Cell growth inhibition of SFN loaded LPHNPs prepared by bulk nanoprecipitation and MF techniques in comparison with pure SFN solution after incubation period of 72 h with MDA-MB-231 (**a**) and PC3-MM2 (**b**) cells at 37 ÅãC (Tahir et al. 2020) (reproduced from Tahir et al. 2020)

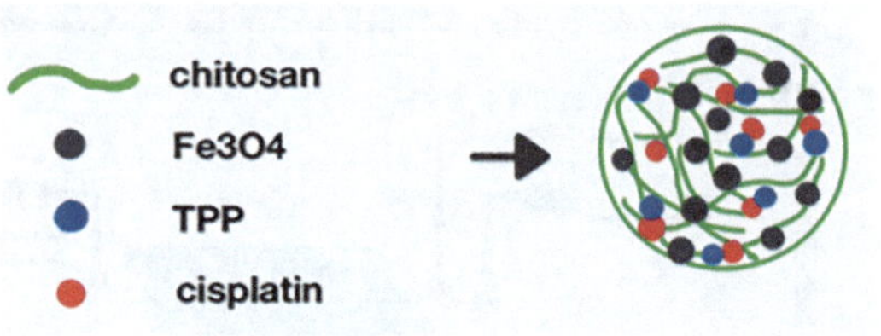

Fig. 6.39 The chitosan-coated magnetic NPs (Siavashy et al. 2021) (reproduced from Siavashy et al. 2021)

were polydisperse, with a size of 423 ± 53.33 nm ($n = 3$). Furthermore, the microfluidic-synthesized NPs demonstrated higher encapsulation efficiency, more controlled drug release, and lower IC50 values when exposed to A2780 cell lines. These findings indicate that the microfluidic approach enables more efficient drug uptake and release in cells compared to the conventional method.

Zhang et al. (2023) developed a "Trojan Horse"-like metal-organic framework (MOF) delivery system, FeTPt@CCM, using droplet microfluidic technology for synergistic cancer chemotherapy, ferroptosis therapy, and photodynamic therapy. Upon entering the cytoplasm, the nanoparticles (NPs) aggregate and target the tumor, where platinum is reduced, releasing iron ions that generate hydroxyl (OH) radicals, ultimately inducing ferroptosis (Fig. 6.40). Fluorescence images of tumor-bearing mice and organs (Figs. 6.40h, i) following injection of FeTPt and FeTPt@CCM NPs show peak distribution in the tumor, kidney, and liver. Notably, the FeTPt@CCM NPs exhibited significantly higher accumulation in these areas compared to the pure FeTPt group, which the authors attribute to the camouflage properties of the CCM coating. Figure 6.40j illustrates the phototherapy strategy used in this approach. After the NPs accumulate in the tumor, a 671 nm red light laser is applied, within the "therapeutic window"—the wavelength range that allows for deep tissue penetration (Sun et al. 2017). The tumor size changes over time, as shown in Fig. 6.40k, demonstrate that FeTPt@CCM NPs exhibited the most potent antitumor effects, even without laser irradiation, outperforming the other treatment strategies.

Wei et al. (2020) created core–shell lipid/PCL-PEI/siRNA nanoparticles (LPS NPs) for siRNA delivery using microfluidic technology. They designed a multi-inlet microfluidic device in which aqueous siRNA was pumped through the central inlet and encapsulated by PEI-PCL polymer micelles, which were dissolved in DMSO and introduced from the side inlets. These particles were then mixed with water from a second pair of side inlets and ultimately merged with a lipophilic solution

Fig. 6.40 (Continued) activated by 670 nm light irradiation to produce $^1O_2$ for PDT. Thus, the FeTPt@CCM exhibits synergistic therapeutic effects by combining chemotherapy, ferroptosis therapy, and PDT. (**d–g**) TEM image and elemental mapping of (**d**, **e**) FeT and (**f**, **g**) FeTPt. (**h**) In vivo fluorescence images of tumor-bearing mice ($n = 3$) after intravenous injection of FeTPt and FeTPt@CCM. (**i**) Fluorescence images of major organs and tumors at 12 h after intravenous injection of FeTPt and FeTPt@CCM. (**j**) Schematic illustration of the therapeutic schedule. (**k**) Tumor growth curves after various treatments (Zhang et al. 2023) (reproduced from Zhang et al. 2023)

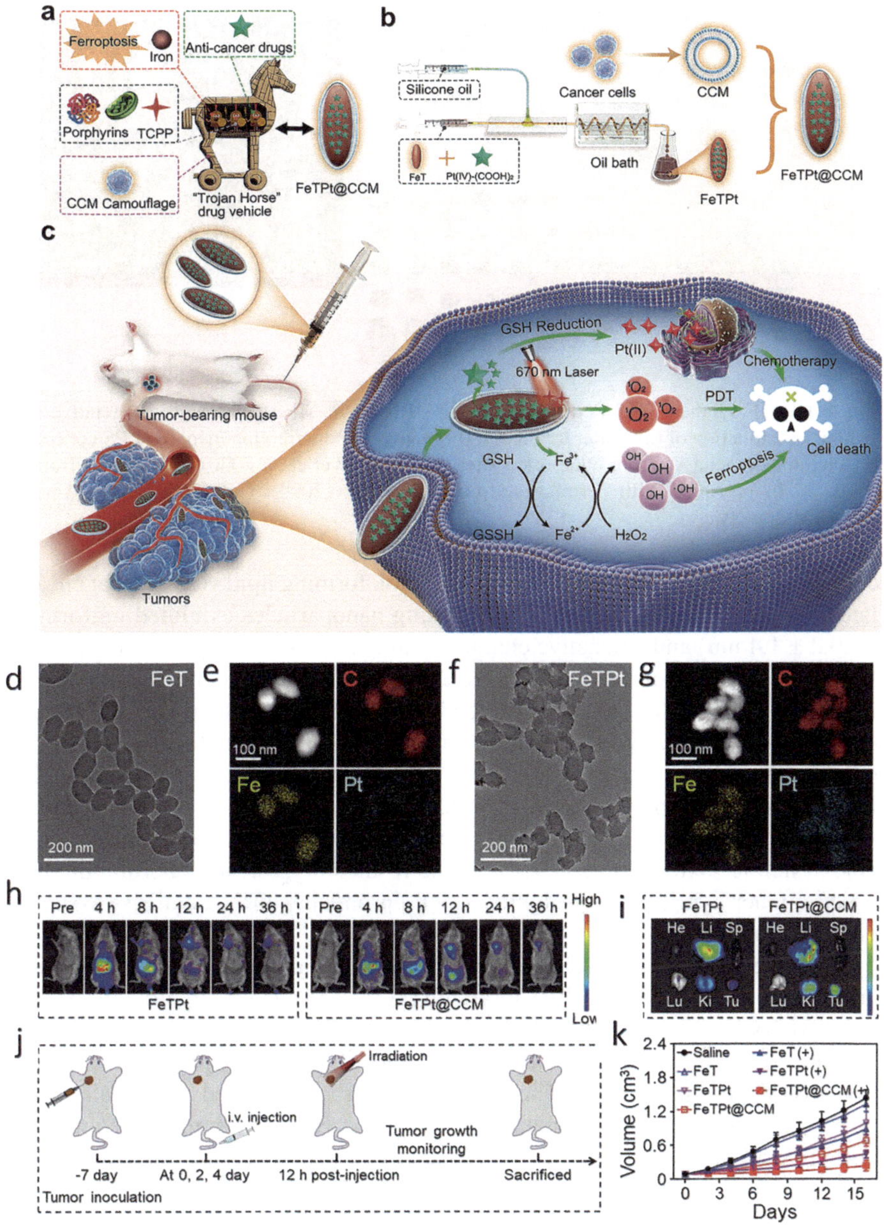

Fig. 6.40 (**a**) The design of the multi-bioinspired "Trojan Horse" drug vehicle. (**b**) The preparation processes of FeTPt through microfluidics, and the generation of FeTPt@CCM through CCM coating. (**c**) After intravenous injection, FeTPt@CCM can efficiently accumulate at the tumor site and enter cancer cells by endocytosis. Subsequently, the Pt(IV)-$(COOH)_2$ was gradually released to the cytoplasm and reduced to high cytotoxic Pt(II) by GSH for chemotherapy. The released Fe^{3+} reacted with GSH and H_2O_2 through the redox reaction, which could promote GSH depletion and •OH generation for initiating ferroptosis. In addition, the TCPP in the FeTPt@CCM could be

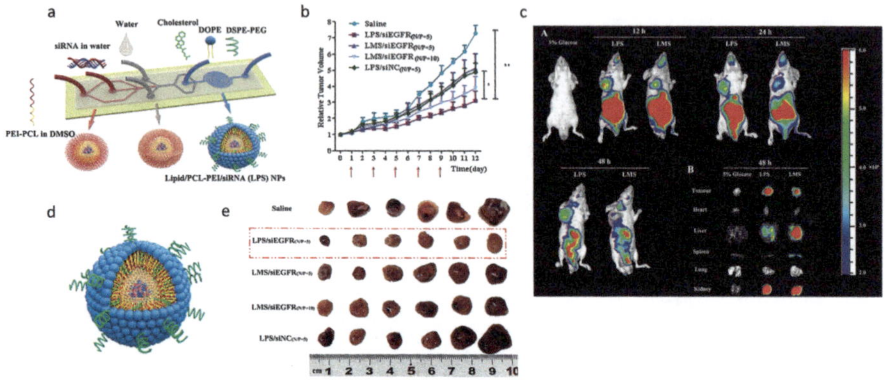

Fig. 6.41 MF synthesis of LPNP-encapsulated siRNA: (**a**) MF HFF device; (**b**) relative tumor volume as a function of time after treatment; (**c**) In vivo cell distribution map; (**d**) LPS concept; (**e**) Ex-vivo tumors excised from nude mice after treatment (Wei et al. 2020) (reproduced from Wei et al. 2020) reprinted with permission from Wei et al. (2020). Copyright (2020) American Chemical Society

containing DOPE, DSPE-PEG, and cholesterol, forming lipid vesicles that encapsulated the particles (Fig. 6.41a). The resulting nanoparticles exhibited uniform size (120.2 ± 1.4 nm) and a negative charge (−8.8 ± 1.6 mV). Compared to bulk-mixed LMS NPs, the microfluidic-based LPS NPs use a lower amount of cationic PCL-PEI material and offer enhanced protection for siRNA in serum. LPS NPs demonstrated significantly greater tumor growth inhibition relative to LMS NPs (Fig. 6.41b). Moreover, the study showed that LPS NPs effectively downregulated EGFR mRNA and protein expression levels both in vitro and in vivo, leading to significant tumor growth inhibition after systemic administration, without noticeable systemic toxicity (Fig. 6.41c). These results suggest that microfluidic-based lipid/polymer hybrid nanoassemblies offer a promising siRNA delivery system for clinical applications.

Younis et al. (2021) developed ultra-small lipid nanoparticles (usLNPs) for the co-delivery of the cytotoxic drug sorafenib (SOR) and siRNA targeting the Midkine gene (MK-siRNA) for hepatocellular carcinoma (HCC) treatment in mice. The usLNPs are composed of a novel pH-sensitive lipid, multiple phospholipids, and highly selective targeting peptides. The particle size was optimized using microfluidic devices to improve tumor penetration efficiency. By fine-tuning the composition and physicochemical properties of the usLNPs, the researchers enhanced tumor accumulation, selectivity, and gene silencing in vivo. This novel combination therapy achieved a remarkable ~85% tumor eradication in mice at an unexpectedly low dose of SOR (2.5 mg/kg), a result that could not be achieved with monotherapy (Fig. 6.42). This approach shows great promise for potential clinical applications in HCC treatment.

Caputo et al. (2023) presented biodegradable and biocompatible polymeric Poly(D,L-lactide-co-glycolide) (PLGA) particles loaded with sorafenib (SOR) (PS) using the emulsion-solvent evaporation method (Fig. 6.43). The resulting particles,

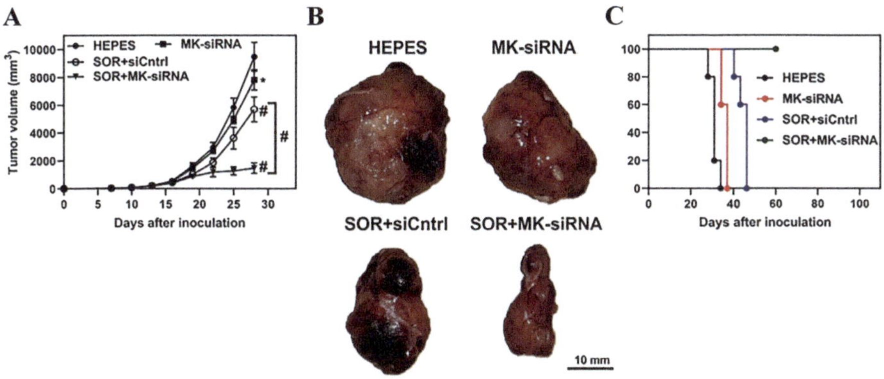

Fig. 6.42 Therapeutic effect of usLNPs with different payloads on HCC-bearing mice. (**a**) Effect on tumor growth. $*P < 0.05$ versus HEPES, $\#P < 0.0001$ for SOR + siCntrl and SOR + MK-siRNA versus HEPES and for SOR + MK-siRNA versus SOR + siCntrl. Two-way ANOVA was applied to verify the interaction between SOR and MK-siRNA, P(SOR*MK-siRNA) < 0.001. (**b**) Representative photos showing tumor tissues from different treatment groups after the end of treatment. (**c**) Survival rates of mice in different treatment groups followed-up until 60-days post-tumor inoculation (Younis et al. 2021) (reproduced from Younis et al. 2021)

with a size of 250 nm, a spherical shape, tunable surface charge, and good stability in culture medium, show promise as a carrier for hydrophobic drug delivery. The PS particles, with a drug loading of 9%, exhibited a sustained release profile lasting over 220 h and fast cellular uptake within 1 h. Their stability and rapid internalization led to significantly increased toxicity to hepatocellular carcinoma cells, with only 17% cell viability after 168 h at a concentration of 7.5 μM. The authors also introduced a novel light-triggered drug delivery system, based on side-emitting optical fibers (seOF) integrated with PLGA particles via a UV-photocleavable linker. This light-triggered release mechanism was further tested in a microfluidic device, which can be easily adapted to various needle models. This study paves the way for the development of a new type of light-triggered drug delivery system that enables on-demand, site-specific drug release.

Roffo et al. (2022) proposed biocompatible lipid-polymer nanoparticles (LiPoNs) fabricated using coupled hydrodynamic flow focusing (cHFF) for theranostic and multi-modal imaging applications. The microfluidic platform enables precise manipulation of species interdiffusion and manages nanoprecipitation and self-assembly, guiding the lipid-polymer coupling process. The resulting LiPoNs have a core–shell structure, with a polymeric chitosan core encapsulated by a lipid bilayer, and are capable of simultaneously co-encapsulating Gd-DTPA and irinotecan/Atto 633 compounds. The LiPoNs demonstrated enhanced delivery and therapeutic efficacy of the irinotecan-loaded hybrid formulation against U87 MG cancer cells.

Disulfiram (DSF) and Cu ions form diethyldithiocarbamate copper complex [$Cu(DDC)_2$] which exhibits potent antitumor activity. In this study, Chang et al. (2020) improved stabilized metal ion ligand nanocomplex (SMILE) technology to

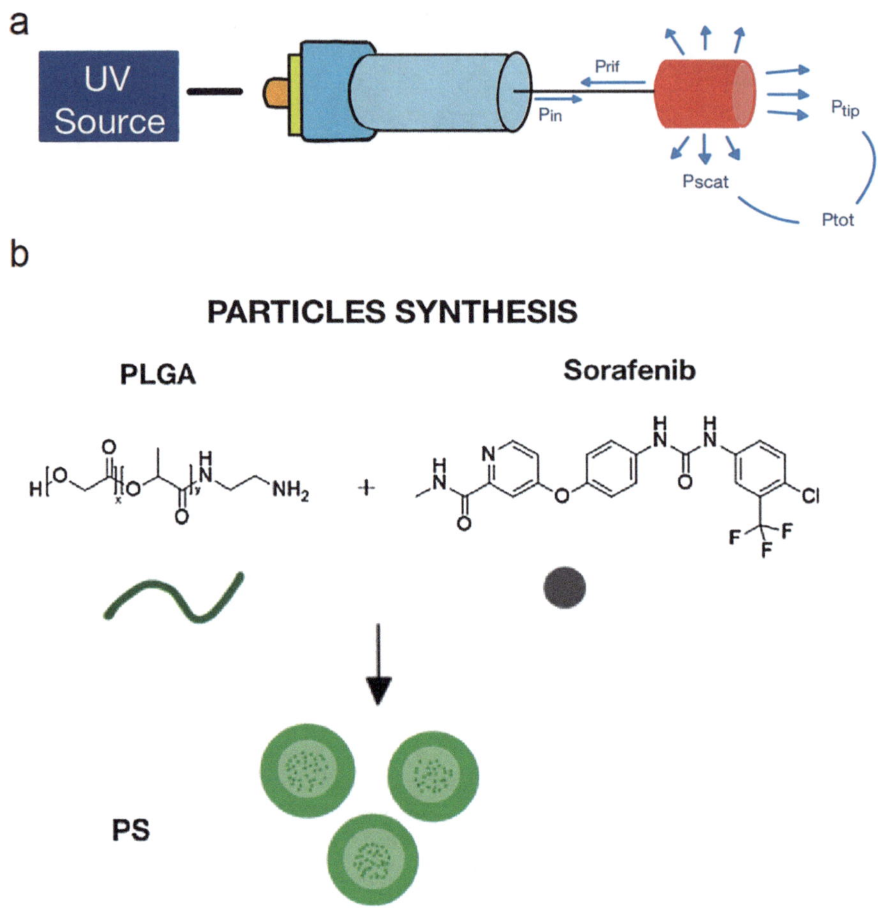

Fig. 6.43 (**a**) Schematic of characterization setup. (**b**) The synthesis of biodegradable polymeric Poly (D,L-Lactide-co-glycolide) (PLGA) particles loaded with SOR (PS) by emulsion-solvent evaporation process (Caputo et al. 2023) (reproduced from Caputo et al. 2023)

achieve scalable and continuous production of $Cu(DDC)_2$ metal-organic nanoparticles (MONs) using a 3D-printed microfluidic device for cancer treatment. The microfluidic platform allows for precise control over the mixing process, enabling the production of nanoparticles with a small size (<100 nm) and a narrow size distribution. The Cy5-labeled BSA/$Cu(DDC)_2$ MON effectively inhibited tumor growth in an orthotopic 4 T1 breast tumor model (Fig. 6.44).

Nanoliposomes produced using microfluidic technology have been widely explored as effective drug carriers for various applications, particularly in tumor chemotherapy. For instance, nanoliposomes loaded with doxorubicin of varying sizes were fabricated using microfluidics for breast cancer treatment (Dong et al. 2019). The results demonstrated that the biodistribution and clearance of these

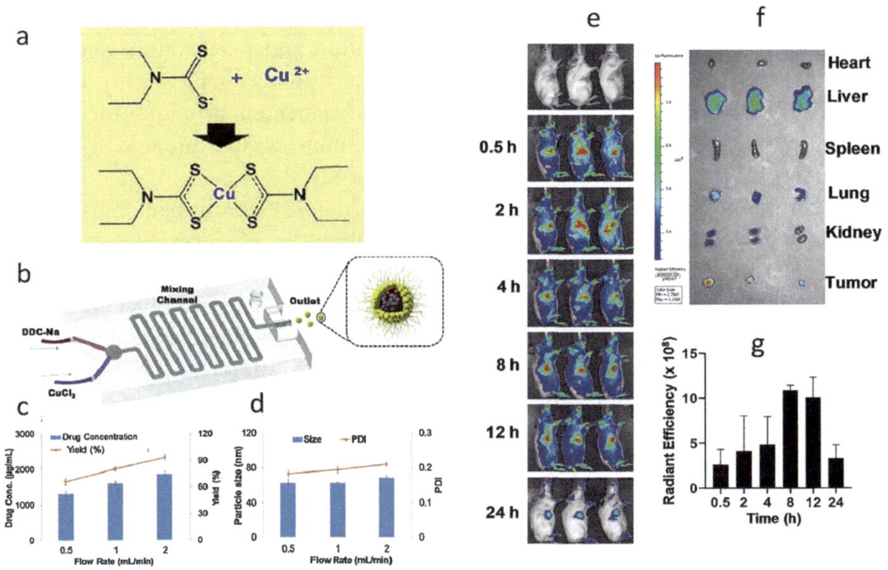

Fig. 6.44 (**a**) Schematic illustration to show the Cu(DDC)2 complex formation. (**b**) Schematic illustration to show the preparation of BSA Cu(DDC)2MONs. The effects of flow rate on (**c**) drug concentration and yield, and (**d**) particle size and PDI. (**e–g**) In vivo biodistribution of Cy5-labeled BSA Cu(DDC)2MONs in mice with 4T1 orthotropic tumors determined with bioimaging. (**e**) The whole-body imaging at different time post injection. (**f**) Ex vivo imaging of isolated organs (heart, liver, spleen, lung, kidneys) and tumor. (**g**) The in vivo radiant efficiency of the tumor sites (Chang et al. 2020) (reproduced from Chang et al. 2020)

nanoliposomes from the blood and organs could be precisely controlled by adjusting their size. Specifically, larger nanoliposomes exhibited greater distribution in the liver and spleen, while smaller nanoliposomes were cleared more rapidly from the bloodstream and organs. Additionally, the nanoliposomes showed enhanced tumor penetration compared to the free drug.

6.4 Conclusions and Perspectives

The versatility of microfluidics in precise operation and fine assembly provides robust technical support for the research of liquid biopsy, nano-drug delivery systems (NDDS), and tumor diagnosis and treatment. A well-designed microfluidic system can achieve many functions that traditional methods cannot, including the following three key aspects:

1. Efficient Capture and Separation of Tumor Markers: Microfluidic systems can efficiently capture and separate trace tumor markers, such as CTCs, EVs, and circulating tumor DNA (ctDNA), from complex biological fluids like blood, urine, or sweat.

2. Integration of Multifunctional Modules: By integrating multiple functions, including sample pretreatment, target separation and purification, and in situ detection, into a single microfluidic chip, these systems enable highly sensitive, quantitative analysis of tumor markers. This advancement promotes the research and application of liquid biopsy, offering a non-invasive method for cancer diagnosis.
3. Controllable Synthesis of Functional Nanoparticles: Microfluidic chips provide a powerful platform for the controllable synthesis of functional composite nanoparticles, including polymer nanoparticles, lipid core-shell nanoparticles, biomimetic lipid nanoparticles, ligand-edited exosomes, and other multifunctional NDDS. Specifically, they offer a reproducible, standardized, and structurally complete process for preparing biomimetic lipid nanoparticles with biological properties. Using microfluidics, researchers can achieve a diverse range of NDDS, including complex nanostructures, varied drug delivery types, multiple targeted delivery strategies, and integrated approaches for diagnosis and treatment.

While current research on microfluidic liquid biopsy and NDDS is largely in the preclinical stage, it is still far from clinical application. This presents significant opportunities for further in-depth and systematic investigation. For example, more precise control over the microstructure, size, and drug distribution in NDDS needs to be explored further. Emerging microfluidic technologies, such as nanofluidics, offer potential for studying the complex control of NDDS properties. It is essential for researchers to continue optimizing and breaking through existing methods and technologies, conducting more studies to evaluate the effectiveness of nanoparticles in cancer treatment.

References

Ahmad I, Al-Dolaimy F, Kzar MH et al (2024) Microfluidic-based nanoemulsion of Ocimum basilicum extract: constituents, stability, characterization, and potential biomedical applications for improved antimicrobial and anticancer properties. Microsc Res Tech 87(3):411–423

Ahmed MG, Fasil AM, Yanling S et al (2017) Isolation, detection, and antigen-based profiling of circulating tumor cells using a size-dictated immunocapture chip. Angew Chem Int Ed 56(36):10681–10685

Alix-Panabières C, Pantel K (2014) Challenges in circulating tumour cell research. Nat Rev Cancer 14(9):623–631

Alix-Panabières C, Pantel K (2021) Liquid biopsy: from discovery to clinical application. Cancer Discov 11(4):858–873

Arya SK, Lim B, Rahman AR (2013) Enrichment, detection and clinical significance of circulating tumor cells. Lab Chip 13(11):1995–2027

Ayala MS, Perez-Gonzalez VH, Mata-Gómez MA et al (2019) Electrokinetically driven exosome separation and concentration using dielectrophoretic-enhanced PDMS-based microfluidics. Anal Chem 91(23):14975–14982

Azadi SA, Vasheghani-Farahani E, Hashemi-Najafbabadi S et al (2016) Co-encapsulation of pancreatic islets and pentoxifylline in alginate-based microcapsules with enhanced immunosuppressive effects. Prog Biomater 5:101–109

Baroud CN, Delville JP, Gallaire F et al (2007) Thermocapillary valve for droplet production and sorting. Phys Rev E Stat Nonlinear Soft Matter Phys 75(4 Pt 2):046302
Bawazer LA, Mcnally CS, Empson CJ et al (2016) Combinatorial microfluidic droplet engineering for biomimetic material synthesis. Sci Adv 2(10):e1600567
Beekman P, Agustin E-M, Suk RH et al (2019) Immuno-capture of extracellular vesicles for individual multi-modal characterization using AFM, SEM and Raman spectroscopy. Lab Chip 19(15):2526–2536
Bhagat AAS, Hansen B, Wei HH et al (2010) Microfluidics for cell separation. Med Biol Eng Comput 48:999–1014
Bishop JM (1987) The molecular genetics of cancer. Science 235(4786):305–311
Broman A, Andreas L, Mikael E et al (2021) Multinodal acoustic trapping enables high capacity and high throughput enrichment of extracellular vesicles and microparticles in miRNA and MS proteomics studies. Anal Chem 93(8):3929–3937
Cai J, Jinyun J, Feng G et al (2017) Rapid prototyping of cyclic olefin copolymer based microfluidic system with CO2 laser ablation. Microsyst Technol 23:5063–5069
Caputo TM, Maria CA, Sofia P et al (2023) Sorafenib-loaded PLGA carriers for enhanced drug delivery and cellular uptake in liver cancer cells. Int J Nanomedicine 18(null):4121–4142
Carlborg CF, Tommy H, Kim Ö et al (2011) Beyond PDMS: off-stoichiometry thiol–ene (OSTE) based soft lithography for rapid prototyping of microfluidic devices. Lab Chip 11(18):3136–3147
Castro-Hernandez E, Pablo G-S, Hwa TS et al (2015) Breakup length of AC electrified jets in a microfluidic flowfocusing junction. Microfluid Nanofluid 19(4):787–794
Chang Y, Jizong J, Chen W et al (2020) Biomimetic metal-organic nanoparticles prepared with a 3D-printed microfluidic device as a novel formulation for disulfiram-based therapy against breast cancer. Appl Mater Today 18:100492
Chen J, Zefan Y, Wenhu X et al (2021) Sorting of circulating tumor cells based on the microfluidic device of a biomimetic splenic interendothelial slit array. Microfluid Nanofluid 25(7):57
Chen Y-S, Chihchen C, Pin-Kuang LC et al (2022) Isolation and digital counting of extracellular vesicles from blood via membrane-integrated microfluidics. Sensors Actuators B Chem 358:131473
Cheng SB, Chen MM, Wang YK et al (2021) A three-dimensional conductive scaffold microchip for effective capture and recovery of circulating tumor cells with high purity. Anal Chem 93(18):7102–7109
Cheung LS, Sarah S, Ajymurat O et al (2018) Rapid detection and trapping of extracellular vesicles by Electrokinetic concentration for liquid biopsy on Chip. Micromachines 9(6):306
Choi CH, Weitz DA, Lee CS (2013) One step formation of controllable complex emulsions: from functional particles to simultaneous encapsulation of hydrophilic and hydrophobic agents into desired position. Adv Mater 25(18):2536–2541
Churski K, Michalski J, Garstecki P (2010) Droplet on demand system utilizing a computer controlled microvalve integrated into a stiff polymeric microfluidic device. Lab Chip 10(4):512–518
Colombo M, Raposo G, Théry C (2014) Biogenesis, secretion, and intercellular interactions of exosomes and other extracellular vesicles. Annu Rev Cell Dev Biol 30:255–289
Dong YD, Tchung E, Nowell C et al (2019) Microfluidic preparation of drug-loaded PEGylated liposomes, and the impact of liposome size on tumour retention and penetration. J Liposome Res 29(1):1–9
Durukan O, Ilhan K, Patricia P et al (2016) Microfluidization, time-effective and solvent free processing of nanoparticle containing thermosetting matrix resin suspensions for producing composites with enhanced thermal properties. Eur Polym J 85:575–587
Emrich CA, Medintz Igor L, Chu Wai K et al (2007) Microfabricated two-dimensional electrophoresis device for differential protein expression profiling. Anal Chem 79(19):7360–7366
Feng X, Ying Y, Xu Y et al (2010) Generation of water–ionic liquid droplet pairs in soybean oil on microfluidic chip. Lab Chip 10(3):313–319
Garstecki P, Stone Howard A, Whitesides George M (2005) Mechanism for flow-rate controlled breakup in confined geometries: a route to monodisperse emulsions. Phys Rev Lett 94(16):164501

Gholizadeh S, Shehata DM, Zarghooni M et al (2017) Microfluidic approaches for isolation, detection, and characterization of extracellular vesicles: current status and future directions. Biosens Bioelectron 91:588–605
Gorges TM, Tinhofer I, Drosch M et al (2012) Circulating tumour cells escape from EpCAM-based detection due to epithelial-to-mesenchymal transition. BMC Cancer 12:178
Gu H, Duits MH, Mugele F (2010) A hybrid microfluidic chip with electrowetting functionality using ultraviolet (UV)-curable polymer. Lab Chip 10(12):1550–1556
Habibi R, Neild A (2019) Sound wave activated nano-sieve (SWANS) for enrichment of nanoparticles. Lab Chip 19(18):3032–3044
Han G, Zhao J, Zhang R et al (2019) Membrane-penetrating carbon quantum dots for imaging nucleic acid structures in live organisms. Angew Chem Int Ed Engl 58(21):7087–7091
Hannes B, Julien V, Bou CE et al (2008) The etching of glass patterned by microcontact printing with application to microfluidics and electrophoresis. Sensors Actuators B Chem 129(1):255–262
Hashimoto M, Francis B, Feng X et al (2007) Serial processing of biological reactions using flow-through microfluidic devices: coupled PCR/LDR for the detection of low-abundant DNA point mutations. Analyst 132(9):913–921
Hilton SH, White IM (2021) Advances in the analysis of single extracellular vesicles: a critical review. Sens Actuators Rep 3:100052
Iqbal M, Zafar N, Fessi H et al (2015) Double emulsion solvent evaporation techniques used for drug encapsulation. Int J Pharm 496(2):173–190
Iwai K, Cheng SK, Xiao L et al (2014) Finger-powered microfluidic systems using multilayer soft lithography and injection molding processes. Lab Chip 14(19):3790–3799
Jeong H-S, Eunseo K, Changwoo N et al (2021) Hydrogel microcapsules with a thin oil layer: smart triggered release via diverse stimuli. Adv Funct Mater 31(18):2009553
Kangning R, Yin C, Hongkai W (2014) New materials for microfluidics in biology. Curr Opin Biotechnol 25:78–85
Kim SM, Burns Mark A, Hasselbrink Ernest F (2006) Electrokinetic protein preconcentration using a simple glass/poly (dimethylsiloxane) microfluidic chip. Anal Chem 78(14):4779–4785
Kim MS, Sim TS, Kim YJ et al (2012) SSA-MOA: a novel CTC isolation platform using selective size amplification (SSA) and a multi-obstacle architecture (MOA) filter. Lab Chip 12(16):2874–2880
Kim D, Woo HK, Lee C et al (2020) EV-ident: identifying tumor-specific extracellular vesicles by size fractionation and single-vesicle analysis. Anal Chem 92(8):6010–6018
Ko J, Yongcheng W, Carlson Jonathan CT et al (2020) Single extracellular vesicle protein analysis using Immuno-droplet digital polymerase chain reaction amplification. Adv Biosyst 4(12):1900307
Ko J, Wang Y, Sheng K et al (2021) Sequencing-based protein analysis of single extracellular vesicles. ACS Nano 15(3):5631–5638
Kulasinghe A, Jian Z, Liz K et al (2019) Capture of circulating tumour cell clusters using straight microfluidic chips. Cancer 11(1):89
Lee C-P, Ting-Sheng L, Mei-Feng L (2014) Fabrication of two-dimensional ferrofluid microdroplet lattices in a microfluidic channel. J Appl Phys 115(17):527
Li L, Ismagilov RF (2010) Protein crystallization using microfluidic technologies based on valves, droplets, and SlipChip. Annu Rev Biophys 39(1):139–158
Li P, Stratton ZS, Dao M et al (2013) Probing circulating tumor cells in microfluidics. Lab Chip 13(4):602–609
Liang LG, Kong MQ, Zhou S et al (2017) An integrated double-filtration microfluidic device for isolation, enrichment and quantification of urinary extracellular vesicles for detection of bladder cancer. Sci Rep 7:46224
Liga A, Vliegenthart AD, Oosthuyzen W et al (2015) Exosome isolation: a microfluidic road-map. Lab Chip 15(11):2388–2394
Lim EK, Kim T, Paik S et al (2015) Nanomaterials for theranostics: recent advances and future challenges. Chem Rev 115(1):327–394

Lim SB, Trifanny Y, Di LW et al (2019) Addressing cellular heterogeneity in tumor and circulation for refined prognostication. Proc Natl Acad Sci 116(36):17957–17962

Linares R, Tan S, Gounou C et al (2015) High-speed centrifugation induces aggregation of extracellular vesicles. J Extracell Vesicles 4:29509

Link DR, Erwan G-M, Agnes D et al (2006) Electric control of droplets in microfluidic devices. Angew Chem Int Ed 45(16):2556–2560

Liu P, Seok ST, Nathaniel B et al (2007) Integrated portable polymerase chain reaction-capillary electrophoresis microsystem for rapid forensic short tandem repeat typing. Anal Chem 79(5):1881–1889

Liu J, Say-Hwa T, Fatt YY et al (2011) Numerical and experimental investigations of the formation process of ferrofluid droplets. Microfluid Nanofluid 11:177–187

Liu C, Feng Q, Sun J (2019a) Lipid Nanovesicles by microfluidics: manipulation, synthesis, and drug delivery. Adv Mater 31(45):e1804788

Liu C, Zhao J, Tian F et al (2019b) λ-DNA- and aptamer-mediated sorting and analysis of extracellular vesicles. J Am Chem Soc 141(9):3817–3821

Liu Z, Huang Y, Liang W et al (2021) Cascaded filter deterministic lateral displacement microchips for isolation and molecular analysis of circulating tumor cells and fusion cells. Lab Chip 21(15):2881–2891

Lone SN, Sabah N, Tariq M et al (2022) Liquid biopsy: a step closer to transform diagnosis, prognosis and future of cancer treatments. Mol Cancer 21(1):79

Lu Y, Park K (2013) Polymeric micelles and alternative nanonized delivery vehicles for poorly soluble drugs. Int J Pharm 453(1):198–214

Lu C, Jian X, Jintao H et al (2020) A novel microfluidic device integrating focus-separation speed reduction design and trap arrays for high-throughput capture of circulating tumor cells. Lab Chip 20(22):4094–4105

Lu Y, Ling Y, Xiaoyu J et al (2022) Integrated microfluidic system for isolating exosome and analyzing protein marker PD-L1. Biosens Bioelectron 204:113879

Ma Y, Qi Z, Kexin Z et al (2023) NTRK fusions in thyroid cancer: pathology and clinical aspects. Crit Rev Oncol Hematol 184:103957

Marczak S, Richards K, Ramshani Z et al (2018) Simultaneous isolation and preconcentration of exosomes by ion concentration polarization. Electrophoresis. https://doi.org/10.1002/elps.201700491

Morimoto Y, Maiko O, Shoji T (2017) Mass production of cell-laden calcium alginate particles with centrifugal force. Adv Healthc Mater 6(13):1601375

Murphy TW, Qiang Z, Naler LB et al (2018) Recent advances in the use of microfluidic technologies for single cell analysis. Analyst 143(1):60–80

Myers FB, Lee LP (2008) Innovations in optical microfluidic technologies for point-of-care diagnostics. Lab Chip 8(12):2015–2031

Navi M, Jennifer K, Tsai Scott SH (2020) Magnetic polyelectrolyte microcapsules via water-in-water droplet microfluidics. Lab Chip 20(16):2851–2860

Neves LF, Krais JJ, Van Rite BD et al (2013) Targeting single-walled carbon nanotubes for the treatment of breast cancer using photothermal therapy. Nanotechnology 24(37):375104

Nguyen NT, Hui TT, Fatt YY et al (2007) Thermally mediated droplet formation in microchannels. Appl Phys Lett 91(8):10404

Nie Z, Minsseok S, Shengqing X et al (2008) Emulsification in a microfluidic flow-focusing device: effect of the viscosities of the liquids. Microfluid Nanofluid 5:585–594

Nisisako T, Toru T, Toshiro H (2002) Droplet formation in a microchannel network. Lab Chip 2(1):24–26

Normanno N, Andres C, Fortunato C et al (2018) The liquid biopsy in the management of colorectal cancer patients: current applications and future scenarios. Cancer Treat Rev 70:1–8

Peng L, Min Y, Shi-Shang G et al (2011) The effect of interfacial tension on droplet formation in flow-focusing microfluidic device. Biomed Microdevices 13:559–564

Pihl J, Mattias K, Chiu DT (2005) Microfluidic technologies in drug discovery. Drug Discov Today 10(20):1377–1383
Ramazani F, Chen W, Van Nostrum CF et al (2016) Strategies for encapsulation of small hydrophilic and amphiphilic drugs in PLGA microspheres: state-of-the-art and challenges. Int J Pharm 499(1–2):358–367
Ramirez MI, Amorim MG, Catarina G et al (2018) Technical challenges of working with extracellular vesicles. Nanoscale 10(3):881–906
Raposo G, Stoorvogel W (2013) Extracellular vesicles: exosomes, microvesicles, and friends. J Cell Biol 200(4):373–383
Ren K, Chen Y, Wu, H (2014) New materials for microfluidics in biology. Curr Opin Biotechnol 25:78–85
Roffo F, Maria PA, Antonio NP et al (2022) Coupled hydrodynamic flow focusing (cHFF) to engineer lipid–polymer nanoparticles (LiPoNs) for multimodal imaging and theranostic applications. Biomedicine 10(2):438
Rothbauer M, Helene Z, Peter E (2018) Recent advances in microfluidic technologies for cell-to-cell interaction studies. Lab Chip 18(2):249–270
Saitoh T, Yuichi S, Masataka H (2002) Preparation of poly (N-isopropylacrylamide)-modified glass surface for flow control in microfluidics. Anal Sci 18(2):203–205
Sartipzadeh O, Morteza NS, Amir S et al (2020) Controllable size and form of droplets in microfluidic-assisted devices: effects of channel geometry and fluid velocity on droplet size. Mater Sci Eng C 109:110606
Schulte TH, Bardell Ron L, Weigl Bernhard H (2002) Microfluidic technologies in clinical diagnostics. Clin Chim Acta 321(1):1–10
Shakeel F, Shafiq S, Haq N et al (2012) Nanoemulsions as potential vehicles for transdermal and dermal delivery of hydrophobic compounds: an overview. Expert Opin Drug Deliv 9(8):953–974
Shao H, Im H, Castro CM et al (2018) New Technologies for Analysis of extracellular vesicles. Chem Rev 118(4):1917–1950
Siavashy S, Soltani M, Fatemeh G-B et al (2021) Microfluidic platform for synthesis and optimization of chitosan-coated magnetic nanoparticles in cisplatin delivery. Carbohydr Polym 265:118027
Sun C, Dai L, Liu F et al (2016) Dynamic high pressure microfluidization treatment of zein in aqueous ethanol solution. Food Chem 210:388–395
Sun W, Thiramanas R, Slep LD et al (2017) Photoactivation of anticancer Ru complexes in deep tissue: how deep can we go? Chemistry 23(45):10832–10837
Sun M, Xu J, Shamul JG et al (2019a) Creating a capture zone in microfluidic flow greatly enhances the throughput and efficiency of cancer detection. Biomaterials 197:161–170
Sun H, Han L, Yang L et al (2019b) Modular chamber assembled with cell-replicated surface for capture of cancer cells. ACS Biomater Sci Eng 5(5):2647–2656
Tahir N, Madni A, Li W et al (2020) Microfluidic fabrication and characterization of sorafenib-loaded lipid-polymer hybrid nanoparticles for controlled drug delivery. Int J Pharm 581:119275
Tan SJ, Yobas L, Lee GY et al (2009) Microdevice for the isolation and enumeration of cancer cells from blood. Biomed Microdevices 11(4):883–892
Tan S-H, Nam-Trung N, Levent Y et al (2010) Formation and manipulation of ferrofluid droplets at a microfluidic T-junction. J Micromech Microeng 20(4):045004
Tan SL, Stanslas J, Basri M et al (2015) Nanoemulsion-based parenteral drug delivery system of carbamazepine: preparation, characterization, stability evaluation and blood-brain pharmacokinetics. Curr Drug Deliv 12(6):795–804
Tauro BJ, Greening DW, Mathias RA et al (2012) Comparison of ultracentrifugation, density gradient separation, and immunoaffinity capture methods for isolating human colon cancer cell line LIM1863-derived exosomes. Methods 56(2):293–304
Tavakoli H, Wan Z, Lei M et al (2019) Recent advances in microfluidic platforms for single-cell analysis in cancer biology, diagnosis and therapy. TrAC Trends Anal Chem 117:13–26

Teo A, Minghong Y, Jing D et al (2020) Controllable droplet generation at a microfluidic T-junction using AC electric field. Microfluid Nanofluid 24:1–9
Théry C, Zitvogel L, Amigorena S (2002) Exosomes: composition, biogenesis and function. Nat Rev Immunol 2(8):569–579
Thorsen T, Roberts RW, Arnold FH et al (2001) Dynamic pattern formation in a vesicle-generating microfluidic device. Phys Rev Lett 86(18):4163–4166
Tian Y, Chunmiao X, Jiao F et al (2021) Personalized gel-droplet monocyte vaccines for cancer immunotherapy. Lab Chip 21(22):4414–4426
Tian Y, Kai W, Chunmiao X et al (2022) Non-invasive T cells adoptive immunotherapy for solid tumor with gel anti-tumor T-cell injections. Chem Eng J 439:135839
Toepke MW, Kenis J (2005) Multilevel microfluidics via single-exposure photolithography. J Am Chem Soc 127(21):7674–7675
Tse L, Hesketh P, Rosen D et al (2003) Stereolithography on silicon for microfluidics and microsensor packaging. Microsyst Technol 9:319–323
Utada AS, Elise L, Link Darren R et al (2005) Monodisperse double emulsions generated from a microcapillary device. Science 308(5721):537–541
Valente KP, Sultan K, Kolahchi AR et al (2017) Microfluidic technologies for anticancer drug studies. Drug Discov Today 22(11):1654–1670
Van DJ, Mestdagh P, Sormunen R et al (2014) The impact of disparate isolation methods for extracellular vesicles on downstream RNA profiling. J Extracell Vesicles 18:3
Vandijken J, Kaigala Govind V, Jana L et al (2007) Microfluidic chips for detecting the t (4; 14) translocation and monitoring disease during treatment using reverse transcriptase-polymerase chain reaction analysis of IgH-MMSET hybrid transcripts. J Mol Diagn 9(3):358–367
Wang H, Mooney DJ (2018) Biomaterial-assisted targeted modulation of immune cells in cancer treatment. Nat Mater 17(9):761–772
Wang H, Yang R, Yang L et al (2009) Nucleic acid conjugated nanomaterials for enhanced molecular recognition. ACS Nano 3(9):2451–2460
Wang X, Christina L, Ralph L et al (2016) A disposable, roll-to-roll hot-embossed inertial microfluidic device for size-based sorting of microbeads and cells. Lab Chip 16(10):1821–1830
Wang SS, Zhao XP, Liu FF et al (2019) Direct Plasmon-enhanced electrochemistry for enabling ultrasensitive and label-free detection of circulating tumor cells in blood. Anal Chem 91(7):4413–4420
Wei J, Xiao-Jie J, Xiao-Yi Z et al (2014) Multi-stimuli-responsive microcapsules for adjustable controlled-release. Adv Funct Mater 24(22):3312–3323
Wei W, Sun J, Guo XY et al (2020) Microfluidic-based holonomic constraints of siRNA in the kernel of lipid/polymer hybrid Nanoassemblies for improving stable and safe in vivo delivery. ACS Appl Mater Interfaces 12(13):14839–14854
Whitesides GM (2006) The origins and the future of microfluidics. Nature 442(7101):368–373
Woo HK, Sunkara V, Park J et al (2017) Exodisc for rapid, size-selective, and efficient isolation and analysis of nanoscale extracellular vesicles from biological samples. ACS Nano 11(2):1360–1370
Wu M, Ouyang Y, Wang Z et al (2017) Isolation of exosomes from whole blood by integrating acoustics and microfluidics. Proc Natl Acad Sci USA 114(40):10584–10589
Wu D, Yu Y, Zhao C et al (2019) NK-cell-encapsulated porous microspheres via microfluidic electrospray for tumor immunotherapy. ACS Appl Mater Interfaces 11(37):33716–33724
Wu L, Ding H, Qu X et al (2020) Fluidic multivalent membrane nanointerface enables synergetic enrichment of circulating tumor cells with high efficiency and viability. J Am Chem Soc 142(10):4800–4806
Wu Z, Pan Y, Wang Z et al (2021) A PLGA nanofiber microfluidic device for highly efficient isolation and release of different phenotypic circulating tumor cells based on dual aptamers. J Mater Chem B 9(9):2212–2220
Wunsch BH, Smith JT, Gifford SM et al (2016) Nanoscale lateral displacement arrays for the separation of exosomes and colloids down to 20 nm. Nat Nanotechnol 11(11):936–940

Yáñez-Mó M, Siljander PR, Andreu Z et al (2015) Biological properties of extracellular vesicles and their physiological functions. J Extracell Vesicles 4:27066
Yang F, Liao X, Tian Y et al (2017) Exosome separation using microfluidic systems: size-based, immunoaffinity-based and dynamic methodologies. Biotechnol J 12(4):699
Yang C, Fangfang C, Shuyi W et al (2019) Circulating tumor cells in gastrointestinal cancers: current status and future perspectives. Front Oncol 9:1427
Yang Q, Liming C, Liang H et al (2020) An integrative microfluidic device for isolation and ultrasensitive detection of lung cancer-specific exosomes from patient urine. Biosens Bioelectron 163:112290
Yang Z, Atiyas Y, Shen H et al (2022) Ultrasensitive single extracellular vesicle detection using high throughput droplet digital enzyme-linked immunosorbent assay. Nano Lett 22(11):4315–4324
Yeh Y-T, Yijing Z, Donghua Z et al (2020) Rapid size-based isolation of extracellular vesicles by three-dimensional carbon nanotube arrays. ACS Appl Mater Interfaces 12(11):13134–13139
Yobas L, Stefan M, Wee-Liat O et al (2006) High-performance flow-focusing geometry for spontaneous generation of monodispersed droplets. Lab Chip 6(8):1073–1079
Younis MA, Khalil IA, Elewa YH, A. et al (2021) Ultra-small lipid nanoparticles encapsulating sorafenib and midkine-siRNA selectively-eradicate sorafenib-resistant hepatocellular carcinoma in vivo. J Control Release 331:335–349
Yu C-C, Yi-Wen C, Po-Ying Y et al (2019) Random and aligned electrospun PLGA nanofibers embedded in microfluidic chips for cancer cell isolation and integration with air foam technology for cell release. J Nanobiotechnol 17(1):31
Zeng S, Bowei L, Jianhua Q et al (2009) Microvalve-actuated precise control of individual droplets in microfluidic devices. Lab Chip 9(10):1340–1343
Zhang P, Xin Z, Yong Z (2019) Multiplexed immunophenotyping of circulating exosomes on nano-engineered ExoProfile chip towards early diagnosis of cancer. Chem Sci 10(21):5495–5504
Zhang J, Bingqian L, Lingling W et al (2020) DNA nanolithography enables a highly ordered recognition interface in a microfluidic chip for the efficient capture and release of circulating tumor cells. Angew Chem Int Ed 59(33):14115–14119
Zhang Q, Kuang G, Wang H et al (2023) Multi-bioinspired MOF delivery systems from microfluidics for tumor multimodal therapy. Adv Sci 10(33):e2303818
Zheng Y, Ouyang WQ, Wei YP et al (2016) Effects of Carbopol(®) 934 proportion on nanoemulsion gel for topical and transdermal drug delivery: a skin permeation study. Int J Nanomedicine 11:5971–5987
Zhou Z, Zhang Y, Shen Y et al (2018) Molecular engineering of polymeric carbon nitride: advancing applications from photocatalysis to biosensing and more. Chem Soc Rev 47(7):2298–2321
Zhu Z, Dan W, Shuang L et al (2021) A polymer-film inertial microfluidic sorter fabricated by jigsaw puzzle method for precise size-based cell separation. Anal Chim Acta 1143:306–314

GPSR Compliance

The European Union's (EU) General Product Safety Regulation (GPSR) is a set of rules that requires consumer products to be safe and our obligations to ensure this.

If you have any concerns about our products, you can contact us on ProductSafety@springernature.com

In case Publisher is established outside the EU, the EU authorized representative is:

Springer Nature Customer Service Center GmbH
Europaplatz 3
69115 Heidelberg, Germany

Batch number: 10755329

Printed by Printforce, the Netherlands